Machine Learning in Cyber Trust

Security, Privacy, and Reliability

Machine Learning in Cyber Trust

Security, Privacy, and Reliability

Edited by

Jeffrey J.P. Tsai

Philip S. Yu

Editors:

Jeffrey J. P. Tsai
Department of Computer Science
University of Illinois at Chicago
851 S. Morgan St., Rm 1120 SEO
Chicago, IL 60607-7053, USA
tsai@cs.uic.edu

Philip S. Yu
Department of Computer Science
University of Illinois at Chicago
851 S. Morgan St., Rm 1138 SEO
Chicago, IL 60607-7053, USA
psyu@cs.uic.edu

ISBN: 978-1-4419-4698-0 e-ISBN: 978-0-387-88735-7
DOI: 10.1007/978-0-387-88735-7

Printed on acid-free paper.

9 8 7 6 5 4 3 2 1

springer.com

To my parents: Ying-Ren and Shiow-Lien,
and my family: Fuh-Te, Edward, Christina

- J.T.

To my family

- P.Y.

Preface

Networked computers reside at the heart of systems on which people now rely, both in critical national infrastructures and in private enterprises. Today, many of these systems are far too vulnerable to cyber attacks that can inhibit their functioning, corrupt important data, or expose private information. It is extremely important to make the system resistant to and tolerant of these cyber attacks.

Machine learning is critical in the study of how to build computer programs that improve their performance through experience. Machine learning algorithms have proven to be of great practical value in a variety of application domains. They are particularly useful for (a) poorly understood problem domains where little knowledge exists for the humans to develop effective algorithms; (b) domains where there are large databases containing valuable implicit regularities to be discovered; or (c) domains where programs must adapt to changing conditions. Not surprisingly, the field of cyber-based systems turns out to be a fertile ground where many security, reliability, performance, availability, and privacy tasks could be formulated as learning problems and approached in terms of learning algorithms.

This book deals with the subject of machine learning applications in the trust of cyber systems. It includes twelve chapters that are organized into four parts – cyber system, security, privacy, and reliability. Cyber-physical systems are a new and popular research area. In Part I, Chapter 1 introduces the motivation and basic concept of cyber-physical systems and reviews a sample of challenges related to real-time networked embedded systems.

In Part II, Chapter 2 describes how new vulnerabilities occur in security decisions using statistical machine learning. Particularly, authors demonstrate how three new attacks can make the filter unusable and prevent victims from receiving specific email messages. Chapter 3 presents a survey of various approaches that use machine learning/data mining techniques to enhance the traditional security mechanisms of databases. Two database security applications, namely, detection of SQL Injection attacks and anomaly detection for defending against insider threats are discussed. Chapter 4 describes an approach to detecting anomalies in a graph-based representation of the data collected during the monitoring of cyber and other infrastructures. The approach is evaluated using several synthetic and real-world datasets. Results show that the approach has high true-positive rates, low false-positive rates, and is capable of detecting complex structural anomalies in several real-world domains. Chapter 5 shows results from an empirical study of seven online-learning methods on the task of detecting malicious executables. Their study gives readers insights into the performance of online methods of machine learning on the task of detecting malicious executables. Chapter 6 proposes a novel network intrusion detection framework for mining and detecting sequential intrusion patterns is proposed. Experiments show promising results with high detection rates, low processing time, and low false alarm rates in mining and detecting sequential intrusion detections. Chapter 7 presents a solution for extending the capabilities of existing systems while simultaneously maintaining the stability of the current systems. It proposes an externalized survivability management scheme based on the observe-reason-modify paradigm and claims that their approach can be applied to a broad class of observable systems. Chapter 8 discusses an image encryption algorithm based on a chaotic cellular neural network to deal with information security and assurance. The comparison with the most recently reported chaos-based image encryption algorithms indicates that the algorithm proposed has a better security performance.

Over the decades, a variety of privacy threat models and privacy principles have been proposed and implemented. In Part III, Chapter 9 presents an overview of data privacy research by taking a close examination at the achievements with the objective of pinpointing individual research efforts on the grand map of data privacy protection. They also examine the research challenges and opportunities of location privacy protection. Chapter 10 presents an algorithm based on secure multiparty computation primitives to compute the nearest neighbors of records in horizontally distributed data. Authors show how this algorithm can be used in three important data mining algorithms, namely LOF outlier detection, SNN clustering, and kNN classification. They prove the security of these algorithms under the semi-honest adversarial model, and describe methods that can be used to optimize their performance.

Service-oriented architecture (SOA) techniques are being increasingly used for developing network-centric systems. In Part IV, Chapter 11 describes an approach for assessing the reliability of SOA-based systems using AI reasoning techniques. Memory-Based Reasoning technique and Bayesian Belief Networks are verified as the reasoning tools best suited to guide the prediction analysis. They also construct a framework from the above approach to identify the least tested and "high usage" input subdomains of the services. Chapter 12 aims for the models, properties, and applications of context-aware Web services by developing an ontology-based context model, and identifying context-aware applications as well as their properties. They developed an ontology-based context model to enable formal description and acquisition of contextual information pertaining to service requestors and services. They also report three context-aware applications built on top of their context model as a proof-of-concept to demonstrate how the context model can be used to enable and facilitate in finding right ser-vices, right partners and right information.

Finally, we would like to thank Melissa Fearon and Valerie Schofield of Spring for guidance of this project and Han C.W. Hsiao and Peter T.Y. Tsai of Asia University for formatting of the book.

Jeffrey J.P. Tsai
Philip S. Yu

Finally, we would like to thank Melissa Fearon and Valerie Schofield of Springer for guidance of this project and Han C.W. Hsiao and Peter T.Y. Tsai of Asia University for formatting of the book.

Jeffrey J.P. Tsai
Philip S. Yu

Contents

Part I: Cyber System

1 Cyber-Physical Systems: A New Frontier

Lui Sha[1], Sathish Gopalakrishnan[2], Xue Liu[3], and Qixin Wang[1]

Keywords: CPS, real time, robustness, QoS composition

Abstract: The report of the President's Council of Advisors on Science and Technology (PCAST) has placed cyber-physical systems on the top of the priority list for federal research investment in the United States of America in 2008. This article reviews some of the challenges and promises of cyber-physical systems.

1.1 Introduction

The Internet has made the world "flat" by transcending space. We can now interact with people and get useful information around the globe in a fraction of a second. The Internet has transformed how we conduct research, studies, business, services, and entertainment. However, there is still a serious gap between the cyber world, where information is exchanged and transformed, and the physical world in which we live. The emerging cyber-physical systems shall enable a modern grand vision for new societal-level services that transcend space and time at scales never possible before.

Two of the greatest challenges of our time are global warming coupled with energy shortage, and the rapid aging of the world's population with the related chronic diseases that threaten to bankrupt healthcare services, such as Medicare, or to dramatically cut back medical benefits.

During the meeting of the World Business Council for Sustainable Development in Beijing on March 29, 2006, George David[4] noted: *"More than 90*

[1] Lui Sha and Qixin Wang
University of Illinois at Urbana Champaign, lrs@cs.uiuc.edu, qwang4@uiuc.edu

[2] Sathish Gopalakrishnan
University of British Columbia, sathish@ece.ubc.ca

[3] Xue Liu
McGill University, xueliu@cs.mcgill.ca

[4] Chairman and CEO of United Technology Corporation.

J.J.P. Tsai and P.S. Yu (eds.), *Machine Learning in Cyber Trust: Security, Privacy, and Reliability*, DOI: 10.1007/978-0-387-88735-7_1,

percent of the energy coming out of the ground is wasted and doesn't end as useful. This is the measure of what's in front of us and why we should be excited."

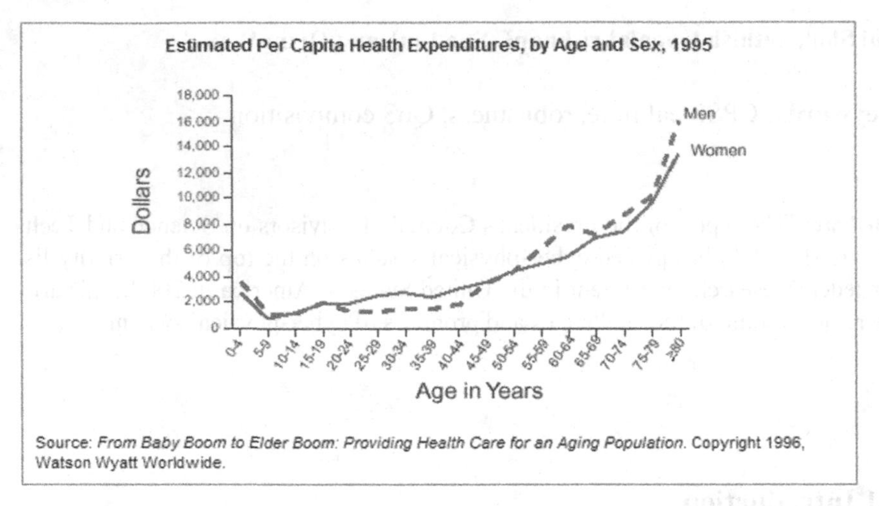

Fig. 1.1 Estimated per Capita Health Expenditures, by Age & Sex, 1995

Buildings and transportation are sectors with heavy energy consumption. During the National Science Foundation's Cyber-Enabled Discovery and Innovation Symposium (September 5-6, 2007) at Rensselaer Polytechnic Institute, Clas A. Jacobson[5] noted that green buildings hold great promises. Approximately 3.3 trillion KWh of energy is used in lighting and temperature control for buildings. Technologically, we can achieve net zero energy buildings, where 60-70% efficiency gains required for reducing demand and balance to be supplied by renewable. To reach the goal of net zero energy buildings we must, however, tightly integrate the cyber world and the physical world. Jacobson noted that in the past the science of computation has systematically abstracted away the physical world and vice versa. It is time to construct a hybrid systems science that is simultaneously computational and physical, providing us with a unified framework for robust CPS co-designs with integrated wired and wireless networking for managing the flows of mass, energy, and information in a coherent way.

According to the Department of Energy, the transportation share of the United States' energy use reached 28.4% in 2006, which is the highest share recorded since 1970[6]. In the United States, passenger and cargo airline operations alone required 19.6 billion gallons of jet fuel in 2006. According to

[5] Chief Scientist, United Technology Research Center

[6] http://cta.ornl.gov/data/new_for_edition26.shtml

Time[7], 88% of all trips in the U.S. are by car. Work related needs including daily work commute and business travel is a significant fraction of the transportation cost. Telepresence research seeks to make all interactions seem local rather than remote. It is one of the three grand challenges of in multimedia research[8] to make interactions with remote people and environments as if interactions with local people and environments. Integrating wired and wireless networks with real-time, interactive, immersive three-dimensional environments and teleoperation hold the promise to greatly reduce travel.

We now turn to the subject of healthcare. The rapidly aging population with age related chronic diseases is another formidable societal challenge. It is alarming to note that the growth of per-capita health cost has been near exponential with an increase in the age of the population. According to the CDC[9], more than 90 million Americans live with chronic illnesses.

- Chronic diseases account for 70% of all deaths in the United States.
- The medical care costs of people with chronic diseases account for more than 75% of the nation's $1.4 trillion medical care costs.
- Chronic diseases account for one-third of the years of potential life lost before age 65.

Advanced cyber physical technology may greatly improve the heath of an aging population. For example, stem-cell biotechnology holds the promise of treatment for many age-related diseases. According to the National Institute of Health[10], "stem cells, directed to differentiate into specific cell types, offer the possibility of a continuous source of replacement cells and tissues to treat diseases including Parkinson's and Alzheimer's diseases, spinal cord injury, stroke, burns, heart disease, diabetes, osteoarthritis, and rheumatoid arthritis." In addition, "Human stem cells could also be used to test new drugs. For example, new medications could be tested for safety on differentiated cells generated from human pluripotent cell lines."

However, much of this potential is not tapped, largely due to insufficient knowledge of the complex and dynamic stem-cell microenvironment, also known as the niche. There is a need to mimic niche conditions precisely in artificial environments to correctly regulate stem cells ex vivo. Indeed, the sensing and the control of the stem cell microenvironment are at the frontier of stem cell research. According to Badri Roysam[11] the stem cell niche has a

---

7 http://www.time.com/time/specials/2007/environment/article/0,28804,1602354_1603074_1603122,00.html

8 http://delivery.acm.org/10.1145/1050000/1047938/p3-rowe.pdf?key1=1047938&key2=8175939811&coll=GUIDE&dl=GUIDE&CFID=15151515&CFTOKEN=6184618

9 http://www.cdc.gov/nccdphp/overview.htm#2

10 http://stemcells.nih.gov/info/basics/basics6.asp

11 Professor, ECSE & Biomedical Engineering, RPI and Associate Director, NSF ERC Center for Subsurface Sensing & Imaging Systems

complex multi-cellular architecture that has many parameters, including multiple cell types related by lineage, preferred spatial locations and orientations of cells relative to blood vessels, soluble factors, insoluble factors related to the extra-cellular matrix, bio-electrical factors, biomechanical factors, and geometrical factors. The combinatorial space of parameter optimization and niche environment control calls are grand challenges in embedded sensing, control and actuation.

A closely related problem is providing care to the elderly population without sending them to expensive nursing homes. In the United States alone, the number of people over age 65 is expected to reach 70 million by 2030, doubling from 35 million in 2000. Expenditure in the United States for health-care will grow to 15.9% of the GDP ($2.6 trillion) by 2010. Unless the cost of healthcare for the elderly can be significantly reduced, financially stressed Social Security and Medicare/Medicaid systems will likely lead to burdensome tax increases and/or benefit reductions

The need for assistance in physical mobility causes many elders moving into nursing homes. Another key factor is cognitive impairment that requires daily supervision of medication and health-condition monitoring. When future CPS infrastructure supports telepresence, persons with one or more minor mobility and/or cognitive impairments can regain their freedom to stay at home. In addition, physiological parameters critical to the medical maintenance of health can be monitored remotely. When the elderly can maintain their independence without loss of privacy, a major financial saving in senior care will result. Furthermore, the elderly will be much happier by living independently at home while staying in contact with friends and family.

To realize the potential of CPS, research needs to be conducted at many frontiers and at different layers of the system of systems to support the coming convergence of the physical and cyber worlds.

1.2 The Challenges of Cyber-Physical System Research

The complex interactions and dynamics of how the cyber and physical subsystem interact with each others form the core of CPS research. The subjects are simultaneously broad and deep. The rest of this paper reviews a sample of challenges related to real-time networked embedded systems.

1.2.1 Real-time System Abstractions

Future distributed sensors, actuators, and mobile devices with both deterministic and stochastic data traffic require a new paradigm for real-time resource management that goes far beyond traditional methods. The interconnection topology of mobile devices is dynamic. The system infrastructure can also be dynamically reconfigured in order to contain system disruptions or optimize system performance. There is a need for novel distributed real-time computing and real-time group communication methods for dynamic topology control in wireless CPS systems with mobile components with dynamic topology control. Understanding and eventually controlling the impact of reconfigurable topologies on real-time performance, safety, security, and robustness will have tremendous impact in distributed CPS system architecture design and control.

Existing hardware design and programming abstractions for computing are largely built on the premise that the principal task of a computer is data transformation. Yet cyber-physical systems are real-time systems. This requires a critical re-examination of existing hardware and software architectures that have been developed over the last several decades. There are opportunities for fundamental research that have the potential to define the landscape of computation and coordination in the cyber-physical world. When computation interacts with the physical world, we need to explicitly deal with events distributed in space and time. Timing and spatial information need to be explicitly captured into programming models. Other physical and logical properties such as physical laws, safety, or power constraints, resources, robustness, and security characteristics should be captured in a composable manner in programming abstractions. Such programming abstractions may necessitate a dramatic rethinking of the traditional split between programming languages and operating systems. Similar changes are required at the software/hardware level given performance, flexibility, and power tradeoffs.

We also need strong real-time concurrent programming abstractions. Such abstractions should be built upon a model of simultaneity: bands in which the system delays are much smaller than the time constant of the physical phenomenon of interest at each band. The programming abstractions that are needed should also capture the ability of software artifacts to execute at multiple capability levels. This is motivated by the need for software components to migrate within a cyber-physical system and execute on devices with different capabilities. Software designers should be capable of expressing the reduced functionality of a software component when it executes on a device with limited resources.

The programming abstractions that we envision will need support at the middleware and operating system layers for:

- Real-time event triggers, both synchronous and asynchronous,

- Isolation and protections in resource sharing for applications with different criticality levels,
- Consistent views of distributed states in real-time within the sphere of influence. This challenge is especially great in mobile devices,
- Topology control and "dynamic real-time groups" in the form of packaged service classes of bounded delay, jitter and loss under precisely specified conditions,
- Interface to access to the same type of controls regardless of the underlying network technology.

1.2.2 Robustness of Cyber-Physical Systems

Uncertainty in the environment, security attacks, and errors in physical devices and in wireless communication pose a critical challenge to ensure overall system robustness, security and safety. Unfortunately, it is also one of the least understood challenges in cyber-physical systems. There is a clear intellectual opportunity in laying the scientific foundations for robustness, security and safety of cyber-physical systems. An immediate aim should be to establish a prototypical CPS model challenge problems and to establish a set of useful and coherent metrics that capture uncertainty, errors, faults, failures and security attacks.

We have long accepted that perfect physical devices are rare. An example is the design of reliable communication protocols that use an inherently error-prone medium, whether wired or wireless. We have also made great advancement in hardware reliability. Random hardware faults can be effectively masked using a combination of innovative circuit design, redundancy and restarts. However, sub-micron scaling of semiconductor devices and increasing complexity in the design of multi-core microprocessors will raise many new challenges. Challenging intermittent errors – that last several milliseconds to a few seconds – may not be uncommon in future generation chip multiprocessors [9].

These trends will, however, make our current efforts of building perfect software even more difficult. Indeed, there has been great advancement in automated theorem proving and model checking in recent years. However, it is important to remember that cyber-physical systems are real-time systems and the complexity of verifying temporal logic specifications is exponential. That is, like the physical counterpart, perfect software components are also rare and will remain that way. This has profound implications. We need to invent a cyber-physical system architecture in which the safety critical services of large and complex CPS can be guaranteed by a small subset of modules and fault tolerant protocols governing their interactions with the rest of the systems [16]. The design of this subset will have to be formally specified and verified. Their assumptions about the physical environments should be fully validated. Furthermore, we need to develop integrated static analysis and testing technologies to ensure that 1) the software code is compliant with the design, and that 2) the assumptions regarding external environment are sound. Finally, cyber-physical systems are deeply embedded and they will evolve in place. The verification and validation of cyber-physical system is not a one-time event; it needs to be a life cycle process that produces an explicit body of evidence for the certification of safety critical services. We also have the great challenge of how to handle residual errors, and security gaps in many useful, but not safety critical components that have not been fully verified and validated.

In physical systems, it is the theory of feedback control that provides the very foundation to achieve robustness and stability despite uncertainty in the environment and errors in sensing and control. The current open loop architecture in software systems may allow a minor error to cascade into system failure. The loops must be closed across both the cyber world and physical world. The system must have the capability to effectively counter-act uncertainties, faults, failures and attacks. The recent development of formal specification based automatic generation of system behavior monitoring, the steering of computation trajectories, and the use of analytically redundant modules based on different principles, while still in infancy, is an encouraging development.

Safety, robustness and security of the composed CPS also require explicit and machine checkable assumptions regarding external environments; formally specified and verifiable reduced complexity critical services and reduced complexity interaction involving safety critical and non-safety critical components; and analytically redundant sensing and control subsystems based on different physical principles and/or algorithms so as to avoid common mode failures due to faults or attacks. We also need theory and tools to design and ensure well-formed dependency relations between components with different criticality as they share resources and interact. By "well-formed dependency", we mean that a less critical component may depend on the service of critical components but not vice versa. The challenge of ensuring well-formed dependency arises when a critical component needs to use but cannot depend on less critical component. For example, after major surgery, a patient is allowed to "operate" an infusion pump with potentially lethal painkillers (patient controlled analgesia (PCA)). When pain is severe, the patient can push a button to get more pain-relieving medication. This is an example of a safety critical device controlled by an error-prone operator (the patient). Nevertheless, the PCA system, as a whole, needs to be certifiably safe despite mistakes made by the patient. Similar challenges are found in modern avionics, where the auto-pilot is certified to be DO 178B Level A (most critical) while the flight guidance system are unable to be certified higher than Level C, owning to its complexity. Nevertheless, the Level A autopilot is guided by the Level C flight guidance system. While solutions for these specific problems are known, the development of a general theory of composing multi-level criticality components remains a challenge.

1.2.3 System QoS Composition Challenge

Large CPS systems will have many different QoS properties encompassing stability, robustness, schedulability, security, each of which will employ a different set of protocols and will need to be analyzed using a different theory. It is important to note that these protocols may not be orthogonal and, sometimes, could have pathological interactions; for example, the well-known problem of unbounded priority inversion when we use synchronization protocols and real-time priority assignments as is [17]. There are also many incidents related to the adverse interactions between certain security, real-time and fault tolerant protocols. Thus, the theory of system composition must address not only the composability at each QoS dimension but also the question of how protocols interact.

The *science* of system composition has clearly emerged as one of the grand themes driving many of our research questions in networking and distributed systems. By *system composition* we mean that the QoS properties and functional correctness of the system can be derived from the architectural structure, subsystem interaction protocols, and the local QoS properties and functional properties of constituent components.

1.2.4 Knowledge Engineering in Cyber-Physical Systems

Knowledge engineering plays an important role in cyber-physical systems. We need a unified framework to represent the myriad types of data and application contexts in different physical domains, and interpret them under the appropriate contexts. For example, under a medical procedure, the real-time monitoring and interpretation of a patient's vital signs depend on the context consisting of a patient's medical history, current condition, medications given, the stage of surgery, and the expected responses. Indeed, stream data mining has already emerged as an important frontier of research [18]. In the coming cyber-physical systems, machine learning and real-time stream data mining will deal with more distributed, dynamic, heterogeneous information sources, including data streams from sensors, distributed events taking place at both physical and cyber-worlds, traditional databases, user inputs, and even records written in natural languages.

1.3 Medical Device Network: An Example Cyber-Physical System

The next generation medical system is envisioned as a ubiquitous system of wired and wireless networked medical devices and medical information systems for a secured, reliable, and privacy-preserving health care. It will be a networked system that improves the quality of life. For example, during a surgical operation, context information such as sensitivity to certain drugs will be automatically routed to relevant devices (such as infusion pumps) to support personalized care and safety management. A patient's reactions – changes in vital signs – to medication and surgical procedures will be correlated with streams of imaging data; streams will be selected and displayed, in the appropriate format and in real-time, to medical personnel according to their needs, e.g., surgeons, nurses, anesthetists and anesthesiologists. During particularly difficult stages of a rare surgical operation, an expert surgeon can remotely carry out key steps using remote displays and robot-assisted surgical machines, sparing the surgeon of the need to fly across the country to perform, say, a fifteen-minute procedure. Furthermore, data recording will be integrated with storage management such that surgeons can review operations and key findings for longitudinal studies for the efficacy of drugs and operational procedures.

While networked medical devices hold many promises, they also raise many challenges. First, from operating rooms to enterprise systems, different devices and subnets have different levels of clinical criticality. Data streams with different time sensitivities and criticality levels may share many hardware and software infrastructure resources. How to maintain safety in an integrated system is a major challenge that consists of many research issues. Indeed, many medical devices are safety critical and must be certified. Thus, it is important to develop a standard-based, certifiable wired and wireless networked medical devices infrastructure to lower the cost of development, approval, and deployment of new technologies/devices. The development of technologies that can formally specify both the application context and the device behaviors is a major challenge for the vision of certifiable plug and play medical devices in the future.

When monitoring devices are being moved from wired networks to wireless networks. It will be a challenge to provide an on-demand, reliable real-time streaming of critical medical information in a wireless network. For example, when an EKG device detects potentially dangerous and abnormal heartbeats, it is of critical importance to ensure that not only the warning but also the real-time EKG streams are reliably displayed at nursing stations. Furthermore, reliable on-demand real-time streaming must coexist with other wireless devices. For example, in an intensive care unit, there are 802.11 wireless networks, cellular phones, wireless PDAs, RFID, two-way radios and other RF emitting devices. This necessitates a network infrastructure to reliably integrate myriad wireless devices, to let them coexist safely, reliably and efficiently. To address these concerns, the FDA has issued an official guideline for medical wireless network development [3].

To design an integrated wired and wireless medical device network, we face all the aforementioned QoS composition challenges. For example, how does one monitor and enforce safe, secure, reliable and real-time sharing of various resources, in particular the wireless spectrum? How does one balance the resources dedicated to reliability, real-time performance and the need for coexistence? What is the programming paradigm and system composition architecture to support safe and secured medical device plug and play [8]?

1.4 Summary

The convergence of computing and networking gave us the Internet, which has profoundly transformed how we conduct research, studies, business, services, and entertainment. The coming convergence of computing, networking and the intelligent sensing and control of the physical world gives us cyber-physical systems that will transform our world again.

Cyber-physical systems hold the promise to help address the great challenges we are facing today: global warming coupled with energy shortage and the aging of the population. As a result, it is a grand challenge that involves not only the computing community but also many of the engineering communities. In this paper, we reviewed a sample of the challenges related to real-time networked and embedded systems.

Acknowledgments Most of the material presented here originated from discussions, presentations, and working group documents from NSF workshops on Real-time GENI and from NSF workshops on Cyber-Physical Systems [1][2]. The authors thank all the workshop participants for their insightful contributions.

References

[1]. Real-time GENI report. http://www.geni.net/GDD/GDD-06-32.pdf

[2]. NSF Workshops on Cyber Physical Systems. http://varma.ece.cmu.edu/cps/

[3]. FDA, Draft Guidance for Industry and FDA Staff – Radio-Frequency Wireless Technology in Medical Devices, Jan. 2007.
http://www.fda.gov:80/cdrh/osel/guidance/1618.html

[4]. Mu Sun, Qixin Wang, and Lui Sha, "Building Reliable MD PnP Systems", Proceedings of the Joint Workshop on High Confidence Medical Devices, Software, and Systems and Medical Device Plug-and-Play Interoperability, Jun. 2007.
[5]. Qixin Wang, et al., "I-Living: An open system architecture for assisted living," Proceedings of the IEEE International Conference on Systems, Man and Cybernetics, Oct. 2006, pp. 4268-4275.
[6]. http://ostp.gov/pdf/nitrd_review.pdf
[7]. Insup Lee, et al., High-confidence medical device software and systems. http://ieeexplore.ieee.org/Xplore/login.jsp?url=/iel5/2/33950/01620992.pdf
[8]. http://www.mdpnp.org/Home_Page.html
[9]. Phillip M. Wells, Koushik Chakraborty, and Gurindar S. Sohi, "Adapting to intermittent faults in future multicore systems," Proceedings of the International Conference on Parallel Architectures and Compilation Techniques, Sept. 2007.
[10]. Jaideep Vaidya and Chris Clifton. "Privacy-preserving data mining", IEEE Security & Privacy Magazine, Vol. 2, No. 6, Nov.-Dec. 2004, pp. 19 – 26.
[11]. Jane W.-S. Liu, et al., "Imprecise computations", Proceedings of the IEEE, Vol. 82, No. 1, Jan. 1994, pp. 83 – 94.
[12]. "Health informatics – Point-of-care medical device communication – Part 10101: Nomenclature," ISO/IEEE 11073-10101, First Edition, Dec. 15, 2004.
[13]. Jungkeun Yoon, Brian D. Noble, Mingyan Liu, Minkyong Kim. "Building realistic mobility models from coarse-grained traces, " In Proceedings of the 4th Annual ACM/USENIX Conference on Mobile Systems Applications, and Services. June 2006.
[14]. Vassilios S. Verykios, Elisa Bertino, Igor Nai Fovino, Loredana Parasiliti Provenza, Yucel Saygin, Yannis Theodoridis, "State-of-the-Art in Privacy Preserving Data Mining," ACM SIGMOD Record, vol. 3, no. 1, pp. 50-57, Mar. 2004.
[15]. Wenbo He, Xue Liu, Hoang Nguyen, Klara Nahrstedt, Tarek Abdelzaher, "PDA: Privacy-preserving Data Aggregation in Wireless Sensor Networks," in Proceedings of the 26th Annual IEEE Conference on Computer Communications (INFOCOM 2007), Anchorage, Alaska, 2007.
[16]. Lui Sha, "Using Simplicity to Control Complexity," *IEEE Software*, Vol. 18, No. 4, pp. 20-28, July/August 2001.
[17]. Lui Sha, Ragunathan Rajkumar, John Lehoczky, "Priority Inheritance Protocols: An Approach to Real-Time Synchronization," *IEEE Transaction on Computers*, Vol. 39, No. 9, pp. 1175-1185, September 1990.
[18]. http://domino.watson.ibm.com/comm/research.nsf/pages/r.kdd.innovation.html

Part II: Security

2 Misleading Learners: Co-opting Your Spam Filter

Blaine Nelson, Marco Barreno, Fuching Jack Chi, Anthony D. Joseph, Benjamin I. P. Rubinstein, Udam Saini, Charles Sutton, J. D. Tygar, Kai Xia[1]

Abstract Using statistical machine learning for making security decisions introduces new vulnerabilities in large scale systems. We show how an adversary can exploit statistical machine learning, as used in the SpamBayes spam filter, to render it useless—even if the adversary's access is limited to only 1% of the spam training messages. We demonstrate three new attacks that successfully make the filter unusable, prevent victims from receiving specific email messages, and cause spam emails to arrive in the victim's inbox.

2.1 Introduction

Applications use statistical machine learning to perform a growing number of critical tasks in virtually all areas of computing. The key strength of machine learning is adaptability; however, this can become a weakness when an adversary manipulates the learner's environment. With the continual growth of malicious activity and electronic crime, the increasingly broad adoption of learning makes assessing the vulnerability of learning systems to attack an essential problem.

The question of robust decision making in systems that rely on machine learning is of interest in its own right. But for security practitioners, it is especially important, as a wide swath of security-sensitive applications build on machine learning technology, including intrusion detection systems, virus and worm detection systems, and spam filters [13, 14, 18, 20, 24].

Past machine learning research has often proceeded under the assumption that learning systems are provided with training data drawn from a natural distribution of inputs. However, in many real applications an attacker might have the ability to provide a machine learning system with maliciously chosen inputs that cause the system to infer poor classification rules. In the spam domain, for example, the adversary can send carefully crafted spam messages

[1] Comp. Sci. Div., Soda Hall #1776, University of California, Berkeley, 94720-1776, USA

J.J.P. Tsai and P.S. Yu (eds.), *Machine Learning in Cyber Trust: Security, Privacy, and Reliability*, DOI: 10.1007/978-0-387-88735-7_2,

that a human user will correctly identify and mark as spam, but which can influence the underlying machine learning system and adversely affect its ability to correctly classify future messages.

We demonstrate how attackers can exploit machine learning to subvert the SpamBayes statistical spam filter. Our attack strategies exhibit two key differences from previous work: traditional attacks modify attack instances to evade a filter, whereas our attacks interfere with the training process of the learning algorithm and *modify the filter itself*; and rather than focusing only on placing spam emails in the victim's inbox, we also present attacks that *remove legitimate emails* from the inbox.

We consider attackers with one of two goals: expose the victim to an advertisement or prevent the victim from seeing a legitimate message. Potential revenue gain for a spammer drives the first goal, while the second goal is motivated, for example, by an organization competing for a contract that wants to prevent competing bids from reaching their intended recipient.

An attacker may have detailed knowledge of a specific email the victim is likely to receive in the future, or the attacker may know particular words or general information about the victim's word distribution. In many cases, the attacker may know nothing beyond which language the emails are likely to use.

When an attacker wants the victim to see spam emails, a broad *dictionary attack* can render the spam filter unusable, causing the victim to disable the filter (Section 2.3.1.1). With more information about the email distribution, the attacker can select a smaller dictionary of high-value features that are still effective. When an attacker wants to prevent a victim from seeing particular emails and has some information about those emails, the attacker can target them with a *focused attack* (Section 2.3.1.2). Furthermore, if an attacker can send email messages that the user will train as non-spam, a *pseudospam attack* can cause the filter to accept spam messages into the user's inbox (Section 2.3.2).

We demonstrate the potency of these attacks and present a potential defense—the *Reject On Negative Impact (RONI) defense* tests the impact of each email on training and doesn't train on messages that have a large negative impact. We show that this defense is effective in preventing some attacks from succeeding.

Our attacks target the learning algorithm used by several spam filters, including SpamBayes (spambayes.sourceforge.net), a similar spam filter called BogoFilter (bogofilter.sourceforge.net), the spam filter in Mozilla's Thunderbird (mozilla.org), and the machine learning component of SpamAssassin (spamassassin.apache.org)—the primary difference between the learning elements of these three filters is in their tokenization methods. We target SpamBayes because it uses a pure machine learning method, it is familiar to the academic community [17], and it is popular with over 700,000 downloads. Although we specifically attack SpamBayes, the widespread use of its statistical learning algorithm suggests that other filters are also vulnerable to similar attacks[2].

Our experimental results confirm that this class of attacks presents a serious concern for statistical spam filters. A dictionary attack makes the spam filter unusable when controlling just 1% of the messages in the training set, and a well-informed focused attack removes the target email from the victim's inbox over 90% of the time. Our pseudospam attack causes the victim to see almost 90% of the target spam messages with control of less than 10% of the training data.

We explore the effect of the *contamination assumption*: the adversary can control some of the user's training data. Novel contributions of our research include:

- A detailed presentation of specific, successful attacks against SpamBayes.
- A discussion of how these attacks fit into a more general framework of attacks against machine learning systems.
- Experimental results showing that our attacks are effective in a realistic setting.
- A potential defense that succeeds empirically against the dictionary attack.

Below, we discuss the background of the training model (Section 2.2); we present three new attacks on SpamBayes (Section 2.3); we give experimental results (Section 2.4); and we propose a defense against these attacks together with further experimental results (Section 2.5).

A preliminary report on this work appeared in the First USENIX Workshop on Large-Scale Exploits and Emergent Threats (LEET) [19].

2.2 Background

SpamBayes counts occurrences of tokens in spam and non-spam emails and learns which tokens are more indicative of each class. To predict whether a new email is spam or not, SpamBayes uses a statistical test to determine whether the email's tokens are sufficiently indicative of one class or the other, and returns its decision or *unsure*. In this section, we detail the statistical method SpamBayes uses to learn token scores and combine them in predic-

---

[2] We note that while some filters, such as SpamAssassin, use the learner only as one of several components of a broader filtering strategy, our attacks would still degrade the performance of SpamAssassin. Since other components of SpamAssassin are fixed rules, the only element that is trained is the learner. For SpamAssassin, our attacks will degrade the performance of this element in their system and thereby diminish its overall accuracy.

tion, but first we discuss realistic models for deploying SpamBayes and our assumption of adversarial control.

2.2.1 Training Model

SpamBayes produces a *classifier* from a *training set* of labeled examples of spam and non-spam messages. This classifier (or *filter*) is subsequently used to label future email messages as *spam* (bad, unsolicited email) or *ham* (good, legitimate email). SpamBayes also has a third label—when it isn't confident one way or the other, it returns *unsure*. We adopt this terminology: the true class of an email can be ham or spam, and a classifier produces the labels *ham*, *spam*, and *unsure*.

There are three natural choices for how to treat *unsure*-labeled messages: they can be placed in the spam folder, they can be left in the user's inbox, or they can be put into a third folder for separate review. Each choice can be problematic because the *unsure* label is likely to appear on both ham and spam messages. If *unsure* messages are placed in the spam folder, the user must sift through all spam periodically or risk missing legitimate messages. If they remain in the inbox, the user will encounter an increased amount of spam messages in their inbox. If they have their own "Unsure" folder, the user still must sift through an increased number of *unsure*-labeled spam messages to locate *unsure*-labeled ham messages. Too much *unsure* email is therefore almost as troublesome as too many false positives (ham labeled as *spam*) or false negatives (spam labeled as *ham*). In the extreme case, if every email is labeled *unsure* then the user must sift through every spam email to find the ham emails and thus obtains no advantage from using the filter.

Consider an organization that uses SpamBayes to filter incoming email for multiple users and periodically retrains on all received email, or an individual who uses SpamBayes as a personal email filter and regularly retrains it with the latest spam and ham. These scenarios serve as our canonical usage examples. We use the terms *user* and *victim* interchangeably for either the organization or individual who is the target of the attack; the meaning will be clear from context.

We assume that the user retrains SpamBayes periodically (*e.g.*, weekly); updating the filter in this way is necessary to keep up with changing trends in the statistical characteristics of both legitimate and spam email. Our attacks are not limited to any particular retraining process; they only require an assumption that we call the *contamination assumption*.

2.2.2 The Contamination Assumption

We assume that the attacker can send emails that the victim will use for training—the *contamination assumption*—but incorporate two significant restrictions: 1) attackers may specify arbitrary email bodies but cannot alter email headers; and 2) attack emails will always be trained as spam, not ham. In our pseudospam attack, however, we investigate the consequences of lifting the second restriction and allowing the attacker to have messages trained as ham.

It is common practice in security research to assume the attacker has as much power as possible, since a determined adversary may find unanticipated methods of attack—if a vulnerability exists, we assume it may be exploited. It is clear that in some cases the attacker can control training data. Here, we discuss realistic scenarios where the contamination assumption is justified; in the later sections, we examine its implications.

Adaptive spam filters must be retrained periodically to cope with the changing nature of both ham and spam. Many users simply train on all email received, using all *spam*-labeled messages as spam training data and all *ham*-labeled messages as ham training data. Generally the user will manually provide true labels for messages labeled *unsure* by the filter, as well as for messages filtered incorrectly as *ham* (false negatives) or *spam* (false positives). In this case, it is trivial for the attacker to control training data: any emails sent to the user are used in training.

The fact that users may manually label emails does not protect against our attacks: the attack messages are unsolicited emails from unknown sources and may contain normal spam marketing content. The *spam* labels manually given to attack emails are correct and yet allow the attack to proceed. When the attack emails can be trained as ham, a different attack is possible; our pseudospam attack explores the case where attack emails are trained as ham (see Section 2.3.2).

2.2.3 SpamBayes Learning Method

SpamBayes is a content-based spam filter that classifies messages based on the tokens (including header tokens) observed in an email. The spam classification model used by SpamBayes comes from Robinson [17, 22], based on ideas by Graham [8] together with Fisher's method for combining independent significance tests [7]. Intuitively, SpamBayes learns how strongly each token indicates *ham* or *spam* by counting the number of each type of email that token appears in. When classifying a new email, SpamBayes looks at all of its tokens and uses a statistical test to decide whether they indicate one label or the other with sufficient confidence; if not, SpamBayes returns *unsure*.

SpamBayes tokenizes each email E based on words, URL components, header elements, and other character sequences that appear in E. Each is treated as a unique token of the email. The SpamBayes algorithm only records whether or not a token occurs in the message, not how many times it occurs. Email E is represented as a binary vector **e** where

$$\mathbf{e}_i = \begin{cases} 1 & \text{the } i^{\text{th}} \text{ token occurs in } E \\ 0 & \text{otherwise} \end{cases} .$$

Below, we use **e** to refer to both the original message E and SpamBayes' representation of it since we are only concerned with the latter.

In training, SpamBayes computes a spam score vector $\mathbf{P}_{(S)}$ where the ith component is a *token spam score* for the ith token given by

$$\mathbf{P}_{(S,i)} = \frac{N_H N_S(i)}{N_H N_S(i) + N_S N_H(i)} \tag{1}$$

where N_S, N_H, $N_S(i)$, and $N_H(i)$ are the number of spam emails, ham emails, spam emails including the ith token and ham emails including the ith token, respectively. The quantity $\mathbf{P}_{(S,i)}$ is an estimate of $\Pr(E \text{ is spam} \mid \mathbf{e}_i)$ if the prior of ham and spam are equal, but for our purposes, it is simply a per-token score for the email. An analogous *token ham score* is given by $\mathbf{P}_{(H,i)} = 1 - \mathbf{P}_{(S,i)}$.

Robinson's method [22] smooths $\mathbf{P}_{(S,i)}$ through a convex combination with a prior belief x (default value of $x = 0.5$), weighting the quantities by $N(i)$ (the number of training emails with the ith token) and s (chosen for strength of prior with a default of $s = 1$), respectively:

$$\mathbf{f}_i = \frac{s}{s + N(i)} x + \frac{N(i)}{s + N(i)} \mathbf{P}_{(S,i)} . \tag{2}$$

Effectively, smoothing reduces the impact that low token counts have on the scores. For rare tokens, the score is dominated by the prior x. However, as more tokens are observed, the smoothed score gradually shifts to the score in Eq. (1). An analogous smoothed ham score is given by 1 - **f**.

After training, the filter computes the overall spam score $S(\mathbf{m})$ of a new message M using Fisher's method [7] for combining the scores of the tokens observed in M. SpamBayes uses at most 150 tokens from M with scores furthest from 0.5 and outside the interval [0.4,0.6]. Let $\delta(\mathbf{m})$ be the binary vector where $\delta(\mathbf{m})_i = 1$ if token i is one of these tokens, and 0 otherwise. The token spam scores are combined into a *message spam score* for M by

$$S(\mathbf{m}) = 1 - \chi^2_{2n}\left(-2(\log \mathbf{f})^{\mathrm{T}} \delta(\mathbf{m})\right) \quad , \tag{3}$$

where n is the number of tokens in M and $\chi^2_{2n}(\bullet)$ denotes the cumulative distribution function of the chi-square distribution with $2n$ degrees of freedom. A ham score $H(\mathbf{e})$ is similarly defined by replacing $\mathbf{f}$ with $1 - \mathbf{f}$ in Eq. (3). Finally, SpamBayes constructs an overall spam score for M by averaging $S(\mathbf{m})$ and $1 - H(\mathbf{m})$ (both being indicators of whether $\mathbf{m}$ is spam) giving the final score

$$I(\mathbf{m}) = \frac{1 + S(\mathbf{m}) - H(\mathbf{m})}{2} \tag{4}$$

for a message; a quantity between *0* (ham) and *1* (spam). SpamBayes predicts by thresholding $I(\mathbf{m})$ against two user-tunable thresholds θ_0 and θ_1, with defaults $\theta_0 = 0.15$ and $\theta_1 = 0.9$. SpamBayes predicts *ham*, *unsure*, or *spam* if $I(\mathbf{m})$ falls into the interval $[0, \theta_0]$, $(\theta_0, \theta_1]$, or $(\theta_1, 1]$, respectively, and filters the message accordingly.

The inclusion of an *unsure* label in addition to *spam* and *ham* prevents us from purely using *ham-as-spam* and *spam-as-ham* misclassification rates (false positives and false negatives, respectively) for evaluation. We must also consider *spam-as-unsure* and *ham-as-unsure* misclassifications. Because of the practical effects on the user's time and effort discussed in Section 2.2.1, *ham-as-unsure* misclassifications are nearly as bad for the user as *ham-as-spam*.

2.3 Attacks

We examine several attacks against the SpamBayes spam filter. Unlike attacks that exploit flaws in an application's implementation or policies, our attacks take advantage of the learner's adaptive nature. Each attack embodies a particular insight about ways in which a machine learning system is vulnerable, which we can better understand in terms of several commonalities between the attacks.

In a previous paper, we categorize attacks against machine learning systems along three axes [1]. The axes of the taxonomy are:

Influence

- *Causative* attacks influence learning with control over training data.
- *Exploratory* attacks exploit misclassifications but do not affect training.

Security violation

- *Integrity* attacks compromise assets via false negatives.
- *Availability* attacks cause denial of service, usually via false positives.

Specificity

- *Targeted* attacks focus on a particular instance.
- *Indiscriminate* attacks encompass a wide class of instances.

The first axis of the taxonomy describes the capability of the attacker: whether (a) the attacker has the ability to influence the training data that is used to construct the classifier (a *Causative* attack) or (b) the attacker does not influence the learned classifier, but can send new emails to the classifier, and observe its decisions on these emails (an *Exploratory* attack).

The second axis indicates the type of security violation caused: (a) false negatives, in which spam slip through the filter (an *Integrity* violation); or (b) false positives, in which ham emails are incorrectly filtered (an *Availability* violation).

The third axis refers to how specific the attacker's intention is: whether (a) the attack is *Targeted* to degrade the classifier's performance on one particular type of email or (b) the attack aims to cause the classifier to fail in an *Indiscriminate* fashion on a broad class of email.

In the remainder of this section, we discuss three novel *Causative* attacks against SpamBayes' learning algorithm in the context of this taxonomy: one is an *Indiscriminate Availability* attack, one is a *Targeted Availability* attack, and the third is a *Targeted Integrity* attack.

A *Causative* attack against a learning spam filter proceeds as follows:

1. The attacker determines the goal for the attack.
2. The attacker sends attack messages to include in the victim's training set.
3. The victim (re-)trains the spam filter, resulting in a tainted filter.
4. The filter's classification performance degrades on incoming messages.

We consider two possible goals for the attacker: to cause spam emails to be seen by the victim or to prevent the victim from seeing legitimate emails. There are at least two motives for the attacker to cause legitimate emails to be filtered as spam. First, a large number of misclassifications will make the spam filter unreliable, causing users to abandon filtering and see more spam. Second, causing legitimate messages to be mislabeled can cause users to miss important messages. For example, an organization competing for a contract could block competing bids by causing them to be filtered as spam, thereby gaining a competitive advantage.

Based on these considerations, we can further divide the attacker's goals into four categories:

1. Cause the victim to *disable* the spam filter, thus letting all spam into the inbox
2. Cause the victim to *miss* a particular ham email filtered away as *spam*
3. Get a *particular* spam into the victim's inbox
4. Get *any* spam into the victim's inbox

In the remainder of this section, we describe attacks that achieve these objectives. Each of the attacks we describe consists of inserting emails into the training set that are drawn from a particular distribution; the properties of these distributions, along with other parameters, determine the nature of the attack. The *dictionary attack* sends email messages with tokens drawn from a broad distribution, essentially including every token with equal probability. The *focused attack* focuses the distribution specifically on one message or a narrow class of messages. If the attacker has the additional ability to send messages that will be trained as ham, a *pseudospam attack* can cause spam messages to reach the user's inbox.

2.3.1 Causative Availability Attacks

We first focus on *Causative Availability* attacks, which manipulate the filter's training data to increase the number of ham messages misclassified. We consider both *Indiscriminate* and *Targeted* attacks. In *Indiscriminate* attacks, enough false positives force the victim to disable the filter or frequently search in *spam/unsure* folders for legitimate messages erroneously filtered away. Hence, the victim is forced to view more spam. In *Targeted* attacks, the attacker does not disable the filter but surreptitiously prevents the victim from receiving certain messages.

Without loss of generality, consider the construction of a single attack message $\mathbf{a}$. The victim adds it to the training set, (re-)trains on the contaminated data, and subsequently uses the tainted model to classify a new message $\mathbf{m}$. The attacker also has some (perhaps limited) knowledge of the next email the victim will receive. This knowledge can be represented as a distribution $\mathbf{p}$—the vector of probabilities that each token will appear in the next message.

The goal of the attacker is to choose the tokens for the attack message $\mathbf{a}$ to maximize the *expected spam score*:

$$\max_{\mathbf{a}} \mathbf{E}_{\mathbf{m}\sim\mathbf{p}}\left[I_{\mathbf{a}}(\mathbf{m})\right] \ . \tag{5}$$

In other words, the attack goal is to maximize the expectation of $I_{\mathbf{a}}(\mathbf{m})$ (Eq. (4) with the attack message **a** added to the spam training set) of the next legitimate email **m** drawn from distribution **p**.

To describe the optimal attack under this criterion, we make two observations, which we detail in Appendix 2.A. First, $I_{\mathbf{a}}(\mathbf{m})$ is monotonically non-decreasing in $\mathbf{f}_i$ for all i. Therefore, increasing the score of any token in the attack message can only increase $I_{\mathbf{a}}$. Second, the token scores of distinct tokens do not interact; that is, adding the ith token to the attack does not change the score $\mathbf{f}_j$ of some different token $j \neq i$. Hence, the attacker can simply choose which tokens will be most beneficial for their purpose. From this, we motivate two attacks, the *dictionary* and *focused* attacks, as instances of a common attack in which the attacker has different amounts of knowledge about the victim's email.

For this, let us consider specific choices for the distribution **p**. First, if the attacker has little knowledge about the tokens in target emails, we give equal probability to each token in **p**. In this case, we can optimize the expected message spam score by including *all possible tokens* in the attack email. Second, if the attacker has specific knowledge of a target email, we can represent this by setting $\mathbf{p}_i$ to 1 if and only if the ith token is in the target email. This attack is also optimal with respect to the target message, but it is much more compact.

In reality, the attacker's knowledge usually falls between these two extremes. If the attacker has relatively little knowledge, such as knowledge that the victim's primary language is English, the attack can include all words in an English dictionary. This reasoning yields the *dictionary attack* (Section 2.3.1.1). On the other hand, the attacker may know *some* of the particular words to appear in a target email, though not all of the words. This scenario is the *focused attack* (Section 2.3.1.2). Between these levels of knowledge, an attacker could use information about the distribution of words in English text to make the attack more efficient, such as characteristic vocabulary or jargon typical of emails the victim receives. Any of these cases result in a distribution **p** over tokens in the victim's email that is more specific than an equal distribution over all tokens but less informative than the true distribution of tokens in the next message. Below, we explore the details of the dictionary and focused attacks, with some exploration of using an additional corpus of common tokens to improve the dictionary attack.

2.3.1.1 Dictionary Attack

The *dictionary attack*, an *Indiscriminate* attack, makes the spam filter unusable by causing it to misclassify a significant portion of ham emails so that the victim disables the spam filter, or at least must frequently search through *spam/unsure* folders to find legitimate messages that were filtered away. In either case, the victim loses confidence in the filter and is forced to view more spam achieving the ultimate goal of the spammer: the victim sees the attacker's spam. The result of this attack is denial of service, a higher rate of ham misclassified as *spam*.

The dictionary attack is an approximation of the optimal attack suggested in Section 3.1, in which the attacker maximizes the expected score by including all possible tokens. Creating messages with every possible token is infeasible in practice. Nevertheless, when the attacker lacks knowledge about the victim's email, this optimal attack can be approximated with an entire dictionary of the victim's native language—the *dictionary attack*. The dictionary attack increases the score of every token in a dictionary; *i.e.*, it makes them more indicative of spam.

The central idea that underlies the dictionary attack is to send attack messages containing a large set of tokens—the attacker's *dictionary*. The dictionary is selected as the set of tokens whose scores maximally increase the expected value of $I_{\mathbf{a}}(\mathbf{m})$ as in Eq. (5). Since the score of a token typically increases when included in an attack message except in unusual circumstances (see Appendix 2.A.2), the attacker simply needs to include any tokens that are likely to occur in future legitimate message. In particular, if the victim's language is known by the attacker, they can use that language's entire lexicon (or at least a large subset) as the attack dictionary. After training on the dictionary message, the victim's spam filter will have a higher spam score for every token in the dictionary, an effect that is amplified for rare tokens. As a result, future legitimate email is more likely to be marked as *spam* since it will contain many tokens from that lexicon.

A refinement uses a token source with a distribution closer to the victim's email distribution. For example, a large pool of *Usenet* newsgroup postings may have colloquialisms, misspellings, and other "words" not found in a proper dictionary; furthermore, using the most frequent tokens in such a corpus may allow the attacker to send smaller emails without losing much effectiveness. However, there is an inherent trade-off in choosing tokens. Rare tokens are the most vulnerable to attack since their scores will shift more towards spam (1.0 in the SpamBayes formulas) with fewer attack emails. However, the rare vulnerable tokens also are less likely to appear in future messages, diluting their usefulness.

2.3.1.2 Focused Attack

Our second *Causative Availability* attack is a *Targeted* attack—the attacker has some knowledge of a specific legitimate email to target for filtering. If the attacker has exact knowledge of the target email, placing all of its tokens in attack emails produces an optimal attack. Realistically, the attacker has partial knowledge about the target email and can guess only some of its tokens to include in attack emails. We model this knowledge by letting the attacker know a certain fraction of tokens from the target email, which are included in the attack message. (The attack email may also include additional tokens added by the attacker to obfuscate the attack message's intent.) When SpamBayes trains on the resulting attack email, the spam scores of the targeted tokens increase, so the target message is more likely to be filtered as *spam*. This is the focused attack.

Consider an example in which the attacker sends spam emails to the victim with tokens such as the names of competing companies, their products, and their employees. The bid messages may even follow a common template known to the attacker, making the attack easier to craft. As a result of the attack, the legitimate bid email may be filtered away as *spam*, causing the victim not to see it.

The focused attack is more concise than the dictionary attack because the attacker has detailed knowledge of the target email and no reason to affect other messages. This conciseness makes the attack both more efficient for the attacker and more difficult to detect for the defender. Further, the focused attack can be more effective because the attacker may know proper nouns and other non-word tokens common in the victim's email that are otherwise uncommon.

An interesting side-effect of the focused attack is that repeatedly sending similar emails tends to not only increase the spam score of tokens in the attack but also *reduce* the spam score of tokens not in the attack. To understand why, recall the estimate of the token posterior in Eq. (1), and suppose that the ith token does not occur in the attack email. Then N_S increases with the addition of the attack email but $N_S(i)$ does not, so $\mathbf{P}_{(S,i)}$ decreases and therefore so does $\mathbf{f}_i$. In Section 2.4.3, we observe empirically that the focused attack can indeed reduce the spam score of tokens not included in the attack emails.

2.3.2 Causative Integrity Attacks—Pseudospam

We also examine *Causative Integrity* attacks, which manipulate the filter's training data to increase false negatives; that is, spam messages misclassified as *ham*. In contrast to the previous attacks, the *pseudospam attack* directly attempts to make the filter misclassify spam messages. If the attacker can choose messages arbitrarily that are trained as ham, the attack is similar to a focused attack with knowledge of 100% of the target email's tokens. However, there is no reason to believe a user would train on arbitrary messages as ham. We introduce the concept of a *pseudospam email*—an email that does not look like spam but that has characteristics (such as headers) that are typical of true spam emails. Not all users consider benign-looking, non-commercial emails offensive enough to mark them as spam.

To create pseudospam emails, we take the message body text from newspaper articles, journals, books, or a corpus of legitimate email. The idea is that in some cases, users may mistake these messages as ham for training, or may not be diligent about correcting false negatives before retraining, if the messages do not have marketing content. In this way, an attacker might be able to gain control of ham training data. This motivation is less compelling than the motivation for the dictionary and focused attacks, but in the cases where it applies, the headers in the pseudospam messages will gain significant weight indicating ham, so when future spam is sent with similar headers (*i.e.*, by the same spammer) it will arrive in the user's inbox.

2.4 Experiments

In this section, we present our empirical results. First, in Section 2.4.1, we outline our experimental setup. Then, we discuss the effect of each of our attacks in the remainder of the section.

2.4.1 Experimental Method

2.4.1.1 Datasets

In our experiments we use the Text Retrieval Conference (TREC) 2005 spam corpus [5], which is based on the Enron email corpus [12] and contains 92,189 emails (52,790 spam and 39,399 ham). This corpus has several strengths: it comes from a real-world source, it has a large number of emails, and its creators took care that the added spam does not have obvious artifacts to differentiate it from the ham.

We use two sources of tokens for attacks. First, we use the GNU Aspell English dictionary version 6.0-0, containing 98,568 words. We also use a corpus of English Usenet postings to generate tokens for our attacks. This corpus is a subset of a Usenet corpus of 140,179 postings compiled by the

University of Alberta's Westbury Lab [23]. An attacker can download such data and build a language model to use in attacks, and we explore how effective this technique is. We build a primary Usenet dictionary by taking the most frequent 90,000 tokens in the corpus (Usenet-90k), and we also experiment with a smaller dictionary of the most frequent 25,000 tokens (Usenet-25k).

The overlap between the Aspell dictionary and the most frequent 90,000 tokens in the Usenet corpus is approximately 26,800 tokens. The overlap between the Aspell dictionary and the TREC corpus is about 16,100 tokens, and the intersection of the TREC corpus and Usenet-90k is around 26,600 tokens.

2.4.1.2 Constructing Message Sets for Experiments

In constructing an experiment, we often need several non-repeating sequences of emails in the form of mailboxes. When we require a mailbox, we sample messages without replacement from the TREC corpus, stratifying our sampling to ensure the necessary proportions of ham and spam. For subsequent messages needed in any part of our experiments (target messages, headers for attack messages, and so on), we again sample emails without replacement from the messages remaining in the TREC corpus. In this way we ensure that no message is repeated in our experiments.

We construct attack messages by splicing elements of several emails together to make messages that are realistic under our model of the adversary's control. We construct our attack email bodies according to the specifications of the attack. We select the header for each attack email by choosing a random spam email from TREC and using its headers, taking care to ensure that the content-type and other Multipurpose Internet Mail Extensions (MIME) headers correctly reflect the composition of the attack message body. Specifically, we discard the entire existing multi- or single-part body and we set relevant headers (such as Content-Type and Content-Transfer-Encoding) to indicate a single plain-text body.

The tokens used in each attack message are selected from our datasets according to the attack method. For the dictionary attack, we use all tokens from the attack dictionary in every attack message (98,568 tokens for the Aspell dictionary and 90,000 or 25,000 tokens for the Usenet dictionary). For the focused and the pseudospam attacks, we select tokens for each attack message based on a fresh message sampled from the TREC dataset. The number of tokens in attack messages for the focused and pseudospam attacks varies, but all such messages are comparable in size to the messages in the TREC dataset.

Finally, to evaluate an attack, we create a control model by training SpamBayes once on the base training set. We incrementally add attack emails to the training set and train new models at each step, giving us a series of models tainted with increasing numbers of attack emails. (Because Spam-

Bayes is order-independent in its training, it arrives at the same model whether training on all messages in one batch or training incrementally on each email in any order.) We evaluate the performance of these models on a fresh set of test messages.

Parameter	Focused Attack	Pseudospam Attack	RONI defense
Training set size	2,000, 10,000	2,000, 10,000	2,000, 10,000
Test set size	200, 1,000	200, 1,000	200, 1,000
Spam percent	50, 75, 90	50, 75, 90	50
Attack percent	0.1, 0.5, 1, 2, 5, 10	0.1, 0.5, 1, 2, 5, 10	10
Validation folds	10	10	-
Target emails	20	-	-

Table 2.1 Parameters used in the experiments

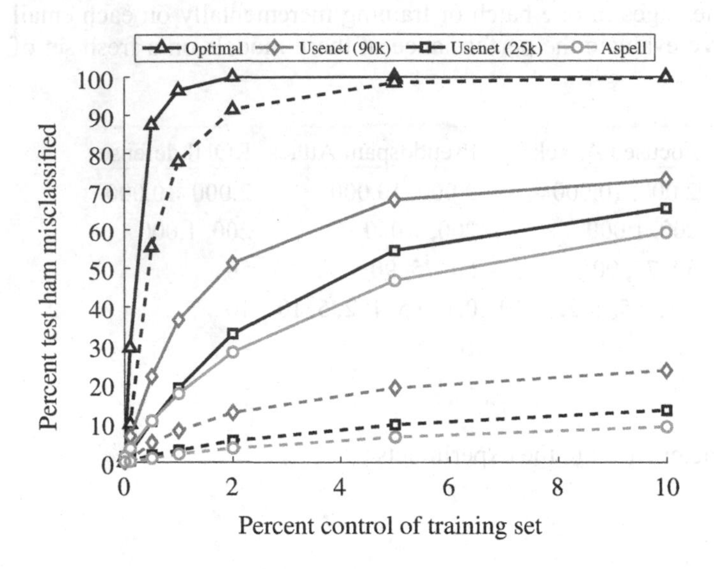

Fig. 2.1 Training on 10,000 messages (50% spam). Figures 2.1-2.4 show the effect of three dictionary attacks on SpamBayes in two settings. Figure 2.1 and Figure 2.2 have an initial training set of 10,000 messages (50% spam) while Figure 2.3 and Figure 2.4 have an initial training set of 2,000 messages (75% spam). Figure 2.2 and Figure 2.4 also depict the standard errors in our experiments for both of the settings. We plot percent of ham classified as *spam* (dashed lines) and as *spam* or *unsure* (solid lines) against the attack as percent of the training set. We show the optimal attack (Δ), the Usenet-90k dictionary attack (◊), the Usenet-25k dictionary attack (□), and the Aspell dictionary attack (○). Each attack renders the filter unusable with adversarial control over as little as 1% of the messages (101 messages).

2.4.1.3 Attack Assessment Method

We measure the effect of each attack by randomly choosing an inbox according to the parameters in Table 2.1 and comparing classification performance of the control and compromised filters using ten-fold cross-validation. In cross-validation, we partition the data into ten subsets and perform ten train-test epochs. During the kth epoch, the kth subset is set aside as a test set and the remaining nine subsets are combined into a training set. In this way, each email from the sample inbox functions independently as both training and test data.

In the following sections, we show the effect of our attacks on test sets of held-out messages. Because our dictionary and focused attacks are designed to cause ham to be misclassified, we only show their effect on ham messages; we found that their effect on spam is marginal. Likewise, for the pseudospam attack, we concentrate on the results for spam messages. Most of our graphs do not include error bars since we observed that the variation on our tests was small compared to the effect of our attacks (see Figure 2.2 and Figure 2.4). See Table 2.1 for our experimental parameters. We found that varying the size of the training set and spam prevalence in the training set had minimal impact on the performance of our attacks (for comparison, see Figure 2.1 and Figure 2.3), so we primarily present the results of 10,000-message training sets at 50% spam prevalence.

2.4.2 Dictionary Attack Results

We examine dictionary attacks as a function of the percent of attack messages in the training set. Figures 2.1-2.4 show the misclassification rates of three dictionary attack variants averaging over ten-fold cross-validation in two settings (Figure 2.1 and Figure 2.2 have an initial training set of 10,000 messages with 50% spam while Figure 2.3 and Figure 2.4 have an initial training set of 2,000 messages with 75% spam). First, we analyze the optimal dictionary attack discussed in Section 2.3.1 by simulating the effect of including every possible token in our attack emails. As shown in the figures, this optimal attack quickly causes the filter to mislabel all ham emails.

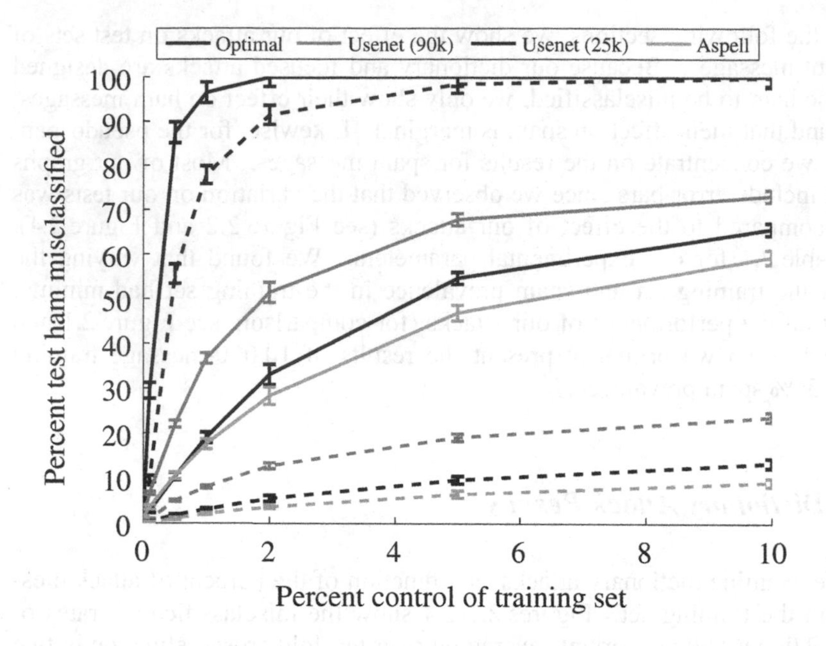

Fig. 2.2 Training on 10,000 messages (50% spam) with error bars. See Figure 2.1.

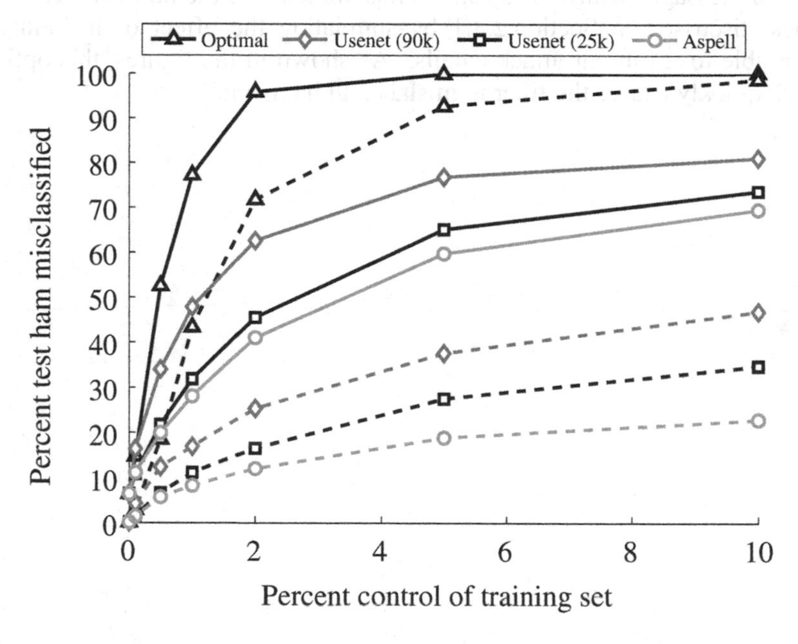

Fig. 2.3 Training on 2,000 messages (75% spam). See Figure 2.1.

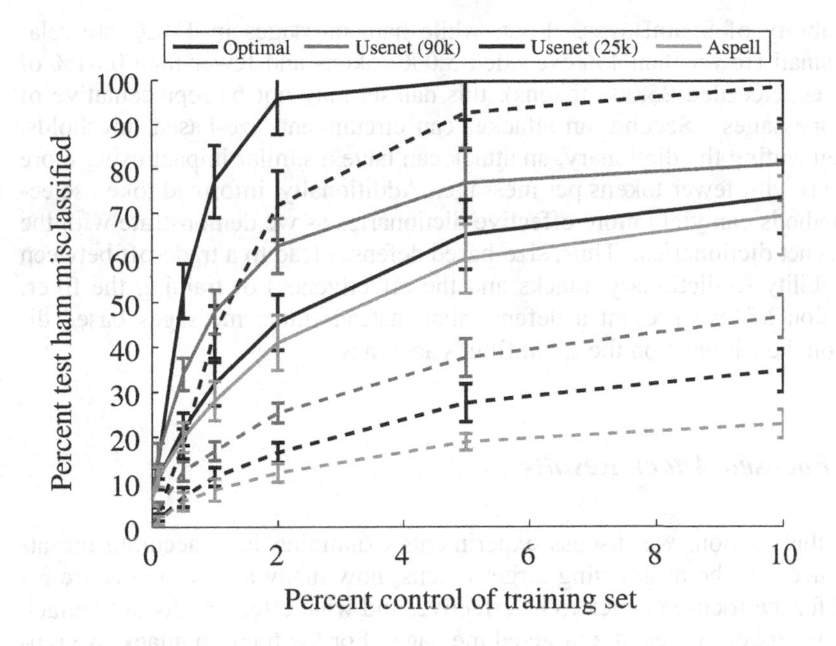

Fig. 2.4 Training on 2,000 messages (75% spam) with error bars. See Figure 2.1.

Dictionary attacks using tokens from the Aspell dictionary are also successful, though not as successful as the optimal attack. Both the Usenet-90k and Usenet-25k dictionary attacks cause more ham emails to be misclassified than the Aspell dictionary attack, since they contains common misspellings and slang terms that are not present in the Aspell dictionary. All of these variations of the attack require relatively few attack emails to significantly degrade SpamBayes' accuracy. After 101 attack emails (1% of 10,000), the accuracy of the filter falls significantly for each attack variation. Overall misclassification rates are 96% for optimal, 37% for Usenet-90k, 19% for Usenet-25k, and 18% for Aspell—at this point most users will gain no advantage from continued use of the filter.

It is of significant interest that so few attack messages can degrade a common filtering algorithm to such a degree. However, while the attack emails make up a small percentage of the *number of messages* in a contaminated inbox, they make up a large percentage of the *number of tokens*. For example, at 204 attack emails (2% of the training messages), the Usenet-25k attack uses approximately 1.8 times as many tokens as the entire pre-attack training dataset, and the Aspell attack includes 7 times as many tokens.

While it seems trivial to prevent dictionary attacks by filtering large messages out of the training set, such strategies fail to completely address this vulnerability of SpamBayes. First, while ham messages in TREC are relatively small (fewer than 1% exceeded 5,000 tokens and fewer than 0.01% of messages exceeded 25,000 tokens), this dataset may not be representative of actual messages. Second, an attacker can circumvent size-based thresholds. By fragmenting the dictionary, an attack can have a similar impact using more messages with fewer tokens per message. Additionally, informed token selection methods can yield more effective dictionaries as we demonstrate with the two Usenet dictionaries. Thus, size-based defenses lead to a trade-off between vulnerability to dictionary attacks and the effectiveness of training the filter. In Section 2.5, we present a defense that instead filters messages based directly on their impact on the spam filter's accuracy.

2.4.3 Focused Attack Results

In this section, we discuss experiments examining how accurate the attacker needs to be at guessing target tokens, how many attack emails are required for the focused attack to be effective, and what effect the focused attack has on the token scores of a targeted message. For the focused attack, we randomly select 20 ham emails from the TREC corpus to serve as the target emails before creating the clean training set. During each fold of cross-validation, we perform 20 focused attacks, one for each email, so our results average over 200 different trials.

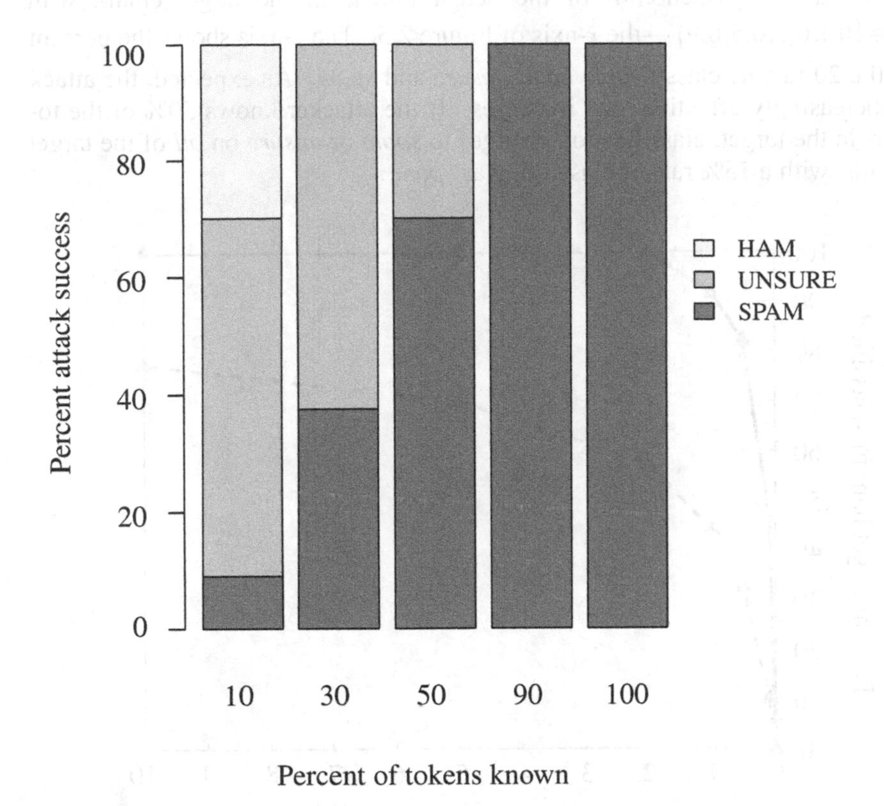

Fig. 2.5 Effect of the focused attack as a function of the percentage of target tokens known by the attacker. Each bar depicts the fraction of target emails classified as *spam*, *ham*, and *unsure* after the attack. The initial inbox contains 10,000 emails (50% spam).

These results differ from our preliminary focused attack experiments [19] in two important ways. First, here we randomly select a fixed percentage of tokens known by the attacker from each message instead of selecting each token with a fixed probability. The later approach causes the percentage of tokens known by the attacker to fluctuate from message to message. Second, here we only select messages with more than 100 tokens to use as target emails. With these changes, our results more accurately represent the behavior of a focused attack. Furthermore, in this more accurate setting, the focused attack is even more effective.

Figure 2.5 shows the effectiveness of the attack when the attacker has increasing knowledge of the target email by simulating the process of the attacker guessing tokens from the target email. We assume that the attacker knows a fixed fraction F of the actual tokens in the target email, with $F \in \{0.1, 0.3, 0.5, 0.9\}$ —the x-axis of Figure 2.5. The y-axis shows the percent of the 20 targets classified as *ham*, *unsure* and *spam*. As expected, the attack is increasingly effective as F increases. If the attacker knows 50% of the tokens in the target, classification changes to *spam* or *unsure* on *all* of the target emails, with a 75% rate of classifying as *spam*.

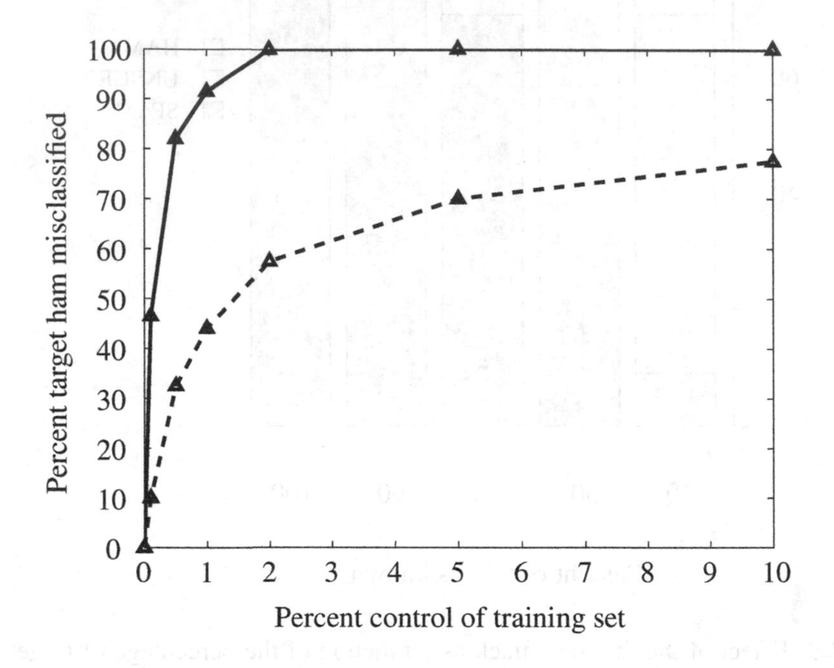

Fig. 2.6 Effect of the focused attack as a function of the number of attack emails with a fixed percentage (50%) of tokens known by the attacker. The dashed line shows the percentage of target ham messages classified as *spam* after the attack, and the solid line the percentage of targets that are *spam* or *unsure* after the attack. The initial inbox contains 10,000 emails (50% spam).

Figure 2.6 shows the attack's effect on misclassifications of the target emails as the number of attack messages increases. We fix the fraction of known tokens at 0.5. The x-axis shows the number of messages in the attack as a fraction of the training set, and the y-axis shows the fraction of target messages misclassified. With 101 attack emails inserted into an initial mailbox size of 10,000 (1%), the target email is misclassified as *spam* or *unsure* over 90% of the time.

Figures 2.7-2.9 show the attack's effect on three representative emails. Each of the figures represents a single target email from each of three attack results: ham misclassified as *spam* (Figure 2.7), ham misclassified as *unsure*

(Figure 2.8), and ham correctly classified as *ham* (Figure 2.9). Each point represents a token in the email. The x-axis is the token's spam score (from Eq. (2)) before the attack, and the y-axis is the token's score after the attack (0 means *ham* and 1 means *spam*). The ×'s are tokens included in the attack (known by the attacker) and the o's are tokens not in the attack. The histograms show the distribution of token scores before the attack (at bottom) and after the attack (at right).

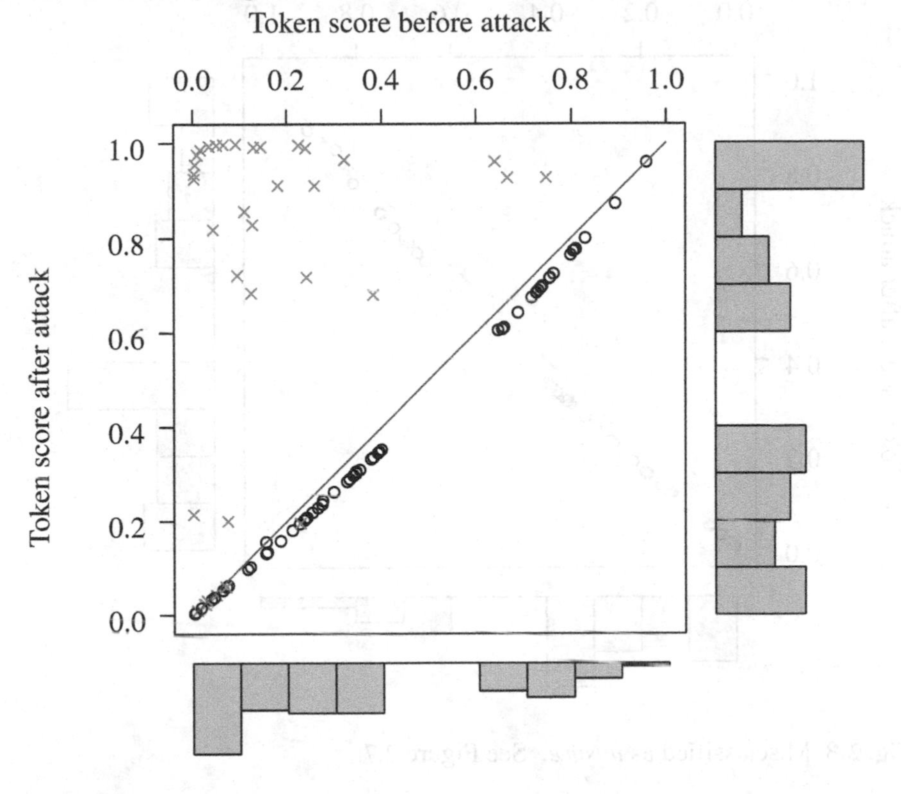

Fig. 2.7 Misclassified as *spam*. Figures 2.7-2.9 show the effect of the focused attack on three representative emails—one graph for each target. Each point is a token in the email. The x-axis is the token's spam score in Eq. (2) before the attack (0 means ham and 1 means spam). The y-axis is the token's spam score after the attack. The ×'s are tokens that were included in the attack and the o's are tokens that were not in the attack. The histograms show the distribution of spam scores before the attack (at bottom) and after the attack (at right).

Any point above the line $y=x$ is a token whose score increased due to the attack and any point below is a decrease. In these graphs we see that the score of the tokens included in the attack typically increase significantly while those not included decrease slightly. Since the increase in score is more significant for included tokens than the decrease in score for excluded tokens, the attack has substantial impact even when the attacker has a low probability of guessing tokens, as seen in Figure 2.5. Further, the before/after histograms in Figures 2.7-2.9 provide a direct indication of the attack's success. By shifting most token scores toward 1, we cause more misclassifications.

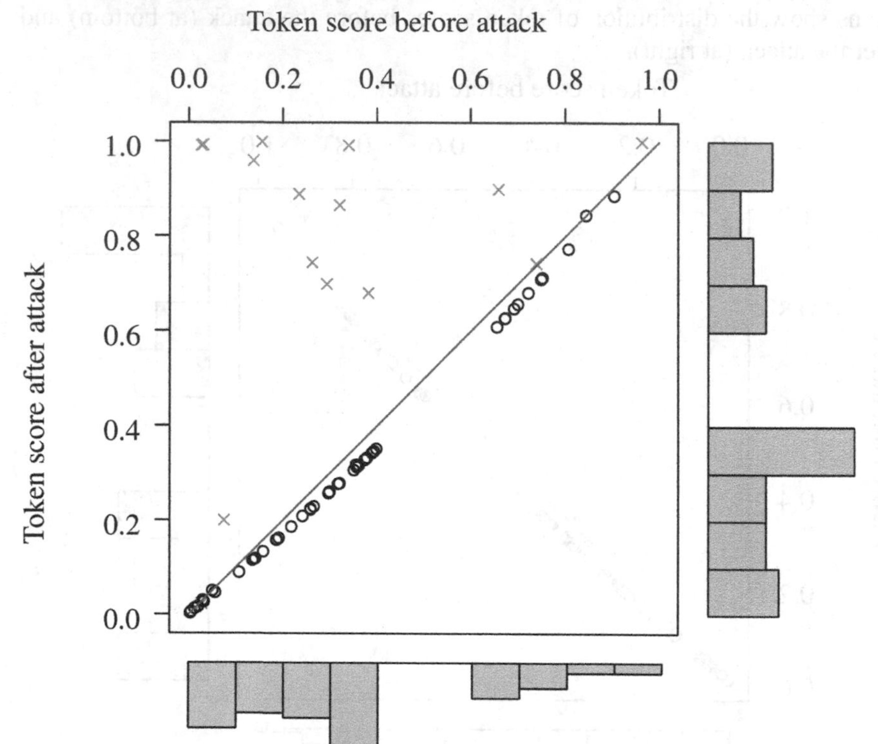

Fig. 2.8 Misclassified as *unsure*. See Figure 2.7.

2.4.4 Pseudospam Attack Experiments

In contrast to the previous attacks, for the pseudospam attack, we created attack emails that may be labeled as ham by a human as the emails are added into the training set. We setup the experiment for the pseudospam attack by first randomly selecting a target spam header to be used as the base header for the attack. We then create the set of attack emails that look similar to ham emails (see Section 2.3.2). To create attack messages, we combine each ham email with the target spam header. This is done so that the attack email has contents similar to other legitimate email messages. Header fields that may modify the interpretation of the body are taken from the ham email to make the attack realistic.

Figure 2.10 demonstrates the effectiveness of the *pseudospam attack*. We plot the percent of attack messages in the training set (x-axis) against the misclassification rates on the test spam email (y-axis). The solid line shows the fraction of target spam classified as *ham* or *unsure* spam while the dashed line shows the fraction of spam classified as *ham*. In the absence of attack, SpamBayes only misclassifies about 10% of the target spam emails (including those labeled *unsure*). If the attacker can insert a few hundred attack emails (1% of the training set), then SpamBayes misclassifies more than 80% of the target spam emails.

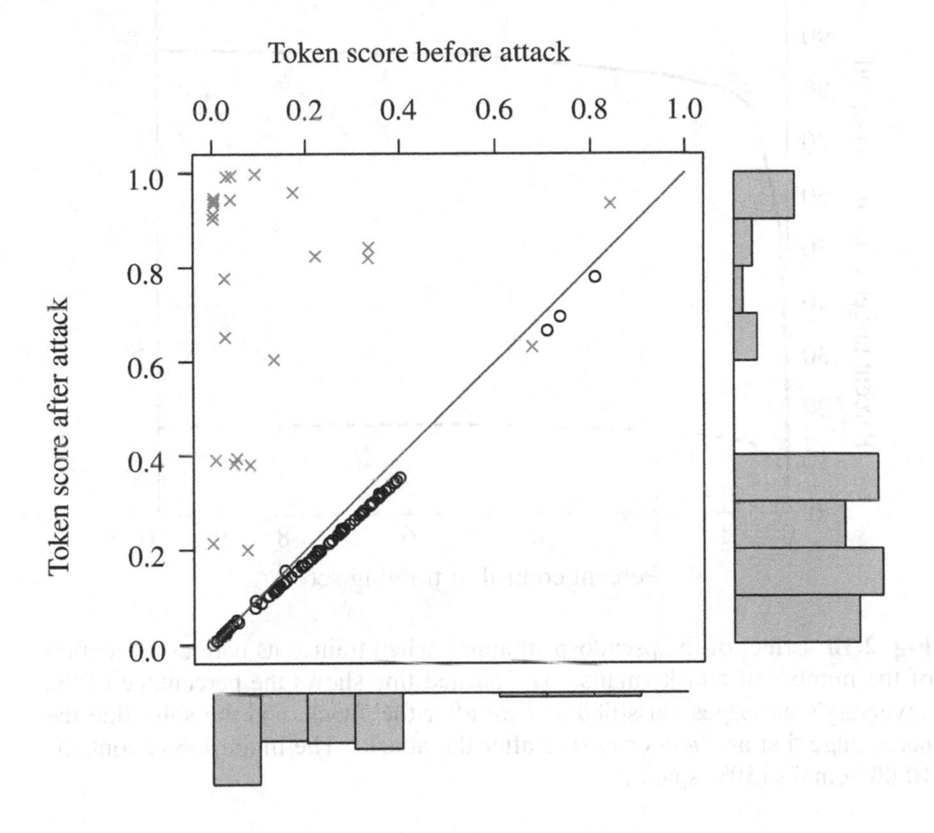

Fig. 2.9 Correctly classified as *ham*. See Figure 2.7.

Further, the attack has a minimal effect on regular ham and spam messages. Other spam email messages are still correctly classified since they do not generally have the same header fields as the adversary's messages. In fact, ham messages may have lower spam scores since they may contain tokens similar to those in the attack emails.

We also explore the scenario in which the pseudospam attack emails are labeled by the user as *spam* to better understand the effect of these attacks if the pseudospam messages fail to fool the user. The result is that, in general, SpamBayes classifies more spam messages incorrectly. As Figure 2.11 indicates, this variant causes an increase in spams mislabeled as either *unsure* or *ham* increases to nearly 15% as the number of attack emails increases. Further, this version of the attack does not cause a substantial impact on normal ham messages.

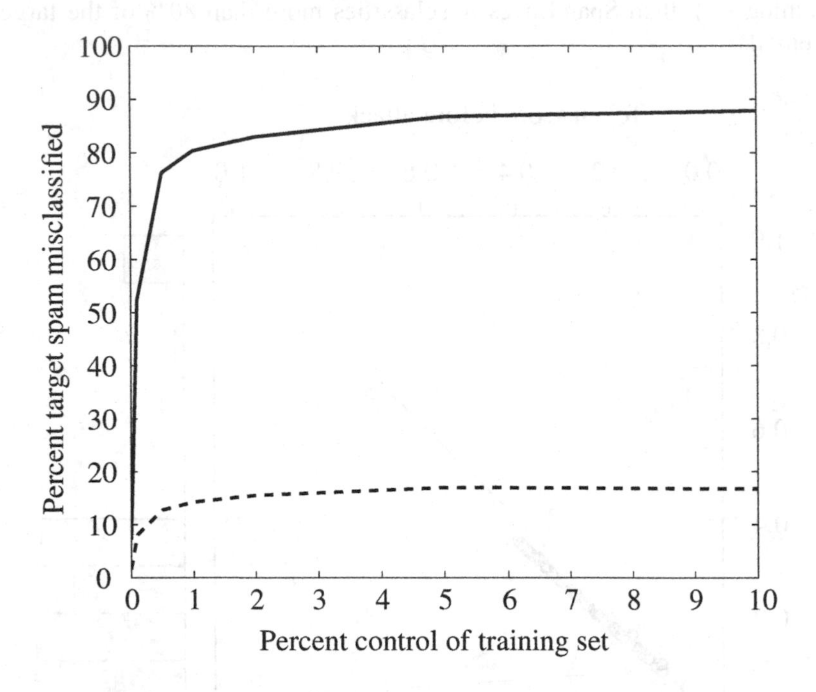

Fig. 2.10 Effect of the pseudospam attack when trained as ham as a function of the number of attack emails. The dashed line shows the percentage of the adversary's messages classified as *ham* after the attack, and the solid line the percentage that are *ham* or *unsure* after the attack. The initial inbox contains 10,000 emails (50% spam).

2.5 A Defense: Reject on Negative Impact (RONI)

Here we describe a potential defense against some of our attacks. Our *Causative* attacks succeed because training on attack emails causes the filter to learn incorrectly and misclassify emails. Each attack email contributes towards the degradation of the filter's performance; if we can measure each email's impact, then we can remove messages with a deleterious effect from the training set.

In the *Reject On Negative Impact* (*RONI*) *defense*, we measure the incremental effect of a *query email* Q by testing the performance difference with and without that email. We independently sample a 20-message training set T and a 50-message validation set V five times from the initial pool of emails given to SpamBayes for training, choosing a small training set so that the effect of a single email will be greater. We train on both T and $T \cup Q$ and measure the impact of Q as the average change in incorrect classifications on V over the five trials. We remove Q from the training pool if its impact is significantly harmful.

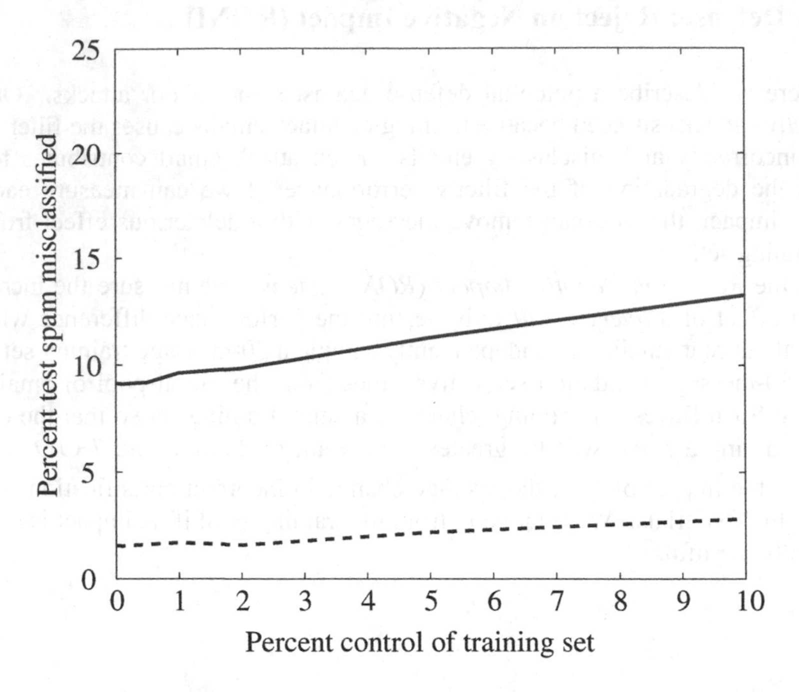

Fig. 2.11 Effect of the pseudospam attack when trained as spam, as a function of the number of attack emails. The dashed line shows the percentage of the normal spam messages classified as *ham* after the attack, and the solid line the percentage that are *unsure* after the attack. Surprisingly, training the attack emails as ham causes an increase in misclassification of normal spam messages. The initial inbox contains 10,000 emails (50% spam).

We test the effectiveness of this defense with 120 random non-attack spam messages and dictionary attack emails using both the Aspell and Usenet dictionaries. Our preliminary experiments show that the RONI defense is extremely successful against dictionary attacks, identifying 100% of the attack emails without flagging any non-attack emails. Each dictionary attack message causes an average decrease of *at least* 6.8 true negatives (ham-as-*ham* messages). In sharp contrast, non-attack spam messages cause *at most* an average decrease of 4.4 true negatives. Hence a simple threshold on this statistic is effective at separating dictionary attack emails from non-attack spam.

However, the RONI defense fails to detect focused attack emails because the focused attack targets a *future* message, so its effect on the training set is minute.

2.6 Related Work

Here we briefly review prior work related to the security of learning systems. A more detailed survey of this literature on appears in a related technical report [2].

Many authors have examined adversarial learning from a more theoretical perspective. For example, within the Probably Approximately Correct framework, Kearns and Li bound the classification error an adversary can cause with control over a fraction □ of the training set [10]. Dalvi et al. apply game theory to the classification problem [6]. They model the interactions between the classifier and attacker as a game and develop an optimal counter-strategy for an optimal classifier playing against an optimal opponent.

We focus on *Causative* attacks. Most existing attacks against content-based spam filters are *Exploratory* attacks that do not influence training but engineer spam messages so they pass through the filter. For example, Lowd and Meek explore reverse-engineering a spam classifier to find high-value messages that the filter does not block [15, 16], Karlberger et al. study the effect of replacing strong spam words with synonyms [9], and Wittel and Wu study the effect of adding common words to spam to get it through a spam filter [25].

Several others have recently developed *Causative* attacks against learning systems. Chung and Mok [3, 4] present a *Causative Availability* attack against the Autograph worm signature generation system [11], which infers blocking rules from the traffic of suspicious nodes. The main idea is that the attack node first sends traffic that causes Autograph to mark it suspicious, then sends traffic similar to legitimate traffic, resulting in rules that cause denial of service.

Newsome, Karp, and Song [21] present attacks against Polygraph [20], a polymorphic virus detector that uses machine learning. They primarily focus on conjunction learners, presenting *Causative Integrity* attacks that exploit certain weaknesses not present in other learning algorithms (such as that used by SpamBayes). They also suggest a *correlated outlier attack*, which attacks a naive-Bayes-like learner by adding spurious features to positive training instances, causing the filter to block benign traffic with those features (a *Causative Availability* attack). They speculate briefly about applying such an attack to spam filters; however, several of their assumptions about the learner are not appropriate in the case of SpamBayes, such as that the learner uses only features indicative of the positive class. Furthermore, although they present insightful analysis, they do not evaluate the correlated outlier attack against a real system. Our attacks use similar ideas, but we develop and test them on a real system. We also explore the value of information to an attacker, and we present and test a defense against the attacks.

2.7 Conclusion

Above, we show that an adversary can effectively disable the SpamBayes spam filter with relatively little system state information and relatively limited control over training data. Using the framework presented in Section 2.3, we have demonstrated the effectiveness of *Causative* attacks against SpamBayes if the adversary is given realistic control over the training process of SpamBayes, by limiting the header fields the attacker can modify and the amount of control over the training set. Our Usenet-25k dictionary attack causes misclassification of 19% of ham messages with only 1% control over the training messages[3], rendering SpamBayes unusable. Our focused attack changes the classification of the target message virtually 100% of the time with knowledge of only 30% of the target's tokens. Our pseudospam attack is able to cause almost 90% of the target spam messages to be labeled as either *unsure* or *ham* with control of less than 10% of the training data.

We also demonstrate a defense against some attacks. We explore the RONI defense, which filters out dictionary attack messages with complete success based on how a message impacts the performance of our classifier. Focused attacks are especially difficult; defending against an attacker with extra knowledge of future events remains an open problem.

Our attacks and defense should also work for other spam filtering systems based on similar learning algorithms, such as BogoFilter, Thunderbird's spam filter and the Bayesian component of SpamAssassin, although their effects may vary (and SpamAssassin uses more components than just the learning algorithm, so the effect of our attacks may be smaller). These techniques may also be effective against other learning systems, such as those used for worm or intrusion detection.

2.A Appendix: Analysis of an Optimal Attack

In this appendix, we justify our claims in Section 2.3.1 about optimal attacks.

[3] While the dictionary attack messages are larger than most typical email, we have demonstrated the effect of these attacks and that one can reduce the attack message size by using a more informed dictionary.

2.A.1 Properties of the Spam Score

The key to understanding heuristic attacks and constructing optimal attacks on SpamBayes is characterizing conditions under which the SpamBayes score $I(\mathbf{m})$ increases when the training corpus is injected with attack spam messages.

Lemma A.1 *The $I(\mathbf{m})$ score defined in Eq. (4) is non-decreasing in $\mathbf{f}_i$ for all i.*

Proof. We show that the derivative of $I(\mathbf{m})$ with respect to $\mathbf{f}_k$ is non-negative. By rewriting, Eq. (3) as $S(\mathbf{m}) = 1 - \chi^2_{2n}\left(-2\log \Pi_{i:\delta(\mathbf{m})_i=1}[\mathbf{f}_i]\right)$, we can use the chain rule. Let $x(\mathbf{f}) = \Pi_{i:\delta(\mathbf{m})_i=1}[\mathbf{f}_i]$ and we have

$$\frac{d}{dx}\left[1-\chi^2_{2n}\left(-2\log(x)\right)\right] = \frac{1}{(n-1)!}\left(-\log(x)\right)^{n-1} ,$$

which is non-negative since $0 \le x \le 1$. We have $\partial x(\mathbf{f})/\partial \mathbf{f}_k$ is $\Pi_{i\neq k:\delta(\mathbf{m})_i=1}[\mathbf{f}_i] \ge 0$ if $\delta(\mathbf{m})_k=1$ or 0 otherwise. Combining these derivatives, we have

$$\frac{\partial S(\mathbf{m})}{\partial \mathbf{f}_k} \ge 0 \quad .$$

By an analogous derivation, replacing $\mathbf{f}_i$ by $1 - \mathbf{f}_i$, we have

$$\frac{\partial H(\mathbf{m})}{\partial \mathbf{f}_k} \le 0 \quad .$$

Finally, we obtain the result

$$\frac{\partial I(\mathbf{m})}{\partial \mathbf{f}_k} = \frac{1}{2}\frac{\partial S(\mathbf{m})}{\partial \mathbf{f}_k} - \frac{1}{2}\frac{\partial H(\mathbf{m})}{\partial \mathbf{f}_k} \ge 0 \quad .$$

Remark A.2 *Given a fixed number of attack spam messages, $\mathbf{f}_j$ is independent of the number of those messages containing the ith token for all $j \neq i$.*

This remark follows from the fact that the inclusion of the ith token in attack spams affects $N_S(i)$ and $N(i)$ but not $N_H(i)$, N_S, N_H, $N_S(j)$, $N_H(j)$, $N(j)$ for all $j \neq i$ (see Eq. (1) and Eq. (2) in Section 2.2.3).

After an attack of a fixed number of spam messages, the score $I(\mathbf{m})$ of an incoming test message **m** can be maximized by maximizing each $\mathbf{f}_i$ separately. This motivates our attacks—intuitively, we increase the $\mathbf{f}_i$ of tokens appearing in **m**.

2.A.2 Effect of Poisoning on Token Scores

We have not yet established how email spam scores change as the result of an attack message in the training set. One might assume that the ith score $\mathbf{f}_i$ should increase when the ith token is added to the attack email. This would be the case, in fact, if the token score in Eq. (1) were computed according to Bayes' Rule. However, the score in Eq. (1) is derived by applying Bayes' Rule with an additional assumption that the prior of spam and ham is equal. As a result we show that $\mathbf{f}_i$ can be smaller if the ith token is included in the attack email.

We consider a single attack spam message, after which the counts become

$$\begin{aligned} N_S &\mapsto N_S + 1 \\ N_H &\mapsto N_H \\ N_S(i) &\mapsto \begin{cases} N_S(i) + 1 & \text{if } \mathbf{a}_i = 1 \\ N_S(i) & \text{otherwise} \end{cases} \\ N_H(i) &\mapsto N_H(i) \quad . \end{aligned}$$

Using these count transformations, we assess the effect on the smoothed SpamBayes score $\mathbf{f}_i$ of training on an attack spam message **a**. If the ith token is included in the attack (*i.e.*, $\mathbf{a}_i = 1$), then the new score for the ith token (from Eq. (1)) is

$$\mathbf{P}^{(1)}_{(S,i)} = \frac{N_H\left(N_S(i)+1\right)}{N_H\left(N_S(i)+1\right) + \left(N_S+1\right)N_H(i)} \quad .$$

If the token is not included in the attack (*i.e.*, $\mathbf{a}_i = 0$), then the new token score is

$$\mathbf{P}_{(S,i)}^{(0)} = \frac{N_H N_S(i)}{N_H N_S(i) + (N_S + 1) N_H(i)} \quad .$$

We use the notation $\mathbf{f}_i^{(1)}$ and $\mathbf{f}_i^{(0)}$ to denote the smoothed spam score after the attack depending on whether or not the ith token was used in the attack message.

We wish to analyze the quantity

$$\Delta\mathbf{f}_i = \mathbf{f}_i^{(1)} - \mathbf{f}_i^{(0)} \quad .$$

One might expect this difference to always be non-negative, but we show that in some scenarios $\Delta\mathbf{f}_i < 0$. After some algebra, we expand $\Delta\mathbf{f}_i$ as follows:

$$\begin{aligned}\Delta\mathbf{f}_i &= \frac{s}{(s + N(i) + 1)(s + N(i))}\left(\mathbf{P}_{(S,i)}^{(1)} - x\right) \\ &\quad + \frac{N_H N(i)}{(s + N(i))(N_H N_S(i) + (N_S + 1) N_H(i))}\mathbf{P}_{(H,i)}^{(1)} \\ &= \frac{1}{(s + N(i) + 1)(s + N(i))} \cdot \alpha(i) \quad ,\end{aligned}$$

where $\mathbf{P}_{(H,i)}^{(1)} = 1 - \mathbf{P}_{(S,i)}^{(1)}$ is the altered ham score of the ith token. The first factor in the above expression is positive and the second is defined as

$$\alpha(i) = s(1 - x) + P_{(H,i)}^{(1)} \frac{N_H N(i)(N(i) + 1) + s N_H N_H(i) - s(N_S + 1) N_H(i)}{N_H N_S(i) + (N_S + 1) N_H(i)} \quad .$$

When we combine these two terms, only the numerator can be negative. Focusing on conditions under which this can happen leads to the (weakest) conditions that $N_S(i) = 0$, $s > 0$, $N_H(i) > 0$ and

$$x > \frac{N_H(N_H(i) + s)(N_H(i) + 1)}{s(N_H(i)(N_S + 1) + N_H)} \quad .$$

If $N_S(i) > 0$, the resulting bound would be larger.

Under SpamBayes' default values of $s = 1$ and $x = 1/2$, we can rewrite the above condition as $N_S(i) = 0$, $N_H(i) > 0$, and

$$N_S > N_H\left[\frac{2\left(N_H\left(i\right)+1\right)^2-1}{N_H\left(i\right)}\right]-1\ .$$

Since $N_H\left(i\right)\geq 1$, we have the weakest possible condition for $\Delta\mathbf{f}_i$ to be negative under default settings: $N_S \geq 7N_H$, $N_H(i) = 1$, and $N_S(i) = 0$. Thus, when the number of spam, N_S, in the training set is sufficiently larger than the number of ham, N_H, it is possible that the score of a token will be lower if it is *included* in the attack message than if it were excluded. This is a direct result of the assumption made by SpamBayes that $N_S = N_H$. Such aberrations will occur most readily in tokens with low initial values of $N_H(i)$ and $N_S(i)$.

Acknowledgments We would like to thank Satish Rao and Carla Brodley for their useful comments and suggestions on this research. This work was supported in part by the Team for Research in Ubiquitous Secure Technology (TRUST), which receives support from the National Science Foundation (NSF award #CCF-0424422), the Air Force Office of Scientific Research (AFOSR #FA9550-06-1-0244), Cisco, British Telecom, ESCHER, Hewlett-Packard, IBM, iCAST, Intel, Microsoft, ORNL, Pirelli, Qualcomm, Sun, Symantec, Telecom Italia, and United Technologies; in part by California state Microelectronics Innovation and Computer Research Opportunities grants (MICRO ID#06-148 and #07-012) and Siemens; and in part by the cyber-DEfense Technology Experimental Research laboratory (DETERlab), which receives support from the Department of Homeland Security Homeland Security Advanced Research Projects Agency (HSARPA award #022412) and AFOSR (#FA9550-07-1-0501). We also gratefully acknowledge the support of the NSF through grants DMS-0434383 and DMS-0707060. The opinions expressed here are solely those of the authors and do not necessarily reflect the opinions of any funding agency, the State of California, or the U.S. government.

References

[1]. Barreno M, Nelson B, Sears R, Joseph AD, Tygar JD (2006) Can machine learning be secure? In: Proceedings of the ACM Symposium on InformAtion, Computer, and Communications Security (ASIACCS), pp 16-25

[2]. Barreno M, Nelson B, Joseph AD, Tygar JD (2008) The security of machine learning. Tech. Rep. UCB/EECS-2008-43, EECS Department, University of California, Berkeley, URL http://www.eecs.berkeley.edu/Pubs/TechRpts/2008/EECS-2008-43.html

[3]. Chung SP, Mok AK (2006) Allergy attack against automatic signature generation. In: Proceedings of the International Symposium on Recent Advances in Intrusion Detection (RAID), pp 61-80

[4]. Chung SP, Mok AK (2007) Advanced allergy attacks: Does a corpus really help? In: Proceedings of the International Symposium on Recent Advances in Intrusion Detection (RAID), pp 236-255

[5]. Cormack G, Lynam T (2005) Spam corpus creation for TREC. In: Proceedings of the Conference on Email and Anti-Spam (CEAS)

[6]. Dalvi N, Domingos P, Mausam, Sanghai S, Verma D (2004) Adversarial classification. In: Proceedings of the ACM SIGKDD International Conference on Knowledge Discovery and Data Mining, pp 99-108
[7]. Fisher RA (1948) Question 14: Combining independent tests of significance. American Statistician 2(5):30-30J
[8]. Graham P (2002) A plan for spam. http://www.paulgraham.com/spam.html
[9]. Karlberger C, Bayler G, Kruegel C, Kirda E (2007) Exploiting redundancy in natural language to penetrate Bayesian spam filters. In: Proceedings of the USENIX Workshop on Offensive Technologies (WOOT), pp 1-7
[10]. Kearns M, Li M (1993) Learning in the presence of malicious errors. SIAM Journal on Computing 22(4):807-837
[11]. Kim HA, Karp B (2004) Autograph: Toward automated, distributed worm signature detection. In: Proceedings of the USENIX Security Symposium, pp 271-286
[12]. Klimt B, Yang Y (2004) Introducing the Enron corpus. In: Proceedings of the Conference on Email and Anti-Spam (CEAS)
[13]. Lazarevic A, Ertöz L, Kumar V, Ozgur A, Srivastava J (2003) A comparative study of anomaly detection schemes in network intrusion detection. In: Barbará D, Kamath C (eds) Proceedings of the SIAM International Conference on Data Mining, pp 25-36
[14]. Liao Y, Vemuri VR (2002) Using text categorization techniques for intrusion detection. In: Proceedings of the USENIX Security Symposium, pp 51-59
[15]. Lowd D, Meek C (2005) Adversarial learning. In: Proceedings of the ACM SIGKDD International Conference on Knowledge Discovery and Data Mining, pp 641-647
[16]. Lowd D, Meek C (2005) Good word attacks on statistical spam filters. In: Proceedings of the Conference on Email and Anti-Spam (CEAS)
[17]. Meyer T, Whateley B (2004) SpamBayes: Effective open-source, Bayesian based, email classification system. In: Proceedings of the Conference on Email and Anti-Spam (CEAS)
[18]. Mukkamala S, Janoski G, Sung A (2002) Intrusion detection using neural networks and support vector machines. In: Proceedings of the International Joint Conference on Neural Networks (IJCNN), pp 1702-1707
[19]. Nelson B, Barreno M, Chi FJ, Joseph AD, Rubinstein BIP, Saini U, Sutton C, Tygar JD, Xia K (2008) Exploiting machine learning to subvert your spam filter. In: Proceedings of the USENIX Workshop on Large-Scale Exploits and Emergent Threats (LEET)
[20]. Newsome J, Karp B, Song D (2005) Polygraph: Automatically generating signatures for polymorphic worms. In: Proceedings of the IEEE Symposium on Security and Privacy, pp 226-241
[21]. Newsome J, Karp B, Song D (2006) Paragraph: Thwarting signature learning by training maliciously. In: Proceedings of the International Symposium on Recent Advances in Intrusion Detection (RAID 2006), pp 81-105
[22]. Robinson G (2003) A statistical approach to the spam problem. Linux Journal
[23]. Shaoul C, Westbury C (2007) A USENET corpus (2005-2007)
[24]. Stolfo SJ, Li WJ, Hershkop S, Wang K, Hu CW, Nimeskern O (2004) Detecting viral propagations using email behavior profiles. ACM Transactions on Internet Technology (TOIT) pp 187-221
[25]. Wittel GL, Wu SF (2004) On attacking statistical spam filters. In: Proceedings of the Conference on Email and Anti-Spam (CEAS)

3 Survey of Machine Learning Methods for Database Security

Ashish Kamra[1] and Elisa Bertino[2]

Abstract Application of machine learning techniques to database security is an emerging area of research. In this chapter, we present a survey of various approaches that use machine learning/data mining techniques to enhance the traditional security mechanisms of databases. There are two key database security areas in which these techniques have found applications, namely, detection of *SQL Injection* attacks and *anomaly detection* for defending against insider threats. Apart from the research prototypes and tools, various third-party commercial products are also available that provide database activity monitoring solutions by profiling database users and applications. We present a survey of such products. We end the chapter with a primer on mechanisms for responding to database anomalies.

3.1 Introduction

Machine learning (ML) techniques have traditionally found applications in the field of computer security for detection of anomalies in the working of an information system. The essential idea is simple and has been applied to different kinds of information systems such as networks, operating system, application software, and so forth[1]. The operation of an anomaly detection (AD) mechanism using ML techniques is typically divided into two phases, a training phase and a detection phase. In the training phase, the AD mechanism learns the *normal* behavior of the information system. The outcome of the training phase is set of *profiles* characterising the system behavior. The profiles are used to logically divide the system behavior in a manner that supports detection of anomalies in the detection phase. For example, the behavior of an

[1] Ashish Kamra Purdue University, West Lafayette IN - 47907 USA, akamra@purdue.edu

[2] Elisa Bertino Purdue University, West Lafayette IN - 47907 USA bertino@cs.purdue.edu

J.J.P. Tsai and P.S. Yu (eds.), *Machine Learning in Cyber Trust: Security, Privacy, and Reliability*, DOI: 10.1007/978-0-387-88735-7_3,

operating system can be modeled as a set of profiles where each profile is associated with the actions of a single user. Such model can then be used for detection of anomalous user actions. Moreover, if the number of users is large and many users exhibit similar behavior, a single profile can be associated with a group of users (a role, such as) that exhibit similar behavior. The information in the learned profile depends on the system events under observation, type of information gathered from these events, and on the learning algorithm used. After the profiles are created, the AD mechanism is operated in a detection mode. In detection mode, a new event under observation is checked against the learned profiles for *fitness*. Many different criterions such as statistical deviation tests, outlier detection tests, posterior probabilities, and so forth can be used to determine the fitness of an event to the learned profiles[1]. If the event does not fit the profiles (according to some threshold), it is flagged off as anomalous. Otherwise, the information from the event is either merged into the profiles or discarded. An AD system has an advantage over a signature-based system in that it can detect potentially novel attacks, while a signature-based system could only go as far as those attacks for which signatures have been created. The downside of an AD system is the number of false alarms it can raise if the profiles learned during the training phase do not fit the normal behavior well.

In this chapter, we present a survey of machine learning techniques for anomaly detection in a Database Management System (DBMS). We first motivate the problem of anomaly detection in context of a DBMS. Data represent today an important asset for companies and organizations. Some of these data are worth millions of dollars and organizations take great care at controlling access to these data, with respect to both internal users, within the organization, and external users, outside the organization. Data security is also crucial when addressing issues related to privacy of data pertaining to individuals; companies and organizations managing such data need to provide strong guarantees about the confidentiality of these data in order to comply with legal regulations and policies[2]. Overall, data security has a central role in the larger context of information systems security. Therefore, the development of DBMSs with high-assurance security (in all its flavors) is a central research issue. The development of such DBMS requires a revision of architectures and techniques adopted by traditional DBMS[3]. An important component of this new generation security-aware DBMS is an anomaly detection mechanism. Even though a DBMS provides access control mechanisms, these mechanisms alone are not enough to guarantee data security. They need to be complemented by suitable AD mechanisms; the use of such mechanisms is crucial for protecting against impersonation attacks and against malicious code embedded in application programs. Also AD mechanisms help in addressing the problem of insider threats, an increasingly important problem in today's organizations for which not many solutions have been devised. However, despite the fact that building AD systems for networks and operating systems has been an active area of research, few AD systems exist that are specifically tailored to a DBMS. There are two main reasons that motivate the necessity of AD systems for a DBMS. The first is that actions deemed malicious for a database application are not necessarily malicious for the network or the operating system; thus AD systems specifically designed for the latter would not be effective for database protection. The second, and more relevant motivation, is that AD systems designed for networks and operating systems are not adequate to protect databases against insider threats, which is an important issue when dealing with privacy. These threats are much more difficult to defend against, because they are from subjects that are legitimate users of the system, and thus may have access rights to data and resources.

We next present a general categorization of threats to databases and identify which threats can be addressed by anomaly detection techniques. The threats can be broadly categorized as follows[4]:

1. **Privilege Elevation.** Elevated privileges may be obtained by an attacker by: *Exploiting software vulnerabilities*, vulnerabilities in the database server code can be exploited to launch arbitrary programs; *SQL Injection*, Web applications and database stored procedures that do not sanitize user input can be exploited to gain elevated privileges.
2. **Privilege Abuse.** This is the insider threat scenario. In this case, a malicious user uses his/her privileges for performing actions that do not conform to its *expected* day-to-day behavior.

To defend against privilege elevation due to software vulnerabilities, the DBMS must be patched regularly[4]. The anomaly detection techniques discussed in this chapter address the problem of SQL injection attacks and the privilege abuse behavior. As is the case with a typical AD mechanism, the goal of all schemes discussed in this chapter is to build profiles representing normal behavior of database users/database applications. Any significant deviation from the normal profiles is then termed as malicious behavior. In what follows, we give some examples of anomalous actions that can be detected by profiling users and applications interacting with a DBMS:

1. User *U* has read access to table *T*. Normally on a daily basis, *U* accesses only a fraction of records in *T*. One day, *U* issues a query to read all the data from all columns of *T*.
2. A DBA logs in from an IP address from which he/she has never logged in before.
3. A backup DBA issues select queries on the application data tables.
4. A database application issues malformed queries that result in database errors on a regular basis. This may be an indication of the information gathering phase of a SQL injection attack.

5. A malicious employee writes code that reads all credit card numbers from a production database table and sends these sensitive data across to a remote development database.

3.1.1 Paper Road Map

The rest of the chapter is as follows. In Section 3.2 we present an overview of two approaches that use ML techniques to detect SQL injection attacks. In Section 3.3, we present a survey of various techniques for detecting the insider threat behavior in context of a DBMS. We present in detail the approach of Kamra et al. that detects anomalous access patterns in relational databases. We conclude in Section 3.4 with emerging research trends in this field.

3.2 Detection of SQL Injection Attacks

With the rapid advancement of information technologies in the recent past, the amount of data stored within an organization's databases has increased manifold. A lot of these data is sensitive, and critical to an organization's day to day operations. Therefore, database vendors have traditionally strived to strengthen the security mechanisms surrounding these databases. One such approach is to install the databases behind an internal firewall, and restrict access to them through only application programs. To access a database, users connect to one such application, and the application submits queries to the database on behalf of the users. The threat to a database arises when these application programs do not behave as intended. Such behavior may often depend on the input provided by users, as is the case of the well known *SQL Injection* attack. As an example of such an attack, consider the following scenario of a database application program that tries to authenticate a user. The application presents the user with an authentication form and the user submits his username and password to the application. The application, at the server side, collects this information, creates a SQL query and submits it to the organization's database in order to obtain the verification results. Consider the following application code fragment:

```
SQLQuery = "SELECT Username FROM Users WHERE Username = '" &
strUsername & "' AND Password = '" & strPassword & "'"
```

In this code, the application receives the input supplied by the user through a form, embeds it directly into a SQL query and submits it to the database. Everything is fine as long as the user provides just the username and password. However, suppose that an attacker submits a login and password that looks like the following:

Login : xyz
Password: abc OR ‘ ’ = ‘

Such input would result in the following SQL query:

SELECT Username FROM Users WHERE Username = ‘xyz’ AND Password = ‘abc’ OR ‘ ’ = ‘ ’

Instead of comparing the user-supplied data with entries in the Users table, the query compares ‘’ (empty string) to ‘’ (empty string). This will always return a true result, and the attacker will then be logged in as the first user returned by the query. The above is a classical example of a *SQL Injection* attack. Essentially, SQL Injection is an attack exploiting applications that construct SQL statements from user-supplied input. When an application fails to properly validate user-supplied input, it is possible for an attacker to alter the construction of backend SQL statements. Several threat scenarios may arise because of these altered SQL queries. As we saw from the above example, SQL Injection may allow an attacker to get unauthorized access to the database. Moreover, the modified SQL queries are executed with the privileges of the application program. An attacker may thus abuse the privileges of the program in a manner unintended by the application program designers. We direct the reader to [5] for an in-depth survey of SQL Injection attacks and countermeasures.

There have been two main approaches [6][7] proposed that make use of machine learning techniques to detect SQL Injection attacks. Both techniques model SQL Injection attacks as an anomaly detection problem. The main differences are in the proposed detection models and the tier at which the schemes operate. In what follows, we discuss these approaches in detail and present their pros and cons.

3.2.1 A Learning-based Approach to the Detection of SQL Attacks

Valeur et al. [6] have proposed an intrusion detection system capable of detecting a variety of SQL injection attacks. Their approach uses multiple statistical models to build profiles of normal access to the database. As with most learning-based anomaly detection techniques, the system requires a training phase prior to detection. The training phase itself is divided into two halves: In the first half, the data fed to the models is used to determine the model parameters. During this phase, the models learn what normal queries look like. The second half of the training phase is the *threshold learning* phase. In this phase, the system, instead of updating the model parameters, calculates an anomaly score based on how well the queries fit the trained models. For each model, the maximum anomaly score observed during this phase is stored and is used as a tunable threshold for flagging anomalous requests in the detection phase.

The system contains a parser that processes each input SQL query and returns a sequence of tokens. Each token has a flag that indicates whether the token is a constant or not. Constants are the only elements that should contain a user supplied input (which may be malicious). Each constant also has a data type attached to it that decides the statistical model to be applied to it. A feature vector is thus created by extracting all tokens marked as constants. A profile is then a collection of statistical models and a mapping that dictates which features are associated with which models. The name of the script generating the query and the skeleton query are used as keys to look up a profile. We now briefly describe the heart of the scheme, that is, the statistical models used to characterize the feature vectors. A detailed description of these statistical models can be found be in [8].

1. **String Length.** The string length model approximates the distribution of the length of the string values in the feature vectors. Features that deviate from the observed normal string length values are characterized as anomalous. The detection of deviation is based on the Chebyshev inequality [9].
2. **String Character Distribution.** The string character distribution model captures the concept of a normal string constant token by looking at its character distribution model. For each observed argument string, its character distribution is stored. The idealized character distribution (ICD) is then approximated by calculating the average of all stored character distributions. In the detection phase, given an ICD, the probability that the character distribution of a given argument is an actual sample drawn from its ICD is determined using a statistical test.
3. **String Structural Inference.** The goal of this model is to generate a grammar capable of producing at least all string values observed during the training phase (for a specific feature). The approach used for the structural inference model generation procedure is to generalize the grammar as much as possible but to stop before too much structural information is lost. The stopping criterion for generalization is specified with the help of Markov models and Bayesian probability. During the

detection phase it is checked whether a string can be generated by the resulting grammar. If not, it is flagged as anomalous.

Other statistical models used in this work are *String Prefix and Suffix Matcher*, and *Token Finder* [6].

One advantage of this scheme is that it uses multiple models to characterize the user input. Thus, the susceptibility of the detection mechanism to *mimicry attacks* [10] is reduced. A mimicry attack is one in which an attacker builds malicious inputs while mimicking the behavior of the model used by the detection mechanism. Using multiple models is one way to protect the detection mechanism against such attacks. A major concern with the scheme is its execution and storage overhead, since multiple statistical models are maintained for each pair of template query and application. In applications with large number of distinct template queries, the overhead due to the detection mechanism may be large.

3.2.2 Profiling Database Applications to Detect SQL Injection Attacks

Bertino et al. [7] have proposed a framework based on anomaly detection techniques to detect malicious behavior of database application programs. The approach is as follows. Profiles of application programs are created that can represent their normal behavior in terms of SQL queries submitted by these programs to the database. Query traces from database logs are used for this purpose. An anomaly detection model based on data mining techniques is then used to detect behavior deviating from normal. The anomalous behavior that is focused on in this work is behavior related to SQL Injection attacks. The authors argue that SQL injection, traditionally, has been considered as an application level vulnerability and solutions have been proposed at that level [5]. However, even though the mechanism of a SQL Injection attack is through an application, the resource the security of which is directly threatened is the database. Therefore, they propose a solution to enhance the traditional security mechanisms of a DBMS to protect against such attacks. The essential idea motivating their approach is that SQL injection can be modeled as an anomaly detection problem. In a typical database application, the input supplied by the users is used to construct the *where* clause of queries at run time. An SQL injection attack typically involves malicious modifications to this input either by adding additional clauses or by changing the structure of an existing clause. The *projection* clause of the query, however, is not modified and remains static because it is not constructed at run time. Driven by this observation, they use association rule mining techniques to derive rules that represent the associative relationships among the various query attributes. Two sets of rules specific to the task are derived. The first set consists of rules binding the *projection* attributes of the query to the attributes used in the *where* clause. The second set of rules represent the relationship among the attributes in the *where* clause of the query. The attributes include the constants in a where clause. These two sets of rules together form the profile of an application. To detect an attack query, it is checked if the relationship among the query (under consideration) attributes can be inferred by the set of rules in the application profile. If not, the system flags the query as an SQL injection query.

There are some fundamental differences between the two schemes. The first difference is the tier at which the systems operate. In the scheme of Bertino et al. [7], the detection mechanism is operated at the database tier with no hooks at the application side. This set-up is more difficult to defend against as no control over the application is assumed. But we believe that it is also a realistic scenario in many real-life DBMS installations. Valeur et al. [6], on the other hand, establish hooks at the application side that allow them to capture user inputs to the application program. The second difference is the way an SQL Injection attack is modeled. Valeur et al. learn models characterizing the user inputs, while Bertino et. al. [7], learn the query structure itself and stores the learned structure as association rules. Both schemes may suffer from high detection overhead in case of large number of distinct template queries. For the former, this will lead to a large number of statistical models to maintain, while for the latter the number of association rules to be maintained will be large.

3.3 Anomaly Detection for Defending Against Insider Threats

In this section, we first briefly review two approaches that use data mining techniques to detect anomalies in a DBMS [11][12]. We then review in detail the work on detecting anomalous database access patterns by Kamra et al. [13][14].

3.3.1 DEMIDS: A Misuse Detection System for Database Systems

DEMIDS is a misuse-detection system, tailored for relational database systems [11]. It uses audit log data to derive profiles describing typical patterns of accesses by database users. Essential to such an approach is the assumption that the access pattern of users typically forms a working scope which comprises sets of attributes that are usually referenced together with some values. The idea of working scopes is captured by mining frequent itemsets which are sets of features with certain values. Based on the data structures and integrity constraints encoded in the system catalogs and the user behavior recorded in the audit logs, DEMIDS describes distance measures that capture the closeness of a set of attributes with respect to the working scopes. These distance measures are then used to guide the search for frequent itemsets in the audit logs. Misuse of data, such as tampering with the data integrity, is detected by comparing the derived profiles against the organization's security policies or new audit information gathered about the users.

The goal of the DEMIDS system is two-fold. The first goal is detection of malicious insider behavior. Since a profile created by the DEMIDS system is based on frequent sets of attributes referenced by user queries, the approach is able to detect an event when a SQL query submitted by an insider does not conform to the attributes in the user profile. The second goal is to serve as a tool for security re-engineering of an organization. The profiles derived in the training stage can help to refine/verify existing security policies or create new policies. The main drawback of the approach presented as in [11] is a lack of implementation and experimentation. The approach has only been described theoretically, and no empirical evidence has been presented of its performance as a detection mechanism.

3.3.2 A Data Mining Approach for Database Intrusion Detection

Hu et al. [12] propose an approach for identifying malicious transactions from the database logs. They propose mechanisms for finding data dependency relationships among transactions and use this information to find hidden anomalies in the database log. The rationale of their approach is the following: if a data item is updated, this update does not happen alone but is accompanied by a set of other events that are also logged in the database log files. For example, due to an update of a given data item, other data items may also be read or written. Therefore, each item update is characterized by three sets: the *read set*, the set of items that have been read because of the update; the *pre-write set*, the set of items that have been written before the update but as consequence of it; and the *post-write set*, the set of items that have been written after the update and as consequence of it. They use data mining techniques to generate dependency rules among the data items. These rules are in the following two forms: before a data item is updated, what other data items are read, and after a data item is updated what other data items are accessed by the same transaction. Once these rules are generated, they are used to detect malicious transactions. The transactions that make modifications to the database without following these rules are termed as malicious.

The approach is novel, but its scope is limited to detecting malicious behavior in user transactions. Within that as well, it is limited to user transactions that conform to the read-write patterns assumed by the authors. Also, the system is not able to detect malicious behavior in individual read-write commands.

3.3.3 Detecting Anomalous Database Access Patterns in Relational Databases

Kamra et al. [14][13] have developed algorithms for detecting anomalous user/role access to a DBMS. In order to identify normal behavior, the system uses the database audit files for extracting information regarding users' actions. The audit records, after being processed, are used to form initial profiles representing acceptable actions. Each entry in the audit file is represented as a separate data unit; these units are then combined to form the desired profiles. The approach considers two different scenarios while addressing this problem. In the first scenario, it is assumed that the database has a Role Based Access Control (RBAC) model in place. Under a RBAC system, permissions are associated with roles, grouping several users, rather than with single users. The intrusion detection (ID) system is able to determine role intruders, that is, individuals that while holding a specific role, behave differently than expected. The existence of roles makes the approach usable even for databases with large user population. When role information does exist, the profile learning problem is transformed into a supervised learning problem. The classification engine used in this case is the naive bayes classifier [15]. A profile for a naive bayes classifier consists of the table and column access probabilities for a specific role. The roles are treated as classes for the classification purpose. The ID task for this setting is as follows: For every new user request, its role (or class) is predicted by the trained classifier. If the predicted role (or class) is different from the original role associated with the query, an anomaly is detected. For benign queries, the classifier can be updated in a straightforward manner by adjusting the probabilities of the relevant features in the user request.

In the second case, the same problem is addressed in the context of DBMS without any role definitions. This is a necessary case to consider because not all organizations are expected to follow a RBAC model for authorizing users of their databases. In such a setting, every transaction is associated with the user that issued it. A naive approach for ID in this setting is to build a different profile for every user. For systems with large user bases such an approach would be extremely inefficient. Moreover, many of the users in those systems are not particularly active and they only occasionally submit queries to the database. In the case of highly active users, profiles would suffer from overfitting, and in the case of inactive users, they would be too general. In the first case we would observe a high number of false alarms, while the second case would result in high number of missed alarms, that is, alarms that should have been raised. These difficulties are overcome by building user-group profiles (clusters of similar behaviors) based solely on the SQL commands users submit to the database. Given such profiles, an *anomaly* is defined as an access pattern that deviates from the profiles. The specific methodology that is used for the ID task is as follows: the training data is partitioned into clusters using standard clustering techniques. In this setting, the clusters obtained after the clustering process represent the profiles. A mapping is maintained for every database user to its representative cluster. The representative cluster for a user is the cluster that contains the maximum number of training records for that user after the clustering phase. For every new query under observation, its representative cluster is determined by examining the user-cluster mapping. Note the assumption that every query is associated with a database user. For the detection phase, two different approaches are outlined. In the first approach, the naive bayes classifier is applied in a manner similar to the supervised case, to determine whether the user associated with the query belongs to its representative cluster or not. In the second approach, a statistical test is used to identify if the query is an outlier in its representative cluster. If the result of the statistical test is positive, the query is marked as an anomaly and an alarm is raised.

In what follows, we first present the system architecture and then describe the information captured in the user/role profiles in this approach.

3.3.3.1 System Architecture

The system's architecture consists of three main components: the conventional DBMS mechanism that handles the query execution process, the database audit log files and the ID mechanism. These components form the new extended DBMS that is enhanced with an independent ID system operating at the database level. The flow of interactions for the ID process is shown in Figure 3.**Error! Reference source not found.**. Every time a query is issued, it is analyzed by the ID mechanism before execution. First, the feature selector converts the raw SQL query into one of the quiplet forms supported by the ID mechanism (see Section 3.3.3.2). The detection engine then checks the quiplet against the existing profiles and submits its assessment of the query (anomalous vs. not anomalous) to the response engine. The response engine consults a policy base of existing response mechanisms to issue a response depending on the assessment of the query submitted by the detection engine. Notice that the fact that a query is anomalous may not necessarily imply an intrusion. Other information and security policies must also be taken into account. For example, if the user logged under the role is performing some special activities to manage an emergency, the ID mechanism may be instructed not to raise alarms in such circumstances. If the response engine decides to raise an alarm, certain actions for handling the alarm can be taken. The most common action is to send an alert to the security administrator. However other actions are possible (Figure 3.**Error! Reference source not found.**), such as disable the role and disconnect the user making the access or drop the query. If by assessment, the query is not anomalous, the response engine simply updates the database audit log and the profiles with the query information. Before the detection phase, the profile creator module creates the initial profiles from a set of intrusion free records from the database audit log.

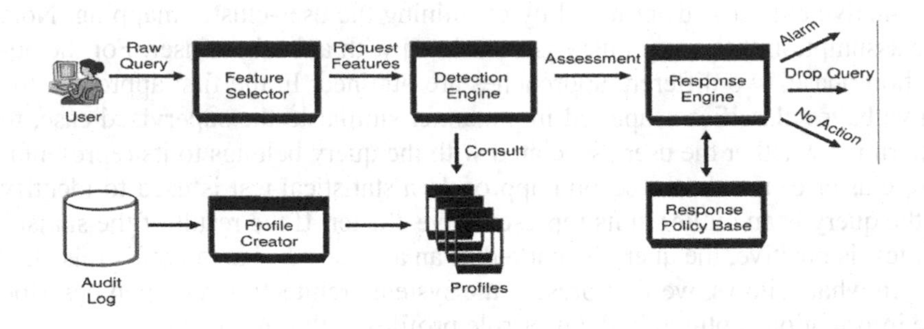

Fig. 3.1 : Overview of the ID process

3.3.3.2 Profile Information

In order to identify user behavior, the database audit files are used for extracting information regarding users' actions. The audit records, after being processed, are used to form initial profiles representing acceptable actions. Each entry in the audit file is represented as a separate data unit; these units are then combined to form the desired profiles.

The authors assume that users interact with the database through commands, where each command is a different entry in the log file, structured according to the SQL language. For example, in the case of *select* queries such commands have the format:

SELECT [DISTINCT] {TARGET-LIST}
FROM {RELATION-LIST}
WHERE {QUALIFICATION}

In order to build profiles, the log-file entries need to be pre-processed and converted into a format that can be analyzed by the detection algorithms. Therefore, each entry in the log file is represented by a basic data unit that contains five fields, and thus it is called a *quiplet*.

Quiplets are the basic unit for viewing the log files and are the basic components for forming profiles. User actions are characterized using sets of such quiplets. Each quiplet contains the following information: the SQL command issued by the user, the set of relations accessed, and for each such relation, the set of referenced attributes. This information is available in the three basic components of the SQL query, namely, the SQL COMMAND, the TARGET-LIST and the RELATION-LIST. The QUALIFICATION component of the query is also processed to extract information on relations and their corresponding attributes that are used in the query predicate. Therefore, the abstract form of such a quiplet consists of five fields (*SQL Command, Projection Relation Information, Projection Attribute Information, Selection Relation Information and Selection Attribute Information*). Depending on the level of detail required in the profile construction phase and in the ID phase, the quiplets are captured from the log file entries using three different representation levels. Each level is characterized by a different amount of recorded information.

The most naive representation of an audit log-file record is termed as, *coarse quiplet* or *c-quiplet*. A c-quiplet records only the number of distinct relations and attributes projected and selected by the SQL query. Therefore, c-quiplets essentially model how many relations and how many attributes are accessed in total, rather than the specific elements that are accessed by the query. Apparently, a large amount of valuable information in the database log is ignored by c-quiplets. It is however useful to consider such a primitive data representation since it is sufficient in the case of a small number of well-separated roles. Moreover, more sophisticated representations of log-file entries are based on the definition of c-quiplets.

The second representation scheme captures more information from the log file records. This representation is termed as, *medium-grain quiplet* or *m-quiplet*. These quiplets extend the coarse quiplets by further exploiting the information present in the log entries. Like a c-quiplet, an m-quiplet represents a single log entry of the database log file. In this case though, each relation is represented separately by the number of its attributes projected (or selected) by the SQL query.

Finally, a third representation level of log-file records is introduced which extracts the maximum information from the log files. This representation is termed as, *fine quiplet* or *f-quiplet*. The structure of an f-quiplet is similar to that of an m-quiplet. In particular, the first, the second and the fourth fields of an f-quiplet are the same as the corresponding fields of the m-quiplets. The f-quiplets and m-quiplets differ only for the third and fifth fields. In the case of f-quiplets, these fields are vector of vectors and are called PROJ-BIN-ATTR[][] and SEL-BIN-ATTR[][] respectively. The *i*-th element of PROJ-BIN-ATTR[][] is a vector corresponding to the *i*-th relation of the database and having size equal to the number of attributes of relation *i*. The *i*-th element of PROJ-BIN-ATTR[][] has binary values indicating which specific attributes of relation *i* are projected in the SQL query. The semantics of SEL-BIN-ATTR[][] are analogous.

SQL Command	c-quiplet	m-quiplet	f-quiplet
Select $R_1.A_1, R_2.C_2$	select<2><2>	select <1,1><1,1>	select <1,1>
From R_1, R_2	<2><2>	<1,1><1,1>	<[1,0,0],[0,0,1]>
Where $R_1.B_1 =$ $R_2.B_2$			<1,1> [0,1,0],[0,1,0]

Table 3.1 Quiplet construction example

Table 3.1 shows a SQL command corresponding to select statement and its representation according to the three different types of quiplets. In the example, a database schema consisting of two relations $R_1 = \{A_1, B_1, C_1\}$ and $R_2 = \{A_2, B_2, C_2\}$, is considered.

The authors have demonstrated experiments that evaluate the performance of the three different representation schemes. For details on the experimental set-up and results, we refer the reader to [14][13].

The approach is useful for three main reasons. First, by modeling the access patterns of users from the SQL command syntax, the insider threat scenario is addressed directly. The approach should be able to capture users/roles that access relations (or columns within relations) not conforming to their normal access behavior. Second, the three different granularity levels of representation proposed in the scheme offer alternatives for space/time/accuracy overhead. The finer the granularity of the information captured, more space it will occupy resulting in larger but more accurate detection assessments. Third, the profiles themselves can be used by the security administrators to refine existing access control policies of the database system or define new ones.

The main drawback of the approach is that it only takes into consideration the information from the SQL query syntax. The semantic information such as constant values in the *where* clause, on which the number of tuples returned/updated by a SQL command depend, is not taken into account.

3.4 Emerging Trends

In the previous sections, we have described the current approaches towards anomaly detection in a DBMS using ML techniques. As we can see from the relatively few number of approaches, this still is a nascent area of research. In the industry as well, none of the DBMS vendors support an anomaly detection mechanism yet. However, recently we have seen an interest in *third-party database activity monitoring* (DAM) products [16]. A DAM product continuously monitors a DBMS externally and reports any relevant suspicious activity [16]. We present an overview of state of the art in DAM products in the next section.

3.4.1 Database Activity Monitoring

Gartner research has identified DAM as one of the top five strategies that are crucial for reducing data leaks in organizations [17], [18]. Such step-up in data vigilance by organizations is partly driven by various government regulations concerning data management such as SOX, PCI, GLBA, HIPAA and so forth [19]. Organizations have also come to realize that current attack techniques are more sophisticated, organized and targeted than the broad-based hacking days of past. Often, it is the sensitive and proprietary data that is the real target of attackers. Also, with greater data integration, aggregation and disclosure, preventing data theft, from both *inside* and *outside* organizations, has become a major challenge. Standard database security mechanisms, such as access control, authentication, and encryption, are not of much help when it comes to preventing data theft from insiders [20]. Such threats have thus forced organizations to re-evaluate security strategies for their internal databases [19]. Today there are several commercial products for database monitoring against intrusions [16]. We categorize them into two broad categories: *network-appliance-based* and *agent-based*.

Network-appliance-based solutions consist of a dedicated hardware appliance that taps into an organization's network, and monitors network traffic to and from the data center. These appliances can either be operated in a *passive* mode in which they observe the traffic and report suspicious behavior, and an *inline* mode where all traffic to the data center passes through them. Agent-based solutions, on the other hand, have a software component installed on the database server that interacts with the database management system (DBMS) in order to monitor accesses to the data.

Each method has its own advantages and disadvantages. A passive network-appliance does not increase network latency but its ability to block an actual attack is limited [16]. Moreover, network appliances, in general, have a major disadvantage in their inability to monitor privileged users who can log into the database server directly. Agent-based solutions, on the other hand, result in some overhead because of the additional software running on the database server and its usage of CPU and memory resources. However, because agent-based solutions are able to directly communicate with the database server, they have a better view of the database structure and are thus able to more effectively monitor the server usage [16].

Many of the DAM products claim to use anomaly detection techniques for detecting exceptional actions of database users [21], [22][23]. But in general, it is hard to comment on their internal details as all of them use proprietary algorithms. One common theme, though, that runs across all the products is that the profiles learnt over a period of time are presented to the administrator as a baseline policy. The baseline policy can be refined by the administrator as per the organization's rules and regulations.

So far we have focused on the *detection* of anomalies in context of DBMS. We now turn our attention to *responding* to a database anomaly. The DAM products are crucial in the line of defense against data theft, but a common shortcoming they all have is their inability to issue a suitable *response* to an ongoing attack. In case of an inline network-appliance, once malicious packets are detected, the system can either be *aggressive* and drop the packets, or take a *conservative* approach by only raising an alarm, but still let the packets go through. At the network level, there is little else that can be done. An agent-based solution, unless it acts as a front-end proxy to the database server, cannot do much either other than raise an alarm for the security management [16]. So what more can a DBMS do to respond to an anomalous request? We discuss this question in the next section.

3.4.2 Responding to Database Anomalies

Standard security violations in the context of a DBMS are dealt with simple and intuitive responses. For example, if a user tries to exercise an unassigned privilege, the access control mechanism will deny the access request. However, it is not trivial to develop a response mechanism capable of automatically taking actions when abnormal database behavior is detected. Let us illustrate this with the following example. Consider a database system with several users, each having the *data-reader* and the *data-writer* role in the database[3]. We consider a database monitoring system in place that builds database user profiles based on SQL queries submitted by the users. Suppose that

---

[3]SQL Server 2000/2005 has the fixed database roles, *db_datareader* and *db_data writer* [24]

a user *U*, who has never accessed table *T*, issues a query that accesses all columns in *T*. The detection mechanism flags this request as anomalous for *U*. The major question is what the system should do next once a request is marked as anomalous by the detection mechanism. Since the anomaly is detected based on the learned profiles, it may well be a false alarm. It is easy to see then there are no simple intuitive response measures that can be defined for such security-related events. The system should take different response actions depending on the details of the anomalous request and the request context. If *T* contains sensitive data, a strong response action is to revoke the privileges corresponding to actions that are flagged as anomalous. In our example, such a response would translate into revoking the *select* privilege on table *T* from *U*. Different and more articulated response actions can be formulated by introducing the notion of "*privilege states*". For example, the privilege corresponding to an anomalous action may be moved into a *suspended* state until a remedial action, such as a 2nd-factor authentication, is executed by user *U*. However, if the user action is a one-time action part of a bulk-load operation, when all objects are expected to be accessed by the request, no response action may be necessary. The key observation here is that a DBMS needs to be instrumented with capabilities to decide *which response measure to take under a given situation*. Therefore, a *response policy* is required by the database security administrator to specify appropriate response actions for different circumstances.

Creation, storage and execution of policies for responding to database anomalies is an exciting new area of research. An important component of such research is once a baseline response policy is defined by the administrator, how can the policy be refined automatically by the response mechanism. Can we create feedback loops such that the response system learns from its own responses (right and wrong both) and gets better over a period of time? We expect to see more work being done in this field in future.

3.4.3 Conclusion

In this chapter we have presented an overview of approaches using machine learning techniques for anomaly detection in databases. These techniques have primarily been used for detection of SQL injection attacks and anomaly detection for addressing the insider threat scenario. This is still a new area of research as we could see from the relatively few number of approaches. We then briefly described two emerging themes in this field. The first theme is database activity monitoring using proprietary third-party products that are external to the DBMS. Some of the products claim to employ anomaly detection techniques for detecting malicious user activity, but there seems to be definite scope for innovation in this area. The second theme is the relatively unexplored field of issuing an appropriate response to an anomalous user action. We expect to see more exciting work being done in these two areas in future.

References

[1]. Animesh Patcha and Jung-Min Park. An overview of anomaly detection techniques: Existing solutions and latest technological trends. *Computer Networks*, 51(12):3448–3470, 2007.

[2]. A. Anton, E.Bertino, N.Li, and T.Yu. A roadmap for comprehensive online privacy policies. In *CERIAS Technical Report*, 2004.

[3]. Rakesh Agrawal, Jerry Kiernan, Ramakrishnan Srikant, and Yiron Xu. Hippocratic databases. In *Proceedings of the 28th International Conference on Very Large Data Bases (VLDB)*, pages 143–154. Morgan-Kaufmann, 2002.

[4]. David Litchfield, Chris Anley, John Heasman, and Bill Grindlay. *The Database Hacker's Handbook: Defending Database Servers*. Wiley, 2005.

[5]. William G.J. Halfond, Jeremy Viegas, and Alessandro Orso. A classification of sql-injection attacks and countermeasures. In *Proceedings of the International Symposium on Secure Software Engineering (ISSSE)*, 2006.

[6]. Fredrik Valeur, Darren Mutz, and Giovanni Vigna. A learning-based approach to the detection of sql attacks. In *Proceedings of the International Conference on detection of intrusions and malware, and vulnerability assessment (DIMVA)*, 2003.

[7]. Elisa Bertino, Ashish Kamra, and James Early. Profiling database application to detect sql injection attacks. *IEEE International Performance, Computing, and Communications Conference (IPCCC) 2007*, pages 449–458, April 2007.

[8]. Darren Mutz, Fredrik Valeur, Giovanni Vigna, and Christopher Kruegel. Anomalous system call detection. *ACM Transactions on Information Systems and Security*, 9(1):61–93, 2006.

[9]. Athanasios Papoulis. *Probability, Random Variables and Stochastic Processes*. McGraw-Hill Companies, 1991.

[10]. David Wagner and Paolo Soto. Mimicry attacks on host-based intrusion detection systems. In *Proceedings of the 9th ACM conference on Computer and Communications Security (CCS)*, pages 255–264, New York, NY, USA, 2002. ACM.

[11]. C.Y. Chung, M. Gertz, and K. Levitt. Demids: a misuse detection system for database systems. In *Integrity and Internal Control in Information Systems: Strategic Views on the Need for Control. IFIP TC11 WG11.5 Third Working Conference*, 2000.

[12]. Yi Hu and Brajendra Panda. A data mining approach for database intrusion detection. In *SAC '04: Proceedings of the 2004 ACM symposium on Applied computing*, pages 711–716, New York, NY, USA, 2004. ACM.

[13]. Elisa Bertino, Ashish Kamra, and Evimaria Terzi. Intrusion detection in rbac-administered databases. In *Proceedings of the Applied Computer Security Applications Conference (ACSAC)*, 2005.

[14]. Ashish Kamra, Elisa Bertino, and Evimaria Terzi. Detecting anomalous access patterns in relational databases. *The International Journal on Very Large Data Bases (VLDB)*, 2008.

[15]. Tom M. Mitchell. *Machine Learning*. McGraw-Hill, 1997.

[16]. Andrew Conry-Murray. The threat from within. Network Computing (Aug 2005), http://www.networkcomputing.com/showArticle.jhtml? articleID=166400792.

[17]. Rich Mogull. Top five steps to prevent data loss and information leaks. Gartner Research (July 2006), http://www.gartner.com.
[18]. Mark Nicolett and Jeffrey Wheatman. Dam technology provides monitoring and analytics with less overhead. Gartner Research (Nov 2007), http://www.gartner.com.
[19]. Ron Ben Natan. *Implementing Database Security and Auditing*. Digital Press, 2005.
[20]. The cyber enemy within ... countering the threat from malicious insiders. In *Proceedings of the 20th Annual Computer Security Applications Conference (ACSAC)*. IEEE Computer Society, 2004.
[21]. Imperva. Imperva securesphere. http://www.imperva.com/products/gateway.html.
[22]. Guardium. Guardium. http://www.guardium.com/.
[23]. Tizor. Tizor mantra. http://www.tizor.com/.
[24]. SQL Server 2005 Books Online. Database-level roles. http://msdn2.microsoft.com/en-us/library/ms189121.aspx.

4 Identifying Threats Using Graph-based Anomaly Detection

William Eberle[1], Lawrence Holder[2], Diane Cook[3]

Abstract Much of the data collected during the monitoring of cyber and other infrastructures is structural in nature, consisting of various types of entities and relationships between them. The detection of threatening anomalies in such data is crucial to protecting these infrastructures. We present an approach to detecting anomalies in a graph-based representation of such data that explicitly represents these entities and relationships. The approach consists of first finding normative patterns in the data using graph-based data mining and then searching for small, unexpected deviations to these normative patterns, assuming illicit behavior tries to mimic legitimate, normative behavior. The approach is evaluated using several synthetic and real-world datasets. Results show that the approach has high true-positive rates, low false-positive rates, and is capable of detecting complex structural anomalies in real-world domains including email communications, cell-phone calls and network traffic.

4.1 Introduction

Maintaining the security of our infrastructure, whether physical or cyber, requires the ability to detect threats to the infrastructure. Modern threats are sophisticated, multifaceted, coordinated and attempt to mimic normal activity. Detecting such threats requires approaches that consider many different types of activities and the relationships between them. We describe an approach that represents the activities and relationships as a graph, mines the graph for normative patterns, and then searches for unexpected deviations to the normative

---

[1] Department of Computer Science, Tennessee Technological University, Box 5101, Cookeville, TN 38505, weberle@tntech.edu

[2] School of Electrical Engineering and Computer Science, Washington State University, Box 642752, Pullman, WA 99164, holder@wsu.edu

[3] School of Electrical Engineering and Computer Science, Washington State University, Box 642752, Pullman, WA 99164, cook@eecs.wsu.edu

J.J.P. Tsai and P.S. Yu (eds.), *Machine Learning in Cyber Trust: Security, Privacy, and Reliability*, DOI: 10.1007/978-0-387-88735-7_4,

patterns. These unexpected deviations may indicate the presence of a threat to the infrastructure being monitored.

The ability to mine data represented as a graph has become important in several domains for detecting various structural patterns (Cook and Holder 2006). One important area of data mining is anomaly detection, particularly for fraud. However, less work has been done in terms of detecting anomalies in graph-based data. While there has been some previous work that has used statistical metrics and conditional entropy measurements, the results have been limited to certain types of anomalies and specific domains.

In this chapter we present graph-based approaches to uncovering anomalies in domains where the anomalies consist of unexpected entity/relationship alterations that closely resemble non-anomalous behavior. We have developed three algorithms for the purpose of detecting anomalies in all three types of possible graph changes: label modifications, vertex/edge insertions and vertex/edge deletions. Each of our algorithms focuses on one of these anomalous types, using the minimum description length principle to first discover the normative pattern. Once the common pattern is known, each algorithm then uses a different approach to discover particular anomalous types. The first algorithm uses the minimum description length to find anomalous patterns that closely compress the graph. The second algorithm uses a probabilistic approach to examine pattern extensions and their likelihood of existence. The third algorithm analyzes patterns that come close to matching the normative pattern, but are unable to make some of the final extensions leading to the construction of the normative pattern.

Using synthetic and real-world data, we evaluate the effectiveness of each of these algorithms in terms of each of the types of anomalies. Each of these algorithms demonstrates the usefulness of examining a graph-based representation of data for the purposes of detecting threats, where some individual or entity is cloaking their illicit activities through an attempt at closely resembling normal behavior.

The next section describes the general area of graph-based learning and the algorithm underlying our ability to find normative patterns in graphs. Section 4.3 defines the problem of graph-based anomaly detection, and Section 4.4 presents the GBAD system, which consists of three variants for detecting different types of graph-based anomalies. Section 4.5 presents experimental results evaluating GBAD on several synthetic and real-world datasets related to threat detection. We discuss related work in section 4.6, and conclude in section 4.7.

4.2 Graph-based Learning

While not specific to anomaly detection, there are several approaches to handling the first stage of detecting anomalies, which is the discovery of the normative pattern in data represented as a graph. One approach called gSpan returns all *frequent* subgraphs in a database that is represented as a graph (Yan and Han 2002). Using a depth-first search (DFS) on the input graphs, the algorithm constructs a hierarchical search tree based upon the DFS code assigned to each graph. Then, from its canonical tree structure, the algorithm performs a pre-order traversal of the tree in order to discover the frequent subgraphs.

Another approach is found in FSG, which is similar to gSpan in that it returns all of the frequent subgraphs in a database of transactions that have been represented as a graph (Kuramochi and Karypis 2004). However, unlike gSpan, FSG uses an Apriori-style breadth-first search. The algorithm takes the input graphs and performs a level-by-level search, growing patterns one edge at a time. The core of the FSG algorithm lies in its candidate generation and counting that are used to determine the frequent subgraphs.

In order to mine large graphs for frequent subgraphs, Huan et al. proposed a maximal frequent subgraphs approach called SPIN as an improvement to gSpan (Huan et al. 2004). By mining only subgraphs that are not part of any other frequent subgraphs, they are able to reduce the number of mined patterns by orders of magnitude. This is accomplished by first mining all frequent trees from a graph, and then reconstructing all maximal subgraphs from the mined trees. Zeng et al. looked at the problem of dense graphs by mining the properties of quasi-cliques (Zeng et al. 2006). Using a system called Cocain, they propose several optimization techniques for pruning the unpromising and redundant search spaces. To help combat the computational complexity of subgraph isomorphism, Gudes et al. proposed a new Apriori-based algorithm using disjoint paths (Gudes et al. 2006). Following a breadth-first enumeration and what they called an "admissible support measure", they are able to prune candidate patterns without checking their support, significantly reducing the search space. MARGIN is another maximal subgraph mining algorithm that focuses on the more promising nodes in a graph (Thomas et al. 2006). This is accomplished by searching for promising nodes in the search space along the "border" of frequent and infrequent subgraphs, thus reducing the number of candidate patterns.

The goal of SUBDUE is to return the substructures that best compress the graph (Holder et al. 1994). Using a beam search (a limited length queue of the best few patterns that have been found so far), the algorithm grows patterns one edge at a time, continually discovering which subgraphs best compress the description length of the input graph. The core of the SUBDUE algorithm is in its compression strategy. After extending each subgraph pattern by one edge, it evaluates each extended subgraph based upon its compression value (the higher the better). A list is maintained of the best substructures, and this process is continually repeated until either there are no more subgraphs that can compress or a user-specified limit is reached.

While each of these approaches is successful at pattern discovery, we will use the SUBDUE compression evaluation technique as the basis for our underlying discovery of the normative patterns. While the gSpan application is not publicly available, there are a few reasons why we found FSG to not be an ideal candidate for our implementation. One reason for our choice of pattern learner lies with the format expected by the FSG application. SUBDUE can effectively discover normative patterns whether it is given all transactions or data as one entire graph, or if each transaction is defined as individual subgraphs. As a graph data miner, FSG shows the frequency of a pattern based upon the number of transactions defined in the graph input file. So, if a graph is not delineated by individual transactions, the frequency of every pattern is 1, and thus very difficult to determine which pattern is the most frequent. However, in some later work by Kuromachi and Karypis, they improve upon this with an approach called Grew that is able to better handle large graphs that consist of connected subgraphs (Kuromachi and Karypis 2004). Another reason we prefer SUBDUE lies in the FSG approach to determining the normative pattern based upon frequency. While tests on various graphs showed SUBDUE and FSG returned the same normative pattern, when the tests involved a graph where the normative pattern is not found across all transactions (e.g., noise), the frequent pattern is not found unless the FSG support percentile is reduced. The issue then is knowing what support percentile should be used for a specific run. Specifying 100% support will result in the normative pattern being lost if the pattern is not found in every transaction, while using a lower percentile may result in other (smaller) normative patterns being found. In short, SUBDUE allows us to find the normative pattern in data that may be less regular or contain some noise. As will be shown in the following section, this is critical to the success of discovering anomalies.

4.3 Graph-based Anomaly Detection

Before we lay the groundwork for our definition of a graph-based anomaly, we need to put forth a framework for the definition of a graph. In general, a graph is a set of nodes and a set of links, where each link connects either two nodes or a node to itself. More formally, we use the following definitions (Gross and Yellen 1999) (West 2001):

Definition: A graph $G = (V, E, L)$ is a mathematical structure consisting of three sets V, E and L. The elements of V are called *vertices* (or nodes), the elements of E are the *edges* (or links) between the vertices, and the elements of L are the string *labels* assigned to each of the elements of V and E.

Definition: A vertex (or node) is an entity (or item) in a graph. For each vertex there is a labeled vertex pair (v, l) where v is a vertex in the set V of vertices and l is a string label in the set L of labels.

Definition: An edge (or link) is a labeled relation between two vertices called its endpoints. For each edge there is a labeled edge pair (e, l) where e is an edge in the set E of edges and l is a string label in the set L of labels.

Definition: An edge can be directed or undirected. A *directed* edge is an edge, one of whose endpoints is designated as the *tail*, and whose other endpoint is designated as the *head*. An *undirected* edge is an edge with two unordered endpoints. A multi-edge is a collection of two or more edges having identical endpoints.

Much research has been done recently using *graph*-based representations of data. Using *vertices* to represent entities such as people, places and things, and *edges* to represent the relationships between the entities, such as friend, lives and owns, allows for a much richer expression of data than is present in the standard textual or tabular representation of information. Representing various data sets, like telecommunications call records, financial information and social networks, in a graph form allows us to discover *structural* properties in data that are not evident using traditional data mining methods.

The idea behind the approach presented in this work is to find anomalies in graph-based data where the anomalous subgraph (at least one edge or vertex) in a graph is part of (or attached to or missing from) a non-anomalous subgraph, or the *normative pattern*. This definition of an anomaly is unique in the arena of graph-based anomaly detection, as well as non-graph-based anomaly detection. The concept of finding a pattern that is "similar" to frequent, or good, patterns, is different from most approaches that are looking for unusual or "bad" patterns. While other non-graph-based data mining approaches may aide in this respect, there does not appear to be any existing approaches that directly deal with this scenario.

Definition: Given a graph G with a normative substructure S, a subgraph S', and a difference d between S and S', let $C(d)$ be the cost of the difference and $P(d)$ be the probability of the difference. Then the subgraph S' is considered anomalous if $0 < A(S') \leq X$, where X is a user-defined threshold and $A(S') = C(d) * P(d)$ is the anomaly score.

The importance of this definition lies in its relationship to fraud detection (i.e., any sort of deceptive practices that are intended to illegally obtain or hide information). If a person or entity is attempting to commit fraud, they will do all they can to hide their illicit behavior. To that end, their approach would be to convey their actions to be as close to legitimate actions as possible. That makes this definition of an anomaly extremely relevant.

For a graph-based anomaly, there are several situations that might occur:

1. *The label on a vertex is different than was expected.*
2. *The label on an edge is different than was expected.*
3. *A vertex exists that is unexpected.*
4. *An edge exists that is unexpected.*
5. *An expected vertex is absent.*
6. *An expected edge between two vertices (or a self-edge to a vertex) is absent.*

These same situations can also be applied to a subgraph (i.e., multiple vertices and edges), and will be addressed as such. In essence, there are three general *categories of anomalies*: modifications, insertions and deletions. Modifications would constitute the first two situations; insertions would consist of the third and fourth situations; and deletions would categorize the last two situations.

4.4 GBAD Approach

Most anomaly detection methods use a supervised approach, which requires a baseline of information from which comparisons or training can be performed. In general, if one has an idea what is normal behavior, deviations from that behavior could constitute an anomaly. However, the issue with those approaches is that one has to have the data in advance in order to train the system, and the data has to already be labeled (i.e., fraudulent versus legitimate).

Our work has resulted in the development of three algorithms, which we have implemented in the GBAD (Graph-based Anomaly Detection) system. GBAD is an *unsupervised* approach, based upon the SUBDUE graph-based knowledge discovery system (Cook and Holder 2000). Using a greedy beam search and Minimum Description Length (MDL) heuristic, each of the three anomaly detection algorithms uses SUBDUE to provide the top substructure (subgraph), or normative pattern, in an input graph. In our implementation, the MDL approach is used to determine the best substructure(s) as the one that minimizes the following:

$$M(S,G) = DL(G \mid S) + DL(S)$$

where G is the entire graph, S is the substructure, $DL(G|S)$ is the description length of G after compressing it using S, and $DL(S)$ is the description length of the substructure.

We have developed three separate algorithms: GBAD-MDL, GBAD-P and GBAD-MPS. Each of these approaches is intended to discover all of the possible graph-based anomaly types as set forth earlier. The GBAD-MDL algorithm uses a Minimum Description Length (MDL) heuristic to discover the best substructure in a graph, and then subsequently examines all of the instances of that substructure that "look similar" to that pattern. The detailed GBAD-MDL algorithm is as follows:

1. Given input graph G:
2. Find the top-k substructures S_i, that minimize $M(S_i,G)$, where S_i is a subgraph of G.
3. Find all close-matching instances I_j of S_i such that the cost $C(I_j,S_i)$ of transforming I_j to match the graph structure of S_i is greater than 0.
4. Determine anomalous value for each I_j by building the substructure definition S_j, finding all *exact* matching instances of S_j such that $F(I_j)$ is the frequency of instances that match I_j, and calculating the value $A(I_j) = F(I_j) * C(I_j,S_i)$, where the lower the value, the more anomalous the instance.
5. Output all I_j minimizing $A(I_j)$.

The cost of transforming a graph A into an isomorphism of a graph B is calculated by adding 1.0 for every vertex, edge and label that would need to be changed in order to make A isomorphic to B. The result will be those instances that are the "closest" (without matching exactly) in structure to the best structure (i.e., compresses the graph the most), where there is a tradeoff in the cost of transforming the instance to match the structure, as well as the frequency with which the instance occurs. Since the cost of transformation and frequency are independent variables, multiplying their values together results in a combinatory value: the lower the value, the more anomalous the structure.

It should be noted that the value of substructure S will include the instances that do not match exactly. It is these inexact matching instances that will be analyzed for anomalousness. It should also be noted that we are only interested in the top substructure (i.e., the one that minimizes the description length of the graph), so *k* will always be 1. However, for extensibility, the *k* can be adjusted if it is felt that anomalous behavior may be found in more than one normative pattern.

The GBAD-P algorithm also uses the MDL evaluation technique to discover the best substructure in a graph, but instead of examining all instances for similarity, this approach examines all *extensions* to the normative substructure (pattern), looking for extensions with the lowest probability. The subtle difference between the two algorithms is that GBAD-MDL is looking at instances of substructures with the same characteristics (i.e., size, degree, etc.), whereas GBAD-P is examining the probability of extensions to the normative pattern to determine if there is an instance that when extended beyond its normative structure is traversing edges and vertices that are probabilistically less likely than other possible extensions.

The detailed GBAD-P algorithm is as follows:

1. **Find the top-*k* substructures** S_i, that minimize $M(S_i,G)$, where S_i is a subgraph of G.
2. **Compress** G by S_i, and **Find the top-k substructures** again.
3. **Find all instances** I_j that match the substructure S_i.
4. **Create extended instances** $I_j^{'}$ that consist of an original instance with an additional extension of an edge and a vertex, such that $I_j \subseteq I_j^{'}$, and $I_j^{'} \subseteq I^{'}$, where $I^{'}$ is the set of all extended instances of S_i.
5. **Determine anomalous value** for each $I_j^{'}$ by finding matching instances of $I_j^{'}$ in set $I^{'}$, and calculating the value of $A(I_j^{'}) = |I_j^{'}| / |I^{'}|$ where $|I_j^{'}|$ is the cardinality of the set of instances that match $I_j^{'}$, and $|I^{'}|$ is the cardinality of the set of extended instances of S_i.
6. **Output** $I_j^{'}$ minimizing $A(I_j^{'})$ and where $A(I_j^{'})$ is less than a user acceptable threshold.
7. **Compress** G by the graph structure of $I_j^{'}$.
8. **Repeat** step 1 and then start again at step 3.

$A(I_j^{'})$ is the *probability* that a given instance should exist given the existence of all of the extended instances. Again, the lower the value, the more anomalous the instance. Given that $\left|I^{'}\right|$ is the total number of possible extended instances, $\left|I_j^{'}\right|$ can never be greater, and thus the value of $A(I_j^{'})$ will never be greater than 1.0.

The GBAD-MPS algorithm again uses the MDL approach to discover the best substructure in a graph, then it examines all of the instances of *parent* (or ancestral) substructures that are missing various edges and vertices. The value associated with the parent instances represents the cost of transformation (i.e., how much change would have to take place for the instance to match the best substructure). Thus, the instance with the lowest cost transformation (if more than one instance have the same value, the frequency of the instance's structure will be used to break the tie if possible) is considered the anomaly, as it is closest (maximum) to the best substructure without being included on the best substructure's instance list.

The detailed GBAD-MPS algorithm is as follows:

1. **Find the top-*k* substructures** S_i, that minimize $M(S_i,G)$, where S_i is a subgraph of G.
2. **Find all ancestor** substructures S' such that $S' \subseteq S_i$.
3. **Find all instances** I' of S'.
4. **Determine the anomalous value** for each instance I' as $A(I') = F(I') * C(I', S_i)$.
5. **Output** I' as an anomalous instance if its anomalous value is less than a user specified threshold.

By allowing the user to specify a threshold, we can control the amount of "anomalousness" that we are willing to accept. By our definition of an anomaly, we are expecting low transformation costs (i.e., few changes for the anomalous instance to match the normative substructure).

4.5 Experimental Results

We have performed several experiments evaluating GBAD on both synthetic and real-world data sets. Results on synthetic data show that GBAD is able to accurately detect different types of target anomalies with low false-positive rates. Results on real-world data show the types of anomalies that can be detected in data describing network intrusions, email communications, cargo shipments, cell-phone communications, and network traffic.

4.5.1 Synthetic Data

We constructed numerous synthetic graphs that were randomly generated based on the following parameters.

- AV is the number of anomalous vertices in an anomalous substructure
- AE is the number of anomalous edges in an anomalous substructure
- V is the number of vertices in the normative pattern
- E is the number of edges in the normative pattern

Each synthetic graph consists of substructures containing the normative pattern (with V number of vertices and E number of edges), connected to each other by one or more random connections, and each test anomaly consists of AV number of vertices and AE number of edges altered.

For *modification* anomalies: an AV number of vertices and AE number of edges, from the same randomly chosen normative instance, have their labels modified to randomly-chosen, non-duplicating labels (e.g., we do not replace a vertex labeled "X" with another vertex labeled "X").

For *insertion* anomalies: randomly inserted AV vertices and AE edges, where the initial connection of one of the AE edges is connected to either an existing vertex in a randomly chosen normative instance, or to one of the already inserted AV vertices.

For *deletion* anomalies: randomly chosen AV vertices and AV edges, from a randomly chosen normative instance, are deleted along with any possible "dangling" edges (i.e., if a vertex is deleted, all adjacent edges are also deleted).

Due to our definition of an anomaly, all tests will be limited to changes that constitute less than 10% of the normative pattern. Again, since anomalies are supposed to represent slight deviations in normal patterns, an excessive change to a pattern is irrelevant. However, in order to analyze the effectiveness of these approaches beyond the upper bounds, we will also perform some tests at deviations above 10%.

Each of the above is repeated for each algorithm, varying sizes of graphs, normative patterns, thresholds, iterations and sizes of anomalies (where the size of the anomaly is $|AV| + |AE|$). Also, due to the random nature in which structures are modified, each test will be repeated multiple times to verify its consistency.

4.5.1.1 Metrics

Each test consists of a single graph from which 30 randomly-altered graphs are generated. The output shown consists of the average of the results of running the algorithms against those 30 graphs for the specified settings. The primary three metrics calculated are:

1. Percentage of runs where the *complete* anomalous substructure was discovered.
2. Percentage of runs where at least *some* of the anomalous substructure was discovered.
3. Percentage of runs containing *false positives*.

After the algorithm has completed, the first metric represents the percentage of success when comparing the results to the known anomalies that were injected into the data. If all of the anomalies are discovered for a particular run, that is counted as a success for that run. For example, if 27 out of the 30 runs found all of their anomalies, the value for this metric would be 90.0.

The second metric represents the percentage of runs where at least one of the injected anomalies was discovered. For example, if the anomaly consisted of 3 vertices and 2 edges that had their labels changed, and the run reported one of the anomalous vertices, then that run would be considered a success. Obviously, this metric will always be at least as high as the first metric.

The last metric represents the percentage of runs that reported at least one anomaly that was not one of the injected anomalies. Since it is possible that multiple reported anomalous instances could have the same anomalous value, some runs may contain both correct anomalies and false ones. Further tuning of these algorithms may enable us to discover other measurements by which we could "break the tie" when it comes to calculating an anomalous score.

4.5.1.2 Information Theoretic Results: GBAD-MDL

Figures 4.1 and 4.2 show the effectiveness of the GBAD-MDL approach on graphs of varying sizes with random anomalous modifications. In these figures, the X axis represents the thresholds, the Y axis is the percentage of anomalies discovered, and the Z axis indicates the sizes of the normative patterns, graphs and anomalies. For example, "10/10/100/100-1v" means a normative pattern with 10 vertices and 10 edges in a graph with 100 vertices and 100 edges and an anomaly consisting of a modification to 1 vertex. (Only a portion of the results are shown for space reasons, and other tests showed similar results.)

In the small synthetic test, when the threshold is high enough, (i.e., the threshold is equal to or higher than the percentage of change), it is clear that this approach was able to find all of the anomalies. The only time false positives are reported is when the threshold is 0.2. For a threshold of 0.2, we are basically saying that we want to analyze patterns that are up to 20% different. Such a huge window results in some noise being considered (along with the actual anomalies, as all of the anomalous instances are discovered). Fortunately, our definition of what is truly an anomaly would steer us towards observing runs with lower thresholds.

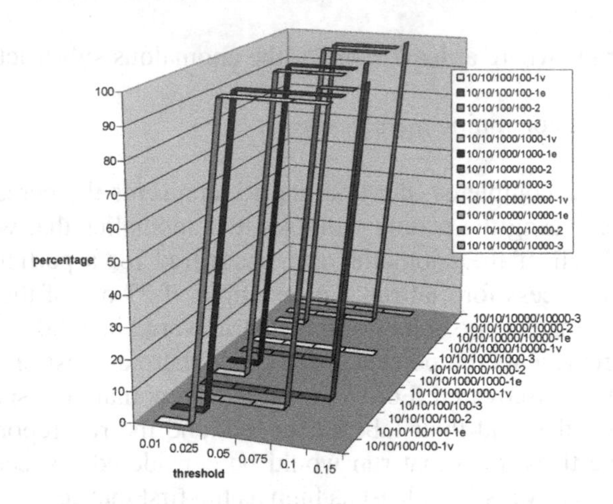

Fig. 4.1 Percentage of GBAD-MDL runs where all anomalies discovered.

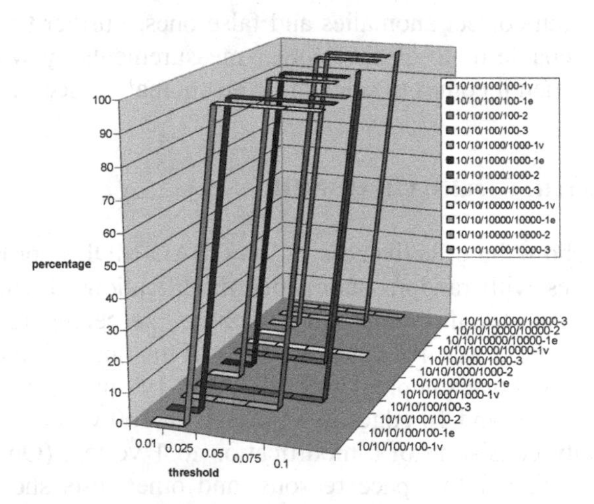

Fig. 4.2 Percentage of GBAD-MDL runs where at least one anomaly is discovered.

4.5.1.3 Probabilistic Results: GBAD-P

Figure 4.3 shows the effectiveness of the GBAD-P approach on graphs of varying sizes with random anomalous insertions. It should be noted in this example that even though unrealistic anomaly sizes were used (representing 20-30% of the normative pattern), this approach is still effective. This same behavior can be observed in larger graphs as well.

As a further experiment, we also tried this approach on different distributions, varying the number of vertices versus the number of edges (e.g., adding more edges than vertices by creating more edges between existing vertices), and also lessening the distribution difference between noise and anomalies. In all cases, the results were relatively the same, with never less than 96.67% of the anomalous instances being found for anomalies of size 8 (40% of the normative pattern) or less, with the lowest discovery rate being 90% for an anomaly of size 10 (50% of the normative pattern).

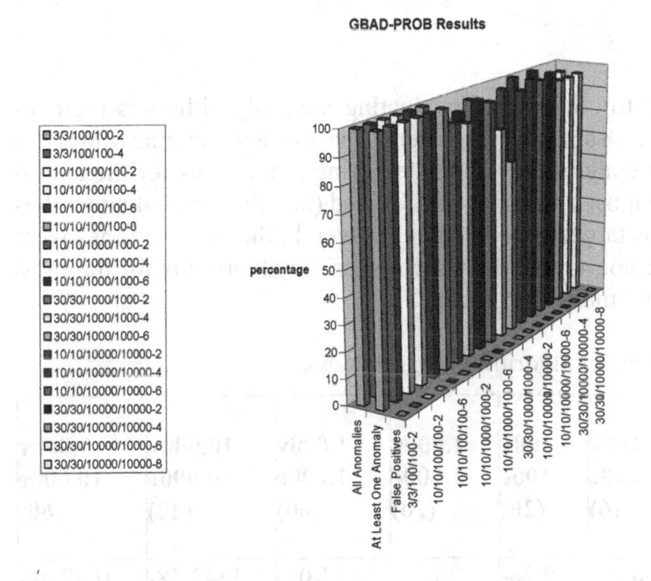

Fig. 4.3 GBAD-P results on anomalous insertions.

4.5.1.4 Maximum Partial Substructure Results: GBAD-MPS

For all tests, across all sizes of graphs and anomalies, the GBAD-MPS algorithm is able to discover all of the anomalous deletions while at the same time reporting no false positives. Initially, the results from the runs on the graph with 1000 vertices and 1000 edges, where the normative pattern consists of 30 vertices and 30 edges, were not good. However, when we increase the number of substructures (to analyze) to 100, and increase the anomalous threshold (i.e., cost of transformation * frequency) to 50.0, the results improve. So, a good rule of thumb is to choose an anomalous threshold based upon the size of the normative pattern. For instance, GBAD could be run first to determine the normative pattern, then based upon the size of the normative pattern, we can determine the maximum size of an anomaly (e.g., around 10%), choose a cost of transformation that would allow for the discovery of an anomaly that size, and then when we rerun the algorithm with the new threshold, the result is complete discovery.

It should be noted that the reason the number of best substructures and the threshold had to be increased is that as the size of the anomaly grows (i.e., the number of vertices and edges deleted increases), the further away the cost of transformation for the anomalous instance is from the normative pattern.

4.5.1.5 Performance

One of the factors to consider in evaluating these algorithms is their respective performances. Table 4.1 represents the average running times (in seconds) for each of the algorithms against varying graph sizes for the anomaly types that were the most effectively discovered (i.e., the types of anomalies that each algorithm was targeted to discover). Overall, the running times were slightly shorter for the non-targeted anomalies, as the algorithms for the most part did not have any anomalies to process.

Table 4.1 Running-times of algorithms (in seconds).

Graph Size (Normative Pattern Size) / Algorithm (Anomaly Type)	100v 100e (6)	100v 100e (20)	1,000v 1,000e (20)	1,000v 1,000e (60)	10,000v 10,000e (20)	10,000v 10,000e (60)
GBAD-MDL (Modification)	0.05-0.08	0.26-15.80	20.25-55.20	31.02-5770.58	1342.58-15109.58	1647.89-45727.09
GBAD-P (Insertion)	1.33	0.95	30.61	18.52	745.45	2118.99
GBAD-MPS (Deletion)	0.14	0.07	4.97	75.59	242.65	813.46

Because the GBAD-MDL algorithm uses a matching threshold, the performance of the algorithm is dependent upon the threshold chosen. The higher the threshold, the longer the algorithm takes to execute, so there is a trade-off associated with the threshold choice. Even on graphs of 10,000 vertices and

10,000 edges, the running times varied anywhere from 1342 seconds to 45,727 seconds, depending upon the threshold chosen. The GBAD-MDL algorithm is tractable given the input parameters. Because of the graph matching that is performed, and the fact that the GBAD-MDL algorithm needs to examine more than just instances of the normative pattern, the user-defined threshold provides a means by which the threat detection analyst can decide how much of the entire graph they want to analyze. The larger the graph, as well as the number of non-matching subgraphs one wants to analyze for anomalous structure, the greater the runtime for this algorithm. Future work on this algorithm will involve a reduction in the number of graph matches by reducing the number of subgraphs that are relevant in terms of matching.

The ability to discover the anomalies is sometimes limited by the *resources* allocated to the algorithm. Given a graph where the anomalous substructure consists of the *minimal* deviation from the normative pattern, if a sufficient amount of processing time and memory is provided, all of these algorithms will discover the anomalous substructure with no false positives. However, the ability to discover anomalies (per our definition) is also hampered by the amount of *noise* present in the graph. The issue is that if noise is a smaller deviation from the normative pattern than the actual anomaly, it may score higher than the targeted anomaly (depending upon the frequency of the noise). Of course, one might say that noise is an anomaly in that it is not normal; however, it is probably not an insider threat, which is the goal of these approaches.

Now, the presence of noise does not eliminate the algorithms' abilities to discover the anomalous substructure. It only results in more false positives being detected if the anomalous score of the noisy structure is better than the desired anomalous substructure. That is where another trade-off is necessary that can be found in most threat detection systems: adjusting thresholds to find a balance of false-positives versus true anomalies. Future work in this area will include not only improved heuristics to reduce the number of graph matches that is performed, but also algorithmic changes to analyze the distribution of vertex and edge labels (especially numeric values) that will aid in the differentiation between seemingly similar labels, thereby reducing the effect of noise.

4.5.2 Real-world Datasets

4.5.2.1 Network Intrusion

One of the more applied areas of research when it comes to anomaly detection can be found in the multiple approaches to intrusion detection. The reasons for this are its relevance to the real-world problem of networks and systems being attacked, and the ability of researchers to gather actual data for testing their models. Perhaps the most used data set for this area of research and experimentation is the 1999 KDD Cup network intrusion dataset (KDD Cup 1999).

In 1998, MIT Lincoln Labs managed the DARPA Intrusion Detection Evaluation Program. The objective was to survey and evaluate research in intrusion detection. The standard data set consisted of a wide variety of intrusions simulated in a military network environment. The 1999 KDD Cup intrusion detection dataset consists of a version of this data. For nine weeks, they simulated a typical U.S. Air Force local-area network, initiated multiple attacks, and dumped the raw TCP data for the competition.

The KDD Cup data consists of connection records, where a connection is a sequence of TCP packets. Each connection record is labeled as either "normal", or one of 37 different attack types. Each record consists of 31 different features (or fields), with features being either continuous (real values) or discrete. The graph representation consisted of a central "record" vertex with 31 outgoing edges, labeled with the feature names, connecting to vertices labeled with the value for that feature. In the 1999 competition, the data was split into two parts: one for training and the other for testing. Groups were allowed to train their solutions using the training data, and were then judged based upon their performance on the test data.

Since the GBAD approach uses unsupervised learning, we ran the algorithms on the test data so that we can judge our performance versus other approaches. Also, because we do not know the possible structural graph changes associated with network intrusions, we have to run all three algorithms to determine which algorithms are most effective for this type of data. Each test contains 50 essentially random records, where 49 are normal records and 1 is an attack record, so the only controlled aspect of the test is that there is only one attack record per data set. This is done because the test data is comprised of mostly attack records, which does not fit our definition of an anomaly, where we are assuming that anomalous substructures are rare. Fortunately, this again is a reasonable assumption, as attacks would be uncommon in most networks.

Not surprisingly, each of the algorithms has a different level of effectiveness when it comes to discovering anomalies in intrusion detection data. Using GBAD-MDL, our ability to discover the attacks is relatively successful. Across all data sets, 100% of the attacks are discovered. However, all but the *apache2* and *worm* attacks produce some false positives. 42.2% of the test runs do not produce any false positives, while runs containing *snmpgetattack*, *snmpguess*, *teardrop* and *udpstorm* attacks contribute the most false positives. False positives are even higher for the GBAD-P algorithm, and the discovery rate of actual attacks decreases to 55.8%. GBAD-MPS shows a similarly bad false positive rate at 67.2%, and an even worse discovery rate at 47.8%.

It is not surprising that GBAD-MDL is the most effective of the algorithms, as the data consists of TCP packets that are structurally similar in size across all records. Thus, the inclusion of additional structure, or the removal of structure, is not as relevant for this type of data, and any structural changes, if they exist, would consist of value modifications.

In order to better understand the effectiveness of the GBAD algorithms on intrusion detection data, we will compare our results with the graph-based approaches of Noble and Cook (Noble and Cook 2003). They proposed two approaches to discovering anomalies in data represented as a graph. Their *anomalous substructure detection* method attempts to find unusual substructures within a graph by finding those substructures that compress the graph the least, compared to our GBAD-MDL approach which uses compression to determine which substructures are closest to the best substructure (i.e., the one that compresses the graph the most). In their results, they use the inverse of the ratio of true anomalies found over the total number of anomalies reported, where the lower the value (i.e., a greater percentage of the reported anomalies are the network attacks), the more effective the approach for discovering anomalies.

The other approach presented is what they call *anomalous subgraph detection.* The objective with this method is to compare separate structures (subgraphs) and determine how anomalous each subgraph is compared to the others. This is similar to our approaches in that the minimum description length is used as a measurement of a substructure's likelihood of existence within a graph. However, in order to implement this approach, the graph must be divided into clearly defined subgraphs so that a proper comparison can be performed. The basic idea is that every subgraph is assigned a value of 1, and the value decreases as portions of the subgraph are compressed away. In the end, the subgraphs are ranked highest to lowest, with the higher the value (i.e., closest to, or equal to, 1), the more anomalous the subgraph. While this works well on intrusion detection data, their approach is restricted to domains where a clear delineation (i.e., subgraphs) must be defined. In other words, the delineation occurs when a graph can be sub-divided into distinct subgraphs, with each subgraph representing a common entity. For example, in domains like terrorist or social networks, this type of delineation may be difficult and subjective.

Using the same set of KDD Cup intrusion detection data as set forth previously, we can compare GBAD-MDL (since it performed the best on this set of data) to both of these approaches, using the same anomalous attack ratios used by Noble and Cook. The ratio used in their work is an inverse fraction of correctly identifying the attacks among all of the attacks reported. For example, if 10 anomalies are reported, but only 1 of them is the actual attack, then the fraction is 1/10, and the inverse is the score of 10, where obviously the lower the score the better. Their anomalous substructure detection method achieves an average anomalous ratio of 8.64, excluding the *snmpgetattack* and *snmpguess* attacks, while using the same scoring ratio, GBAD-MDL generates an average of 7.22 *with snmpgetattack* and *snmpguess* included. In Noble and Cook's paper, both the *snmpgetattack* and *snmpguess* attacks were excluded from the anomalous substructure detection approach results because they had high average attack values of around 2211 and 126 respectively (i.e., too many false positives). However, GBAD-MDL is much more successful at discovering these two attack types, as their respective averages are 8.55 and 7.21. Then, for their anomalous subgraph detection approach, they get an average ranking of 4.75, whereas the GBAD-MDL algorithm is able to achieve a better average ranking of 3.02. Figure 4.4 shows the ranking results for each of the different attack types.

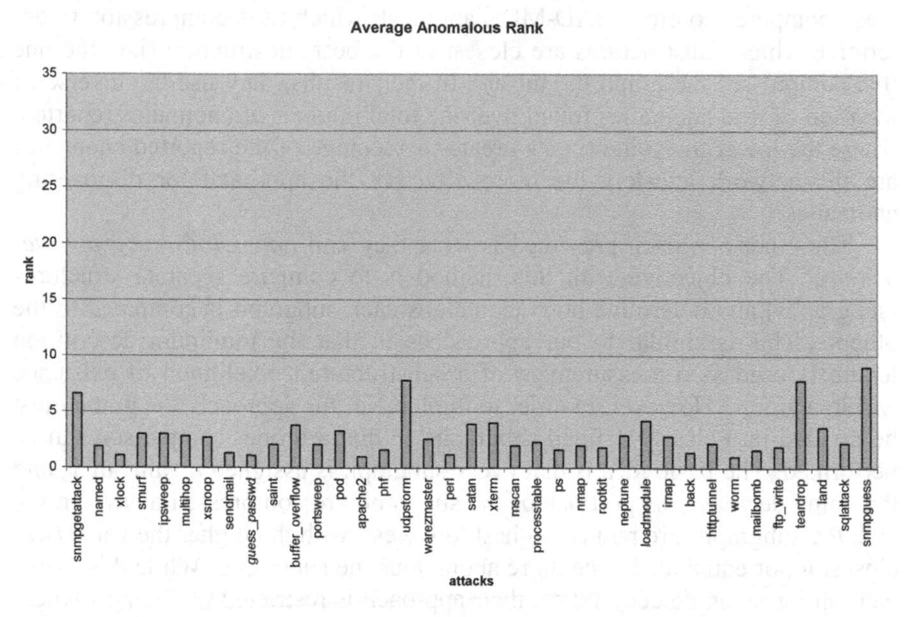

Fig. 4.4 Average anomalous ranking using GBAD-MDL on KDD intrusion detection data.

These results, when compared to the ones presented in Noble and Cook's paper (Noble and Cook 2003), not only show an overall average improvement, but also again show a significant improvement when it comes to effectively discovering the *snmpgetattack* and *snmpguess* attacks, which both had values over 20 using the anomalous subgraph detection approach, whereas the GBAD-MDL algorithm was under 10 for both attack types. It should also be noted that the false positives are mostly due to the fact that we have to increase the anomalous threshold in order to detect some of the anomalous patterns. Unlike our assumption that anomalies are small deviations from the normative pattern, several of the attack records are actually large deviations from the norm.

4.5.2.2 Enron E-mail

One of the more recent domains that has become publicly available is the data set of e-mails between management from the Enron corporation. The Enron e-mail dataset was made public by the Federal Energy Regulatory Commission during its investigation. After subsequent data integrity resolutions, as well as the removal of some e-mails due to requests from affected employees, William Cohen at CMU put the dataset on the web for researchers. From that dataset, Shetty and Adibi further cleaned the dataset by removing duplicate e-mails, and putting the final set into a publicly available database (http://www.isi.edu/~adibi/Enron/Enron.htm). This dataset contains 252,759 messages from 151 employees distributed in approximately 3000 user-defined folders.

This Enron e-mail database consists of messages not only between employees but also from employees to external contacts. In addition to providing the e-mails, the database also consists of employee information such as their name and e-mail address. However, since we do not have information about their external contacts, we decided to limit our graph to the Enron employees and just their correspondences, allowing us to create a more complete "social" structure. In addition, since the body of e-mails consists of many words (and typos), we limited the textual nature of the e-mails to just the subject headers. From these decisions, we created a graph consisting of the structure shown in Figure 4.5. The message vertex can have multiple edges to multiple subject words, and multiple recipient type edges (i.e., TO, CC and BCC) to multiple persons. Running the GBAD algorithms on this data set produced the small normative pattern shown in Figure 4.6.

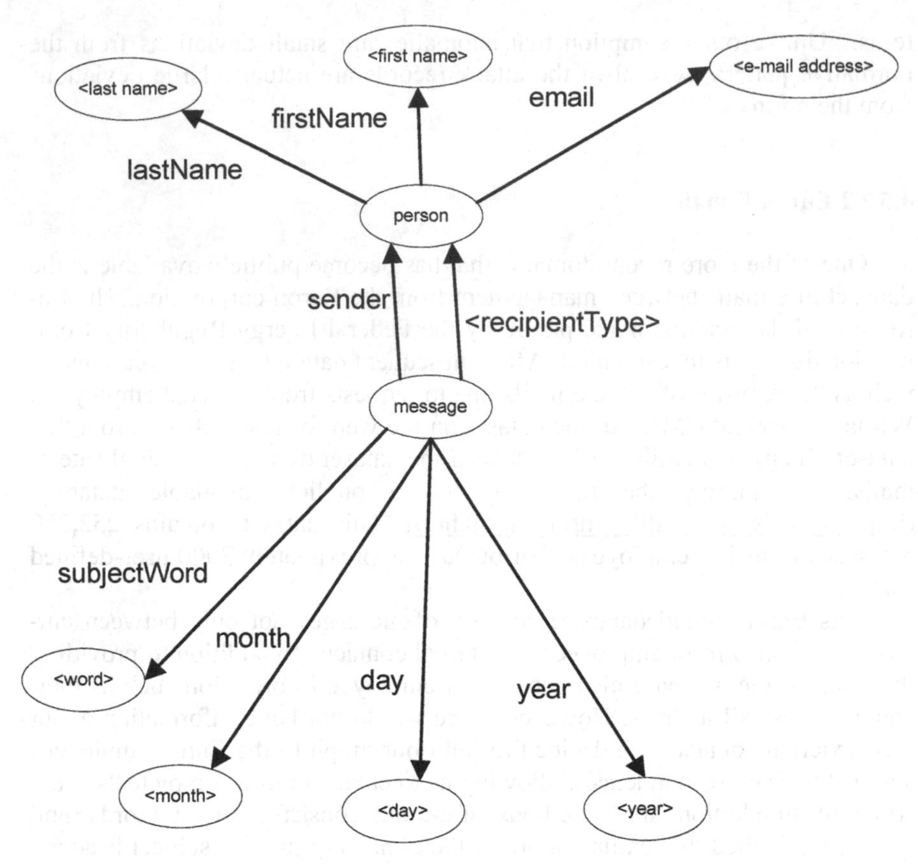

Fig. 4.5 Graphical representation of Enron e-mail.

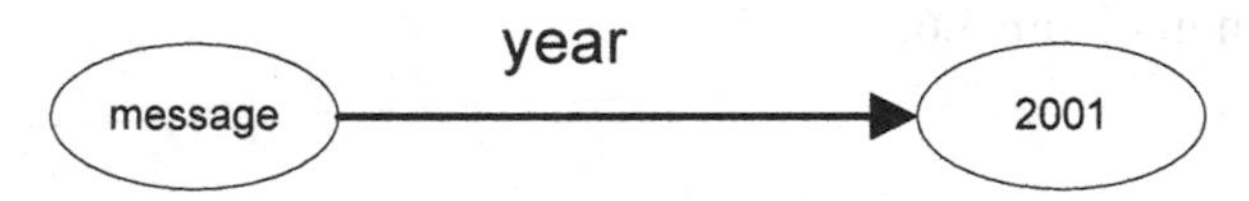

Fig. 4.6 Normative pattern from Enron e-mail data set.

This was an expected normative pattern because most of the e-mail was from the year 2001, with little regularity beyond the fact that messages were sent. Considering the small size of the normative pattern, we did not run the GBAD-MDL and GBAD-MPS algorithms, as clearly nothing of importance would be derived from a modification or deletion to this normative pattern. However, running the GBAD-P algorithm resulted in the substructure shown in Figure 4.7.

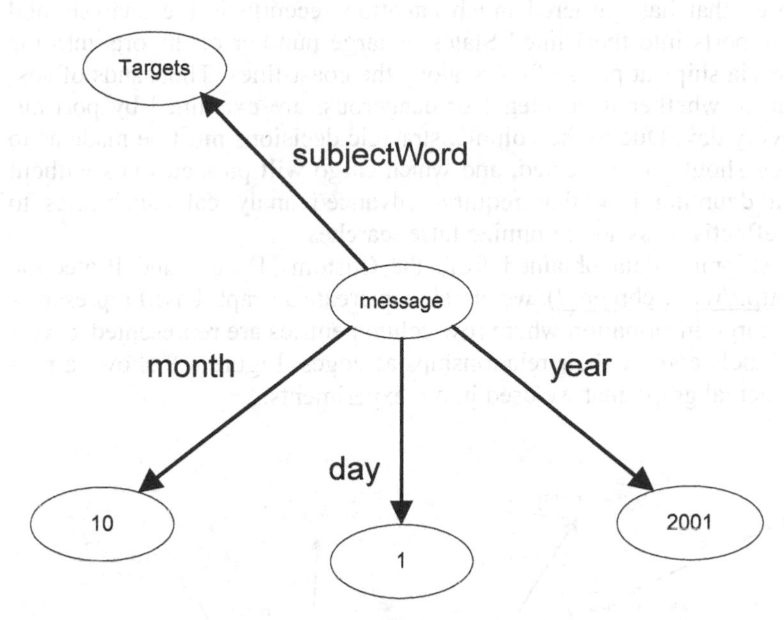

Fig. 4.7 Results from running GBAD-P on Enron e-mail data set.

It is interesting to point out that 859 messages were sent on October 1, 2001, and of all of the messages in the data set, this was the only one with a subject of "Targets", and 1 of only 36 messages in the entire dataset that had the word "Targets" anywhere in its subject line, and no messages anywhere that were a response to this message.

Also, analyzing just the transfer of e-mails between the employees, GBAD has been able to discover various anomalous patterns as they relate to specific individuals. For instance, in the case of Enron employee Mark Taylor (one of the employees with a high volume of e-mail traffic), the *normative* pattern consists of sending e-mails to three other employees. While the employees that are involved in e-mails with Mr. Taylor varies, the GBAD-P algorithm identified one instance as anomalous because it included a link to another employee, Tana Jones, who was not found anywhere else to have been part of a correspondence that contained this normative pattern associated with Mr. Taylor. While Ms. Jones is involved in many e-mails in the Enron dataset, the *structural* anomaly associated with the normative pattern of communication for Mr. Taylor results in GBAD detecting this instance of an anomalous substructure.

4.5.2.3 Cargo Shipments

One area that has garnered much attention recently is the analysis and search of imports into the United States. A large number of imports into the U.S. arrive via ships at ports of entry along the coast-lines. Thousands of suspicious cargo, whether it be illegal or dangerous, are examined by port authorities every day. Due to the volume, strategic decisions must be made as to which cargo should be inspected, and which cargo will pass customs without incident; a daunting task that requires advanced analytical capabilities to maximize effectiveness and minimize false searches.

Using shipping data obtained from the Customs Border and Protection Agency (http://www.cbp.gov/), we are able to create a graph-based representation of the cargo information where row/column entries are represented as vertices, and labels convey their relationships as edges. Figure 4.8 shows a portion of the actual graph that we used in our experiments.

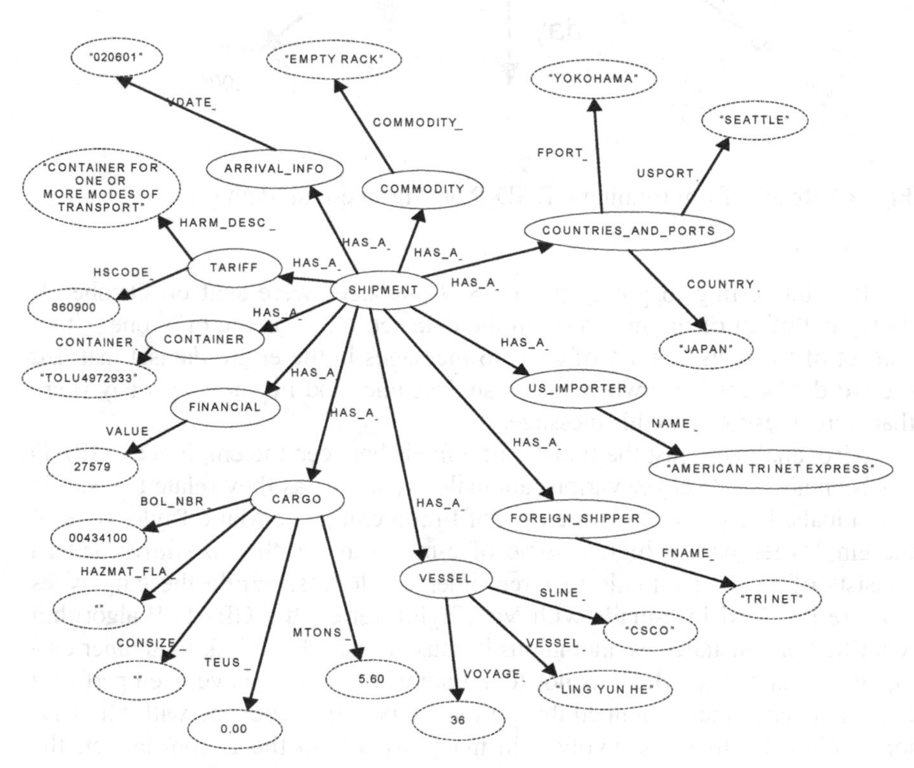

Fig. 4.8 Example graph of cargo information.

While we were not given any labeled data from the CBP, we can draw some results from simulations of publicized incidents. Take for instance the example from a press release issued by the U.S. Customs Service. The situation was that almost a ton of marijuana was seized at a port in Florida (US

Customs Service 2000). In this drug smuggling scenario, the perpetrators attempt to smuggle contraband into the U.S. without disclosing some financial information about the shipment. In addition, an extra port is traversed by the vessel during the voyage. For the most part, the shipment looks like it contains a cargo of toys and bicycles from Jamaica.

When we run all three algorithms on this graph, GBAD-MDL is unable to find any anomalies, which makes sense considering none of the anomalies are modifications. When the graph contains the anomalous insertion of the extra traversed port, the GBAD-P algorithm is able to successfully discover the anomaly. Similarly, when the shipment instance in the graph is missing some financial information, GBAD-MPS reports the instance as anomalous.

There are many different non-graph-based machine learning approaches to anomaly detection. In order to compare our results to a non-graph-based approach, we chose perhaps the most popular approach to anomaly detection - the class of approaches known as *clustering*. The idea behind clustering is the grouping of similar objects, or data. Clustering is an unsupervised approach whose goal is to find all objects that are similar where the class of the example is unknown (Frank and Witten 2005). From an anomaly detection perspective, those objects that fall outside a cluster (*outliers*), perhaps within a specified deviation, are candidate anomalies. Due to the popularity of this approach, and the fact that it is an unsupervised approach (like GBAD), we evaluate the effectiveness of the simple k-Means clustering approach on the cargo shipment data using the WEKA tool (WEKA). For the simple k-Means approach (*SimpleKMeans*), given a set of n data points in d-dimensional space R^d, and an integer k, the problem is to determine a set of k points in R^d, called centers, so as to minimize the mean squared distance from each data point to its nearest center (Kanungo et al. 2000).

First, we randomly select 200 cargo records and generate a graph from the chosen records. Second, we run the graph through SUBDUE to determine the normative pattern in the graph. Then, we generate multiple anomalies of each of the anomaly types (modifications, insertions and deletions), where each of the induced anomalies is similar to the real-world examples mentioned earlier (e.g., change in the name of a port), and the anomaly is part of a normative pattern. Only choosing anomalies that are small deviations (relative to the size of the graph), all three of the GBAD algorithms are able to successfully find all of the anomalies, with only one GBAD-MDL test and one GBAD-P test reporting false positives.

Using these same 200 records, we then convert the data into the appropriate WEKA format for the k-Means algorithm. For each of the tests involving what were vertex modifications in the graph, and are now value or field modifications in the text file, the k-Means algorithm is able to successfully find all of the anomalies. Similar to how we have to adjust GBAD parameters, we have to increase the default WEKA settings for the number of clusters and the seed, as the defaults do not produce any anomalous clusters on the cargo test set which consists of 12 attributes for 200 records. By increasing the number of clusters to 8 and the seed to 31, we are able to discover the anomalous modifications, as shown in the example in Figure 4.9.

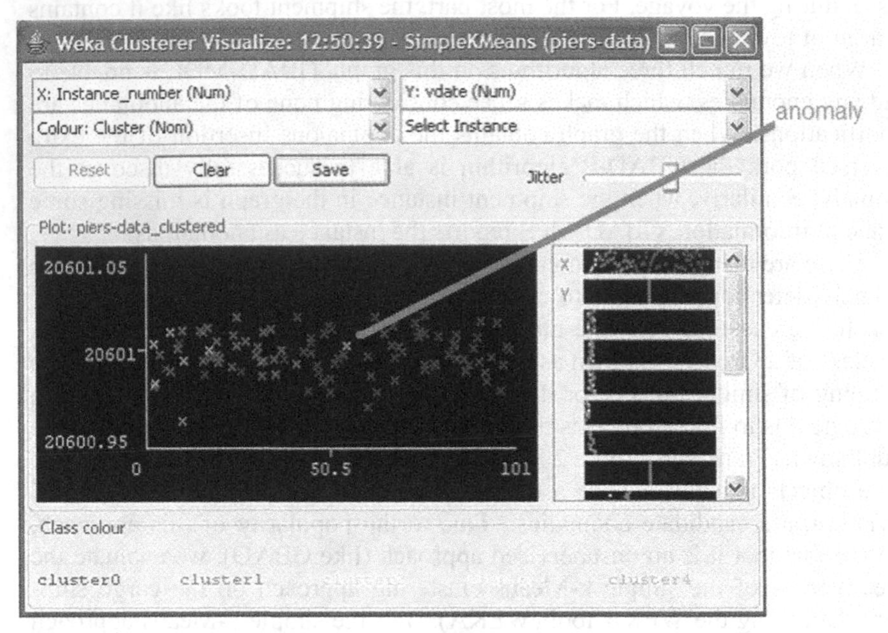

Fig. 4.9 Results from k-Means clustering algorithm on cargo data with anomalous modification.

In these tests, a cluster is considered anomalous if it contains only a single instance.

However, for insertions and deletions, the k-Means approach is not effective, but in some ways, that is to be expected. The k-Means algorithm assumes that every specified record is of the same length. So, in order to simulate anomalous insertions, extra attributes must be added to represent the extra vertices, where the values for those attributes are NULL, unless an anomalous insertion is present. Yet, despite the additional non-NULL attribute when the anomaly is present, the k-Means algorithm never reports an anomalous cluster for any of the tests. When we increase the number of clusters and the seed, it only increases the number of false positives (i.e., clusters of single instances that are not the anomalous instances). This is surprising in that we would have assumed that the unique value for an individual attribute would have been discovered. However, again, we are attempting to simulate a structural change, which is something that the k-Means algorithm (or other traditional clustering algorithms) is not intended to discover.

Similarly, we can only simulate an anomalous deletion by replacing one of the record's attributed values with a NULL value. Again, at first, we would have considered this to be the same as a modification, and clearly identifiable by the k-Means algorithm. But, the algorithm was unable to find the anomalous deletion in any of the tests, which leads us to believe that the presence of a NULL value has an effect on the functionality of the k-Means algorithm. The importance of these tests is to show that for some anomalies (specifically modifications) traditional machine learning approaches like clustering are also effective, and at the same time, the inability to discover anomalous insertions and deletions further justifies the use of an approach like GBAD for structural anomalies. In addition, approaches like k-Means are only able to report the anomalous record – not the specific anomaly within the record.

The use of the k-Means clustering algorithm for anomaly detection and intrusion detection has been reported in other research efforts (Portnoy 1999)(Caruso and Malerba 2004). For more information on how the WEKA tools work, please refer to the WEKA website (WEKA).

We also compared our algorithms against a traditional non-graph-based anomaly detection approach found in the commercially available application called Gritbot, from a company called RuleQuest (http://www.RuleQuest.com/). The objective of the Gritbot tool is to look for anomalous values that would compromise the integrity of data that might be further analyzed by other data modeling tools.

There are two required input files for Gritbot: a .names file that specifies the attributes to be analyzed, and a .data file that supplies the corresponding comma-delimited data. There are several optional parameters for running Gritbot, of which the most important is the "filter level". By default, the filter level is set at 50%. The lower the filter level percentage, the less filtering that occurs, resulting in more possible anomalies being considered.

In order to compare Gritbot to our GBAD algorithms, we gave Gritbot the same cargo data files used in the previous experiments (formatted to the Gritbot specifications). Using the default parameters, no anomalies were reported. We then lowered the filter level to 0 (which specifies that all anomalies are requested). In every case, anomalies were reported, but none of the anomalies reported were the ones we had injected into the data set. So, we increased the number of samples from 200 shipments to ~1000 shipments, so that Gritbot could infer more of a statistical pattern, and then randomly injected a single modification to the country-of-origin attribute. In the cargo data files, all of the country-of-origins were "JAPAN", except for the randomly selected records where the value was changed to "CHINA". Again, Gritbot did not report this anomaly (i.e. 1020 cases of "JAPAN" and one case of "CHINA"), and instead reported a couple of other cases as anomalous.

While we consider the existence of a record with "CHINA" as anomalous, Gritbot does not view that as an anomaly. The issue is that Gritbot (and this is similar to other outlier-based approaches), does not treat discrete attributes the same as numeric attributes. This is because Gritbot views continuous distributions (such as "age") as a much easier attribute to analyze because the distribution of values leads to certain expectations. While discrete distributions are more difficult because there is not a referential norm (statistically), it limits the tool's ability to provide its user with a comprehensive list of anomalies. That is not to say that Gritbot will not discover anomalous discrete values - it will if it can determine a statistical significance. For example, we found (when examining by hand) records that contained a significant number of identical attribute values (e.g., COUNTRY, FPORT, SLINE, VESSEL). In our data set, approximately 250 out of the approximately 1,000 records had identical SLINE values. When we arbitrarily modified the SLINE value of one of these cases from "KLIN" to "PONL" (i.e., another one of the possible SLINE values from this data set), Gritbot did not report the anomaly. When we changed it to "MLSL", Gritbot still did not report it. However, when we changed it to "CSCO", Gritbot reported that case as being anomalous (albeit, not the most anomalous). Why? This behavior is based on what Gritbot can determine to be statistically significant. Of all of the ~1,000 records, only 1 has an SLINE value of "MLSL", and only 3 have a value of "PONL". However, there are 123 records with an SLINE value of "CSCO". Thus, Gritbot was able to determine that a value of "CSCO" among those ~250 records is anomalous because it had enough other records containing the value "CSCO" to determine that its existence in these other records was significant. In short, the behavior depends upon the definition of what is an anomaly.

Gritbot's approach to anomaly detection is common among many other outlier-based data mining approaches. However, in terms of finding what we would consider to be anomalous (small deviations from the norm), Gritbot's approach typically does not find the anomaly.

4.5.2.4 Internet Movie Database (IMDb)

Another common source of data mining research is the Internet Movie Database (http://www.imdb.com/). This database consists of hundreds of thousands of movies and television shows, with all of the credited information such as directors, actors, writer, producers, etc. In their work on semantic graphs, Barthelemy et al. proposed a statistical measure for semantic graphs and illustrated these semantic measures on graphs constructed from terrorism data and data from the IMDb (Barthelemy et al, 2005). While they were not directly looking for anomalies, their research presented a way to measure useful relationships so that a better ontology could be created. Using bipartite graphs, Sun et al. presented a model for scoring the normality of nodes as they related to the other nodes (Sun et al, 2005). Using the IMDb database as one of their datasets, they analyzed the graph for just anomalous nodes.

In order to run our algorithms on the data, due to the voluminous amount of information, we chose to create a graph of the key information (title, director, producer, writer, actor, actress and genre) for the movies from 2006. Running the GBAD algorithms on this data set produced the normative pattern shown in Figure 4.10.

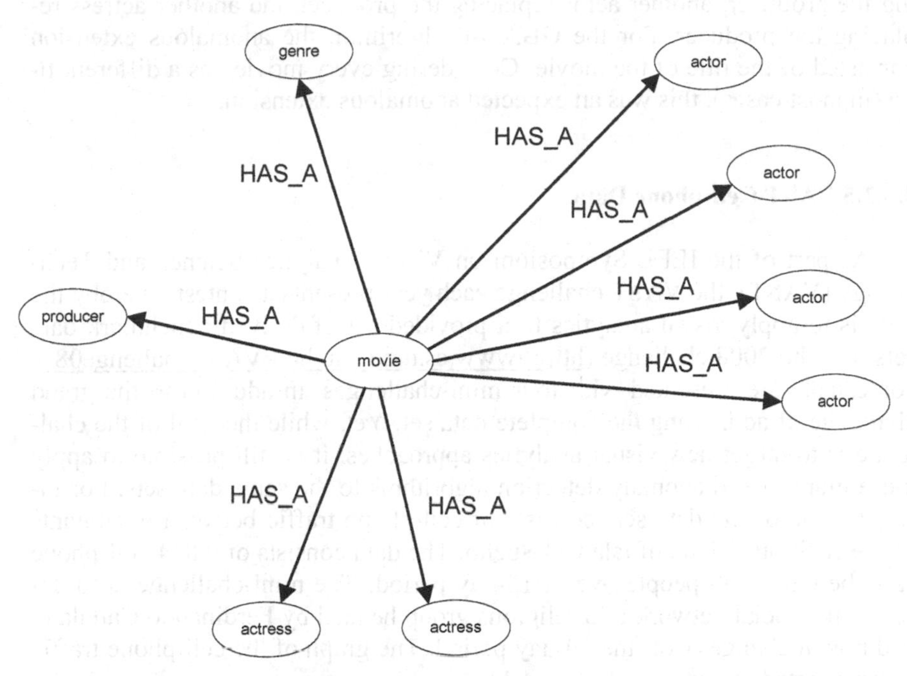

Fig. 4.10 Normative pattern from graph representation of movie data.

This pattern is not surprising, as movies typically consist of multiple actors and actresses. However, because of the size of the database, we chose a beam width that would produce seemingly relevant substructures within a reasonable amount of time. For instance, all of our runs are made with a default beam width of 4, resulting in reasonable running-times of approximately 15 minutes each. In contrast, with a beam width of 7, the running-times are nearly tripled. However, this does result in some additional elements being discovered, like another producer and more actors, so there is always the trade-off of time versus knowledge discovery. One may also wonder why some of the other movie elements like director and writer were not discovered. (Even with the aforementioned beam width increase they are not discovered.) This is due to the fact that several genres are reality shows, which do not require directors and writers, and many documentaries do not have writers credited in the IMDb.

Running the GBAD-MDL and GBAD-MPS algorithms on the IMDb data produced a variety of anomalies that all scored equally. The GBAD-MDL algorithm reported a single anomalous instance where an actor label was replaced by an anomalous actress label. The GBAD-MPS algorithm reported multiple anomalous instances consisting of a writer replacing an actor, a writer replacing the producer, another genre replacing the producer, a director replacing the producer, another actor replacing the producer and another actress replacing the producer. For the GBAD-P algorithm, the anomalous extension consisted of the title of the movie. Considering every movie has a different title (in most cases), this was an expected anomalous extension.

4.5.2.5 VAST Cell-phone Data

As part of the IEEE Symposium on Visual Analytics Science and Technology (VAST), the VAST challenge each year presents a contest whereby the goal is to apply visual analytics to a provided set of derived benchmark data sets. For the 2008 challenge (http://www.cs.umd.edu/hcil/VASTchallenge08/), contestants are presented with four mini-challenges, in addition to the grand challenge of addressing the complete data set. Yet, while the goal of the challenge is to target new visual analytics approaches, it is still possible to apply these graph-based anomaly detection algorithms to the same data sets. For instance, one of the data sets consists of cell-phone traffic between inhabitants of the fictitious island of Isla Del Sueño. The data consists of 9,834 cell-phone calls between 400 people over a 10-day period. The mini-challenge is to describe the social network of a religious group headed by Ferdinando Cattalano and how it changes over the 10-day period. The graph of the cell-phone traffic is represented as shown in Figure 4.11.

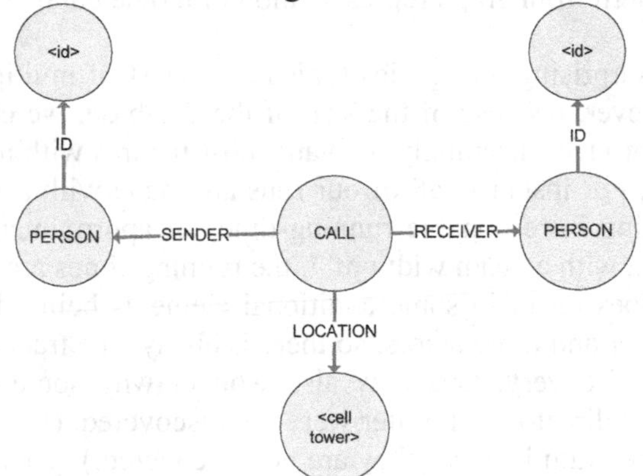

Fig. 4.11 Graph representation of a cell-phone call from the VAST dataset.

Applying the GBAD algorithms to this information results in several structural anomalies within the data, especially when particular individuals are analyzed in terms of their calling patterns. For instance, when we look at the calling patterns of individuals who correspond with Ferdinando Cattalano, identified in the challenge as ID 200, one notices several anomalous substructures, including some who contact Ferdinando on days that are out of the ordinary, and even some individuals who call others outside of their normal chain of cell-phone calls. In addition, a graph of the *social network* of phone usage (i.e., a phone call between two individuals indicating a social interaction), was also applied to the GBAD approaches, yielding additional anomalous behavior between targeted persons. From these results we are able to determine the members of the normative social network surrounding Ferdinando and how the network begins to breakdown after about day 6.

4.5.2.6 CAIDA

The Cooperative Association for Internet Data Analysis (CAIDA), is a publicly available resource for the analysis of IP traffic. Through a variety of workshops, publications, tools, and projects, CAIDA provides a forum for the dissemination of information regarding the interconnections on the internet. One of the core missions of CAIDA is to provide a data repository to the research community that will allow for the analysis of internet traffic and its performance (http://www.caida.org/data/). Using GBAD, we analyzed the CAIDA AS (Autonomous Systems) data set for normative patterns and possible anomalies. The AS data set represents the topology of the internet as the composition of various Autonomous Systems. Each of the AS units represents routing points through the internet. For the purposes of analysis, the data is represented as a graph, composed of 24,013 vertices and 98,664 edges, with each AS depicted as a vertex and an edge indicating a peering relationship between the AS nodes. The normative pattern for this graph is depicted in Figure 4.12. After running GBAD-P on the AS graph, the anomalous substructure discovered is shown in Figure 4.13.

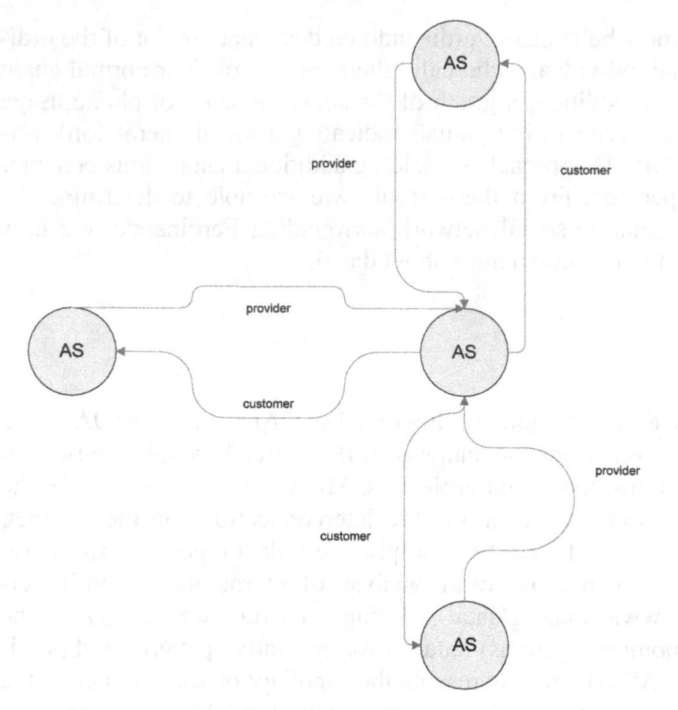

Fig. 4.12 Normative pattern discovered in the CAIDA dataset.

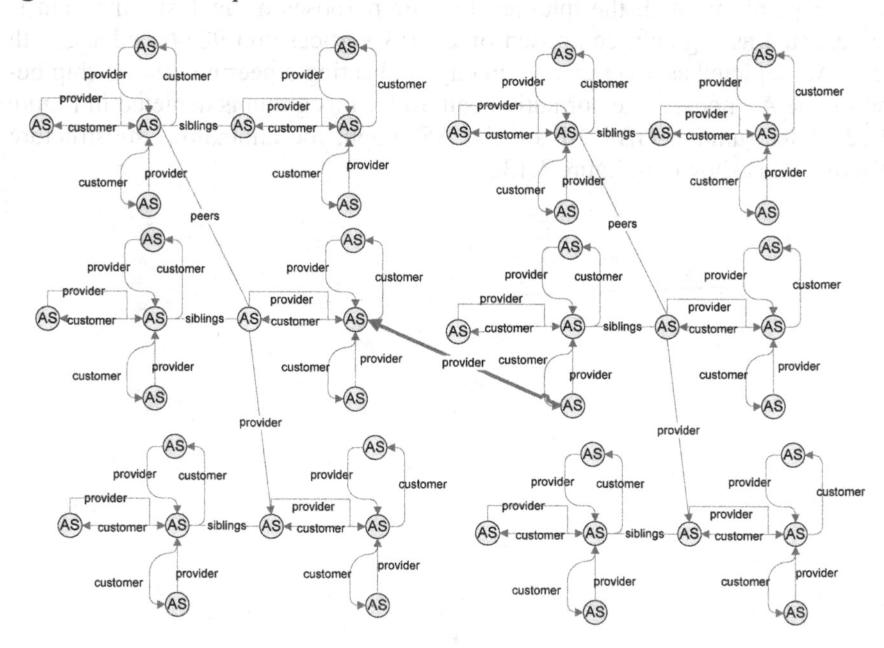

Fig. 4.13 Anomalous pattern discovered in the CAIDA dataset, where the crucial anomalous component is the central "provider" edge (emphasized).

This example shows the advantage of using a graph-based approach on a complex structure. While the data indicates many provider/customer relationships, of which the norm is a particular AS being the provider to three different customers, this single substructure indicates an unusual connection between two ASes. Such a complex structure would probably be missed by a human analyst, and shows the potential of an approach like GBAD to find these complex anomalies in network traffic data.

4.5.3 Other Domains

Since 9/11, one of the more common domains used in data mining consists of terrorist activity and relationships. Organizations such as the Department of Homeland Security use various techniques to discover the inherent patterns in the network representation of known terrorists and their relations (Kamarck 2002). Much research has been applied to not only understanding terrorist networks (Sageman 2004), but also discovering the patterns that discriminate the terrorists from the non-terrorists (Taipale 2003). Much of this area of research has also been applied to what is known as social network analysis, which is a more general term for the measuring and mapping of relationships between people, places and organizations (Mukherjee and Holder 2004).

Through the Evidence Assessment, Grouping, Linking, and Evaluation (EAGLE) program, under the auspices of the Air Force Research Lab (AFRL), we have been able to gather counter-terrorism data. While the data is simulated, it does represent scenarios based on input from various intelligence analysts (Holder et al. 2005). The data represents different terrorist organization activities as they relate to the exploitation of vulnerable targets. Our goal is to use this data as another example of real-world data to further validate the effectiveness of this approach. If a terrorist can be distinguishable from a non-terrorist by a small deviation in a normative pattern, we should be able to discover actual terrorist instances within a network of people and relationships.

Another domain worth investigating is data from the Financial Crimes Enforcement Network (FinCEN, http://www.fincen.gov). The purpose of FinCEN is to analyze financial transactions for possible financial crimes including terrorist financing and money laundering. Again, if illegal transactions consist of small deviations from normal transactions, we should be able to uncover genuine fraudulent activity within a network of people and their related monetary dealings.

Similar techniques could be applied to a myriad of domains, including telecommunications call records and credit card transactions. In short, any data source where transactions and relationships can be represented structurally as a graph, and possible anomalous behavior consists of minor deviations from normal patterns, these approaches to graph-based anomaly detection could prove to be a viable alternative to more traditional anomaly detection methods. In addition, by analyzing the effectiveness of our algorithms against real-world, labeled data sets, we can establish a baseline of comparison that can be used in subsequent anomaly detection endeavors.

4.6 Related Work

Recently there has been an impetus towards analyzing multi-relational data using graph-theoretic methods. Not to be confused with the mechanisms for analyzing "spatial" data, graph-based data mining approaches are an attempt at analyzing data that can be represented as a graph (i.e., vertices and edges). Yet, while there has been much written as it pertains to graph-based intrusion detection (Staniford-Chen et al. 1996), very little research has been accomplished in the area of graph-based anomaly detection.

In 2003, Noble and Cook used the SUBDUE application to look at the problem of anomaly detection from both the anomalous substructure and anomalous subgraph perspective (Noble and Cook 2003). They were able to provide measurements of anomalous behavior as it applied to graphs from two different perspectives. *Anomalous substructure* detection dealt with the unusual substructures that were found in an entire graph. In order to distinguish an anomalous substructure from the other substructures, they created a simple measurement whereby the value associated with a substructure indicated a degree of anomaly. They also presented the idea of *anomalous subgraph* detection which dealt with how anomalous a subgraph (i.e., a substructure that is part of a larger graph) was to other subgraphs. The idea was that subgraphs that contained many common substructures were generally less anomalous than subgraphs that contained few common substructures. In addition, they also explored the idea of conditional entropy and data regularity using network intrusion data as well as some artificially created data.

(Lin and Chalupsky 2003) took a different approach and applied what they called rarity measurements to the discovery of unusual *links* within a graph. Using various metrics to define the commonality of paths between nodes, the user was able to determine whether a path between two nodes was interesting or not, without having any preconceived notions of meaningful patterns. One of the disadvantages of this approach was that while it was domain independent, it assumed that the user was querying the system to find interesting relationships regarding certain nodes. In other words, the unusual patterns had to originate or terminate from a user-specified node.

The AutoPart system presented a non-parametric approach to finding outliers in graph-based data (Chakrabarti 2004). Part of Chakrabarti's approach was to look for outliers by analyzing how *edges* that were removed from the overall structure affected the minimum descriptive length (MDL) of the graph. Representing the graph as an adjacency matrix, and using a compression technique to encode node groupings of the graph, he looked for the groups that reduced the compression cost as much as possible. Nodes were put into groups based upon their entropy.

In 2005, the idea of entropy was also used by (Shetty and Adibi 2005) in their analysis of a real-world data set: the famous Enron scandal. They used what they called "event based graph entropy" to find the most interesting people in an Enron e-mail data set. Using a measure similar to what Noble and Cook had proposed, they hypothesized that the important nodes (or people) were the ones who had the greatest effect on the entropy of the graph when they were removed. Thus, the most interesting node was the one that brought about the maximum change to the graph's entropy. However, in this approach, the idea of important nodes did not necessarily mean that they were anomalous.

In the December 2005 issue of SIGKDD Explorations, two different approaches to graph-based anomaly detection were presented. Using just *bipartite* graphs, (Sun et al. 2005) presented a model for scoring the normality of nodes as they relate to the other nodes. Again, using an adjacency matrix, they assigned what they called a "relevance score" such that every node x had a relevance score to every node y, whereby the higher the score the more related the two nodes. The idea was that the nodes with the lower normality score to x were the more anomalous ones to that node. The two drawbacks with this approach were that it only dealt with bipartite graphs and it only found anomalous nodes, rather than what could be anomalous substructures. In (Rattigan and Jensen 2005), they also went after anomalous links, this time via a *statistical* approach. Using a Katz measurement, they used the link structure to statistically predict the likelihood of a link. While it worked on a small dataset of author-paper pairs, their single measurement just analyzed the links in a graph.

In (Eberle and Holder 2006), we analyzed the use of *graph properties* as a method for uncovering anomalies in data represented as a graph. While our initial research examined many of the basic graph properties, only a few of them proved to be insightful as to the structure of a graph for anomaly detection purposes: average shortest path length, density and connectedness. For a measurement of *density*, we chose to use a definition that is commonly used when defining social networks (Scott 2000). For *connectedness*, we used a definition from (Broder et al. 2000). Then, for some of the more complex graph properties, we investigated two measurements. First, there is the maximum *eigenvalue* of a graph (Chung et al. 2003). Another, which was used in identifying e-mail "spammers", is the *graph clustering coefficient* (Boykin and Roychowdhury 2005). We tested these approaches on the aforementioned cargo data when injecting one of two real-world anomalies related to drugs and arms smuggling. No significant deviations are seen using the average shortest path or eigenvalue metrics. However, there are visible differences for the density, connectedness and clustering coefficient measurements. One issue with this approach is that while graphs are indicated as anomalous, this does not identify the specific anomaly within what could be a very large graph. However, the algorithms presented in this work rectify that problem by not only indicating a graph contains an anomaly, but more importantly, they identify the specific anomaly and its pertinent structure within the graph.

4.7 Conclusions

The purpose of this chapter was to present an approach for discovering the three possible graph anomalies: modifications, insertions and deletions. Using a practical definition of fraud, we designed algorithms to specifically handle the scenario where the anomalies are small deviations to a normative pattern. We have described three novel algorithms, each with the goal of uncovering one of the specified anomalous types. We have validated all three approaches using synthetic data. The tests verified each of the algorithms on graphs and anomalies of varying sizes, with the results showing very high detection rates with minimal false positives. We further validated the algorithms using real-world cargo data and actual fraud scenarios injected into the data set. Despite a less regular set of data, normative patterns did exist, and changes to those prevalent substructures were detected with 100% accuracy and no false positives. We also compared our algorithms against a graph-based approach using intrusion detection data, again with better discovery rates and lower false positives. We also looked at other real-world datasets where we were able to show unusual patterns in diverse domains. In short, the GBAD algorithms presented here are able to consistently discover graph-based anomalies that are comprised of the smallest deviation of the normative pattern, with minimal false positives.

There have been many approaches to anomaly detection over the years, most of which have been based on statistical methods for determining outliers. As was shown here, recent research in graph-based approaches to data mining have resulted in new methods of anomaly detection. This work shows promising approaches to this problem, particularly as it is applied to threat detection. However, there are still many avenues to be explored. One avenue is to detect anomalies in weighted graphs, where the relational edges are associated with weights reflecting the extent of the relationship (e.g., trust). Another avenue is to detect anomalies in graphs changing over time both in terms of their attributes and structure. Techniques for detecting anomalies in real-time within such data are crucial for securing our cyber-infrastructure against modern, sophisticated threats.

References

[1]. Barthélemy, M., Chow, E. and Eliassi-Rad, T, *Knowledge Representation Issues in Semantic Graphs for Relationship Detection.* AI Technologies for Homeland Security: Papers from the 2005 AAAI Spring Symposium, AAAI Press, 2005, pp. 91-98.

[2]. Boykin, P. and Roychowdhury, V. *Leveraging Social Networks to Fight Spam.* IEEE Computer, April 2005, 38(4), 61-67, 2005.

[3]. Broder, A., Kumar, R., Maghoul, F., Raghavan, P., Rajagopalan, S., Stata, R., Tomkins, A. and Wiener, J. *Graph Structure in the Web.* Computer Networks, Vol. 33, 309-320, 2000.

[4]. Caruso, C. and Malerba, D. *Clustering as an add-on for firewalls.* Data Mining, WIT Press, 2004.

[5]. Chakrabarti, D. *AutoPart: Parameter-Free Graph Partitioning and Outlier Detection.* Knowledge Discovery in Databases: PKDD 2004, 8th European Conference on Principles and Practice of Knowledge Discovery in Databases, 112-124, 2004.

[6]. Chung, F., Lu, L., Vu, V. *Eigenvalues of Random Power Law Graphs.* Annals of Combinatorics, 7, 21-33, 2003.

[7]. Cook, D. and Holder, L. *Graph-based data mining.* IEEE Intelligent Systems 15(2), 32-41, 2000.

[8]. Cook, D. and Holder, L. *Mining Graph Data.* John Wiley and Sons, 2006.

[9]. Eberle, W. and Holder, L. *Detecting Anomalies in Cargo Shipments Using Graph Properties.* Proceedings of the IEEE Intelligence and Security Informatics Conference, 2006.

[10]. Frank, E. and Witten, I. *Data Mining: Practical Machine Learning Tools and Techniques.* Morgan Kaufman, Second Edition, 2005.

[11]. Gross, J, and Yellen, J. *Graph Theory and Its Applications.* CRC Press. 1999.

[12]. Gudes, E. and Shimony, S. *Discovering Frequent Graph Patterns Using Disjoint Paths.* IEEE Transactions of Knowledge and Data Engineering, 18(11), November 2006.

[13]. Holder, L., Cook, D. and Djoko, S. *Substructure Discovery in the SUBDUE System.* Proceedings of the AAAI Workshop on Knowledge Discover in Databases, pp. 169-180, 1994.

[14]. Holder, L., Cook, D., Coble, J., and Mukherjee, M. *Graph-based Relational Learning with Application to Security.* Fundamenta Informaticae Special Issue on Mining Graphs, Trees and Sequences, 66(1-2):83-101, March 2005.

[15]. Huan, J., Wang, W. and Prins, J. *SPIN: Mining Maximal Frequent Subgraphs from Graph Databases.* Knowledge Discovery and Data Mining, KDD '04, 2004.

[16]. KDD Cup 1999. Knowledge Discovery and Data Mining Tools Competition. http://kdd.ics.uci.edu/databases/kddcup99/kddcup99.html. 1999.

[17]. Kamarck, E. *Applying 21st Century Government to the Challenge of Homeland Security.* Harvard University, PriceWaterhouseCoopers, 2002.

[18]. Kanungo, T, Mount, D., Netanyahu, N., Piatko, C., Silverman, R. and Wu, A. *The Analysis of a Simple k-Means Clustering Algorithm.* Proceedings on the 16th Annual Symposium on Computational Geometry, 100-109, 2000.

[19]. Kuramochi, M. and Karypis, G. *An Efficient Algorithm for Discovering Frequent Subgraphs.* IEEE Transactions on Knowledge and Data Engineering, pp. 1038-1051, 2004.

[20]. Kuramochi, M. and Karypis, G. *Grew – A Scalable Frequent Subgraph Discovery Algorithm.* IEEE International Conference on Data Mining (ICDM '04), 2004.

[21]. Lin S. and Chalupsky, H. *Unsupervised Link Discovery in Multi-relational Data via Rarity Analysis*. Proceedings of the Third IEEE ICDM International Conference on Data Mining, 171-178, 2003.
[22]. Mukherjee, M. and Holder, L. *Graph-based Data Mining on Social Networks*. Workshop on Link Analysis and Group Detection, KDD, 2004.
[23]. Noble, C. and Cook, D. *Graph-Based Anomaly Detection*. Proceedings of the 9th ACM SIGKDD International Conference on Knowledge Discovery and Data Mining, 631-636, 2003.
[24]. Portnoy, L., Eskin, E. and Stolfo, S. *Intrusion detection with unlabeled data using clustering*. Proceedings of ACM CSS Workshop on Data Mining Applied to Security, 2001.
[25]. Rattigan, M. and Jensen, D. *The case for anomalous link discovery*. ACM SIGKDD Explor. Newsl., 7(2):41-47, 2005.
[26]. Sageman, M. *Understanding Terror Networks*. University of Pennsylvania Press, 2004.
[27]. Scott, J. *Social Network Analysis: A Handbook*. SAGE Publications, Second Edition, 72-78, 2000.
[28]. Shetty, J. and Adibi, J. *Discovering Important Nodes through Graph Entropy: The Case of Enron Email Database*. KDD, Proceedings of the 3rd international workshop on Link discovery, 74-81, 2005.
[29]. Staniford-Chen, S., Cheung, S., Crawford, R., Dilger, M., Frank, J., Hoagland, J. Levitt, K., Wee, C., Yip, R. and Zerkle, D. *GrIDS – A Graph Based Intrusion Detection System for Large Networks*. Proceedings of the 19th National Information Systems Security Conference, 1996.
[30]. Sun, J, Qu, H., Chakrabarti, D. and Faloutsos, C. *Relevance search and anomaly detection in bipartite graphs*. SIGKDD Explorations 7(2), 48-55, 2005.
[31]. Taipale, K. *Data Mining and Domestic Security: Connecting the Dots to Make Sense of Data*. Columbia Science and Technology Law Review, 2003.
[32]. Thomas, L., Valluri, S. and Karlapalem, K. *MARGIN: Maximal Frequent Subgraph Mining*. Sixth International Conference on Data Mining (ICMD '06), 109-1101, 2006.
[33]. U.S. Customs Service: *1,754 Pounds of Marijuana Seized in Cargo Container at Port Everglades*. November 6, 2000. (http://www.cbp.gov/hot-new/pressrel/2000/1106-01.htm)
[34]. WEKA, *http://www.cs.waikato.ac.nz/~ml/index.html*.
[35]. West, D. *Introduction to Graph Theory*. Prentice-Hall International. Second Edition. 2001.
[36]. Yan, X. and Han, J. *gSpan: Graph-Based Substructure Pattern Mining*. Proceedings of International Conference on Data Mining, ICDM, pp. 51-58, 2002.
[37]. Zeng, Z., Wang, J., Zhou, L. and Karypis, G. *Coherent closed quasi-clique discovery from large dense graph databases*. Conference on Knowledge Discovery in Data, SIGKDD, 797-802, 2006.

5 On the Performance of Online Learning Methods for Detecting Malicious Executables

Marcus A. Maloof

Abstract We present results from an empirical study of seven online-learning methods on the task of detecting previously unseen malicious executables. Malicious software has disrupted computer and network operation and has compromised or destroyed sensitive information. Methods of machine learning, which build predictive models that generalize training data, have proven useful for detecting previously unseen malware. In previous studies, batch methods detected malicious and benign executables with high true-positive and true-negative rates, but doing so required significant time and space, which may limit applicability. Online methods of learning can update models quickly with only a single example, but potential trade-offs in performance are not well-understood for this task. Accuracy of the best performing online methods was 93%, which was 3-4% lower than that of batch methods. For applications that require immediate updates of models, this may be an acceptable trade-off. Our study characterizes these trade-offs, thereby giving researchers and practitioners insights into the performance of online methods of machine learning on the task of detecting malicious executables.

5.1 Introduction

For people to trust systems, they must be secure, usable, dependable, and available. They must also preserve user privacy. Methods of machine learning and data mining can be useful for achieving all of these objectives. Indeed, researchers have developed and applied methods to tasks important for cyber trust, such as detecting insider threats (e.g., [28]), detecting intrusions (e.g., [45]), tuning systems that detect intrusions (e.g., [44]), and detecting anomalies in network flows (e.g., [11]) and in web applications (e.g., [7]).

Marcus A. Maloof
Department of Computer Science, Georgetown University, Washington, DC 20057-1232, USA
e-mail: maloof@cs.georgetown.edu

J.J.P. Tsai and P.S. Yu (eds.), *Machine Learning in Cyber Trust: Security, Privacy, and Reliability*, DOI: 10.1007/978-0-387-88735-7_5,

In previous work, we investigated how one might use methods of machine learning to detect and classify malicious executables [17, 18]. As we discuss further in Sect. 5.3, we considered the task of detecting previously unseen malicious executables and analyzed them as they would appear on a user's hard drive (i.e., "in the wild"). A system reliably performing this detection task could be the basis of a tool to help computer-forensic experts analyze hard drives or could supplement existing techniques for detecting malicious software, such as those that use signatures extracted from previously detected malware.

In this chapter, we examine the performance of seven online methods of machine learning on the task of detecting malicious executables. Results from our previous studies suggest that batch methods of machine learning detect malicious executables with high true-positive and true-negative rates, but these methods can be expensive in time and space [17, 18]. Online methods of machine learning, some of which can learn from a single new example, may help mitigate the computational complexity of batch methods. However, there will likely be trade-offs in performance. For example, without the benefit of processing all available training examples at once, online methods may not detect malicious executables as well as batch methods. Unfortunately, these trade-offs are not well-understood because there has been little or no work focused on characterizing online learning methods for tasks in computer security. Indeed, to the best of our knowledge, ours is the first study that evaluates a variety of online-learning methods on the task of detecting malicious executables. The main contribution of this chapter is our discussion and analysis of a rigorous evaluation of seven online-learning methods on the task of detecting malicious executables. All of these methods have appeared in the literature; some have been applied to real-world problems, some have not. None has been applied previously to the online task we consider herein. Although the detect rates and accuracies of these online methods were not higher than those of batch methods [18], they were high enough to warrant further study, especially if one takes into account the expense of running batch methods on a large collection of executables to accommodate only a few new examples. For example – and we discuss our previous results more thoroughly in Sect. 5.3 – the best-performing online method achieved a detect rate of .95, whereas the best-performing batch method achieved a detect rate of .98. While this difference may seem small, one must take into account the number of executables one might need to classify and the cost of not detecting a malicious executable, which usually will be quite high, since malicious software can disrupt computer and network operation and destroy or compromise sensitive information. The organization of this chapter is as follows. In an attempt to make the article self-contained, Sect. 5.2 is a brief overview of machine learning. In Sect. 5.3, we discuss and motivate the task of detecting malicious executables and the role of online methods of machine learning. In Sect. 5.4, we describe the seven online-learning methods included in our study. We then explain how we converted executables into training data for learning algorithms (Sect. 5.5). Section 5.6 is where we detail the experimental design and the results from the empirical evaluation of the learning methods. Finally, in Sect. 5.7, we discuss some of the implications of our findings and analysis, and suggest avenues for future work.

5.2 Preliminaries

In an effort to make the chapter self-contained, in this section, we briefly discuss basic concepts of machine learning and evaluation. Readers may wish to supplement this presentation with material from textbooks on artificial intelligence (e.g., [33]) and on machine learning (e.g., [4, 21, 30, 42]). Discussion in these books is oriented toward batch learning, so for additional information about online learning, readers may need to consult the papers cited herein. One can also find introductory chapters on machine learning and information assurance in [26] (e.g., [27, 38])

5.2.1 Machine Learning

A method of supervised machine learning builds a predictive model from training data. One can then use the model to make predictions about new observations. Let $S = \{\mathbf{x}, y\}_{t=1}^{T}$ be training data such that $y_t = f(\mathbf{x}_t)$ and f is unknown. We refer to $\mathbf{x}$ as an *observation*, y as the *class label*, and the pair $\langle \mathbf{x}, y \rangle$ as an *example*.

Each dimension of $\mathbf{x}$ corresponds to an *attribute*, and we refer to a value of x_i as an attribute value. y is also an attribute, but since it is the one we wish to predict, it is traditionally treated separately. The domains of y and the components of $\mathbf{x}$ can be numeric or categorical. For categorical attributes, let v_{ik} be the *k*th value of the *i*th attribute's domain. When we refer to x_i, it is understood that $x_i \equiv [x_i = v_k] \equiv [x_i = v_{ik}]$, for some value of *k*. For the purposes of this article, we will define the domains of the components of $\mathbf{x}$ to be Boolean (i.e., $x_i \in \{T, F\}$), and we will define the domain of y to be either a positive or negative example of the class of interest (i.e., $y \in \{-, +\}$).

A learning method consists of three components: the concept description language, the learning element, and the performance element. The concept description language is the language the method uses to represent a model. Popular representations include probabilities, trees, rules, logic (propositional and first-order), and equations (linear and non-linear). Herein, we will use $\hat{f}$ to refer to the model, which is an approximation to the function f that generated the training data. The learning element takes as input the training examples and produces a model expressed in the concept description language. That is, it takes S and produces $\hat{f}$. The performance element uses the model to make a prediction about a given observation, meaning it takes $\hat{f}$ and $\mathbf{x}$, and returns $\hat{y} \in \{-, +\}$.

Batch learning is a setting in which one provides all pertinent or available training data to a learning method in a single batch. After building a suitable model, one can use it to make predictions about new observations. *Online learning* is a setting in which examples and observations are distributed over time. Learning and prediction must occur simultaneously. One can always use a batch-learning method in an online setting by saving training examples, adding to them any new training examples, and reapplying the method, which may be costly in time and space.

The advantage of batch learning is the learner uses all available examples to build its model. The main disadvantage is running time, since it depends on the number of examples and the number of attributes (and potentially the number of values as well). In an online setting, the cost of using batch methods to rebuild a model every time new examples arrive could be impractical, depending on the application. An alternative is to use online-learning methods, which can learn and update a model with a single new example. This process can be much more efficient in time and space than is rebuilding a model from scratch. A disadvantage is that performance may suffer because of *ordering effects*. For example, a new example may cause the learner to adjust its model in a way that it now misclassifies observations that it previously classified correctly.

The methods discussed thus far build and use a single model, but drawing from the adage that two heads are better than one, researchers have recently begun investigating *ensemble methods*, which maintain and use multiple models. Generally, a given learning method will produce the same model for a given a training set, so ensemble methods must have a mechanism to achieve *diversity*. Researchers have sampled examples of the training set, sampled attributes of the training set, used different learning methods, and weighted models based on their performance.

Boosting [34] is an ensemble method, and it produced the best performing classifier in our study of batch-learning methods for detecting malicious executables [18]. Briefly, the algorithm for adaptive boosting (AdaBoost) [12] incrementally creates a collection of weighted classifiers. It begins with a uniformly weighted set of training examples and builds a model, which it uses to calculate a weight based on the model's error on the examples. The method then uses this weight to calculate new weights for the examples in the training set. Over a series of iterations, examples that are "difficult" to classify will have larger weights than those that are "easy." The learning method takes into account the weights of the examples, and the process repeats until producing a prescribed number of weighted models. To classify an observation, AdaBoost obtains a prediction from the models in the ensemble, and returns the weighted-majority vote as the overall prediction. Ensemble methods in general and boosting in particular have received much attention recently because they tend to improve accuracy over methods that use only a single model for prediction.

Finally, concept drift [25, 36] is an online setting in which the process generating examples changes over time. Although we do not explicitly deal with this issue in this article, such considerations are important for applications of machine learning to problems in computer security. Moreover, a number of the algorithms we included in this study were originally designed to cope with concept drift, so we mention it for context.

5.2.2 Evaluation

Evaluation is critically important. Learning algorithms carry out inductive inference, reasoning from facts (i.e., training examples) to hypotheses (i.e., models). Such inference in all but the trivial case is uncertain. Consequently, it is important to estimate this uncertainty, to estimate the true error rate of the model.

Researchers have developed a number of evaluation methods. In an offline setting, all generally make use of the currently available examples by randomly splitting them into training and testing sets. One applies a learning method to the examples in the training set and then evaluates the resulting model on the examples in the testing set. Since the examples in the testing set have labels, one can compute a variety of performance measures, such as accuracy, which is the proportion of the testing examples the model correctly classifies.

For batch learning, a popular choice is k-fold cross-validation, in which one uses random selection to create k partitions of the data set. For $i = 1, \ldots, k$, one uses the ith partition as the testing set, forms a training set from the union of the other partitions, applies the learning method to the training set, and evaluates the resulting model on the testing set. To obtain an overall measure of performance, one averages the performance measures from the k runs. As a measure of variability, one can compute a 95% confidence interval.

For online learning, an analog of k-fold cross-validation is again to use random selection to create k partitions of the data set. As before, one uses the ith partition as the testing set, but one applies the online-learning method to the other partitions in succession and evaluates the resulting model on the testing set after each application. One produces a measure of performance for each application, which forms a *performance curve* that shows how performance changes with increasing numbers of training examples. After k runs, one can average the performance metrics at each point of the performance curve and compute a measure of dispersion, such as the 95% confidence interval.

A second method of evaluating methods of online learning is, again, to use random selection to create k partitions of the data set. Rather than selecting one to be the testing set, one processes the partitions in some order. For $i = 1,\ldots,k - 1$, one applies the learning method to the ith partition and evaluates the resulting model on the i+1th partition. This again produces a performance curve. One repeats either by generating new partitions or by reordering the current partitions. One then averages the performance measures at each point on the performance curve. One can evaluate performance on the first partition by using a default rule, such as predicting the majority class or randomly selecting a class to predict.

For this study, we selected this second method of evaluation. We tested and trained each online method on individual examples, meaning that each partition contained a single example. This design best represents how online methods would be used to detect malicious executables in a real-world application, as discussed in the next section.

5.3 Detecting Malicious Executables

Detecting malicious executables is one of the many tasks important for securing computers. Early research concentrated on detecting malicious code in the form of computer viruses (e.g., [15, 24, 37, 40]). More recently, researchers have extended their investigation and analysis from malicious code to malicious executables (e.g., [3, 17, 18, 37]).

In previous work, we evaluated the ability of batch methods of machine learning to detect and classify malicious executables [17, 18]. We obtained an initial collection of 1,651 malicious and 1,971 benign executables in July of 2003, which we discuss further in Sect. 5.5. Based on area under the receiver operating characteristic (ROC) curve, we concluded that such learning methods showed promise for detecting malicious executables at high true-positive and true-negative rates. Indeed, boosted decision trees outperformed all of the other methods included in the study, achieving an area of .996±.002.

We also considered a classification task in which we used batch methods of machine learning to predict the function of the malicious executables [18]; that is, whether the executable opened a backdoor, was a virus loader, or contained a mass-mailer. Accuracy on this task was less than that on the detection task, but boosted decision trees achieved areas under the ROC curve of .889±.015 for detecting mass-mailing functions, .870±.016 for detecting backdoor functions, and .991±.017 for detecting virus-loader functions.

One issue with these experiments is we evaluated the methods using ten-fold cross-validation. This is a standard method of evaluating learning algorithms, but it presents a problem for this task, because there were occasions when we trained on newer malicious executables and tested on older ones. Unfortunately, we had little choice, since we had no information for the executables in our collection about when they initially appeared in the wild.

In an attempt to address this issue, we conducted an evaluation of online performance (not online learning) of the methods by collecting the malicious executables that appeared after we obtained our initial collection. In August of 2004, we gathered an additional 3,082 executables and subsequently verified that 291 of these executables were indeed malicious and had appeared in the wild between July of 2003 and August of 2004.

We trained the learning methods on the initial collection of executables and evaluated them on each of the new malicious executables in the order they appeared in the wild. For three different decision thresholds, boosted decision trees outperformed the other methods and performed almost as well in this setting as it did in an offline setting with ten-fold cross-validation. For example, for a desired true-negative rate of .95, boosted decision trees achieved a true-positive rate of .99 in the offline setting and .98 in the online setting. Overall, boosted decision trees had the most consistent performance in the offline and online settings, as compared to the other methods included in the study [18].

We found these results encouraging, that the batch method of boosted decision trees achieved high rates of detection in both the offline and online settings, especially since the malicious executables that comprised the online evaluation had not been part of our initial experimental study [16]. Indeed, these results suggest that machine learning methods could be used in a commercial system for detecting previously unseen malicious executables.

If one were to construct such a system, then to use batch methods of machine learning, one would have to maintain a collection of all known malicious executables and a large collection of benign executables, including those of different versions. One might attempt to decrease the burden of this requirement by extracting and storing the relevant features from executables as they are discovered, but it is quite possible that features relevant for prediction will change over time, especially as new technologies emerge and as those who write malicious software change and increase the sophistication of their strategies.

With such a system, the most computationally intensive procedures will be selecting the most relevant features (see Sect. 5.5) and building, evaluating, and selecting the best-performing model. In this chapter, we attend to the latter process.

As new executables appear, whether benign or malicious, one will have to monitor the performance of the current model and decide when to rebuild it. However, it is unfortunately not only a decision to rebuild it, for new executables and new methods of writing malicious software may necessitate doing so for all methods that show promise of performing well on this task. One will potentially have thousands of executables, thousands of attributes, a dozen or so learning methods, each with parameters, and hundreds if not thousands of experiments to run.

Online learning methods, some of which can update their models using a single new example, may help reduce the computational burden of building, evaluating, and selecting models as new executables arrive. However, there will likely be trade-offs in performance. One may gain the ability to update models quickly with the arrival of each new executable, but such updates may not significantly improve accuracy on previously unseen malicious executables or, because of ordering effects, may even decrease accuracy on known executables. The results we present in Sect. 5.2 provide insight into these trade-offs, but in the next section, we describe the online methods included in our evaluation.

5.4 Online Learning Methods

We included seven online-learning methods in our study, which we discuss in the following sections. These seven, while not exhaustive, are representative of a range of online methods. We selected both generative (e.g., Naive Bayes) and discriminative methods (e.g., Winnow). They have different inductive biases, memory requirements, and running times. We chose methods that use both single (e.g.,

Stagger) and multiple models (e.g., Dynamic Weighted Majority) to make predictions.

5.4.1 Naive Bayes

An online implementation of Naive Bayes (NB) stores as its concept description counts of examples and attribute values for each class. Learning, therefore, entails using a new example to update these counts. To classify an observation, the method uses these counts to compute estimates of the prior probability of each class, $\hat{p}(y = y_i)$, and the conditional probabilities of each attribute value given the class, $\hat{p}(x_j = v_k \mid y = y_i)$. That is, following the notation in [31], for all i class labels, j attributes, and k attribute values,

$$\hat{p}(y = y_i) = \frac{\#s\{y = y_i\}}{|S|},$$

$$\hat{p}(x_j = v_k \mid y = y_i) = \frac{\#s\{x_j = v_k\}}{\#s\{y = y_i\}},$$

where $\#s\{\cdot\}$ counts the number of occurrences of an event in S. Then, assuming attributes are conditionally independent, Naive Bayes uses Bayes' rule to predict the most probable class $\hat{y}$:

$$\hat{y} = \underset{y_i}{\operatorname{argmax}}\, \hat{p}(y = y_i) \prod_j \hat{p}(x_j = v_k \mid y = y_i).$$

Naive Bayes can be sensitive to irrelevant and conditionally dependent attributes (see, e.g., [10, 13, 20]). Nonetheless, it performs well on a variety of problems, is easy to implement, and is efficient in time and space, so it is a method that researchers and practitioners often use. For this study, we used the implementation of Naive Bayes in WEKA with its default parameter settings [42].

5.4.2 Stagger

Stagger [35, 36] is a probabilistic method that maintains four counts of attribute values given the class, which corresponded to positive and negative confirming and infirming evidence [6]. From these counts, it computes probabilistic measures of necessity and sufficiency for each attribute value given the positive class. To classify an observation, Stagger uses the observation's attribute values and the measures of necessity and sufficiency to compute the probability that the observation is a member of the positive class. Stagger predicts the most probable class. Stagger also uses *chunking* to assert and retract conjunctions, disjunctions, and negations of features. To cope with concept drift, Stagger decays its counts using a parameterized Gaussian function so that the decrease is 0% when Stagger's prior probability of the positive class is 0.5 and is 10% when its prior is either 0.0 or 1.0. For this study, we used our implementation of Stagger, which we implemented using WEKA [42]. We have not yet implemented chunking, and we did not set Stagger to decay its weights, which for this problem produced the best performance, as we discuss further in Sect. 5.2.

5.4.3 Winnow

Winnow represents concepts as coefficients of a linear equation [22]. We refer to the coefficients as weights, and there is one non-negative, real-valued weight, w_i, for each attribute of the data set. Given the observation $\mathbf{x}$, Winnow uses the weight vector to compute a weighted sum. If that sum surpasses the threshold θ, then Winnow predicts the positive class; otherwise, it predicts the negative class. That is,

$$\hat{y} = \begin{cases} + & \text{if } \sum_{i=1}^{n} w_i x_i > \theta \ ; \\ - & \text{otherwise} \ . \end{cases}$$

Learning, given a set of examples, involves setting the weight for each attribute. Winnow sets the weights and threshold to initial default values (e.g., $w_i = 1.0,\ \theta = n/2$) and processes examples one by one. It first makes a prediction for the example, and if the prediction is correct, then Winnow does nothing. If the prediction is incorrect, then Winnow either promotes or demotes each weight based on the label of the example. If it is a negative example, then Winnow promotes each weight by multiplying it by α (e.g., α = 2.0). If it is a positive example, then Winnow demotes each weight by multiplying it by β (e.g., β = 0.5). For this study, we used the implementation of Winnow in WEKA with its default parameter settings [42].

5.4.4 Hoeffding Tree

A decision tree is a rooted tree with internal nodes that correspond to attributes, branches that correspond to attribute values, and leaf nodes that correspond to class labels. A Hoeffding tree (HT) is a decision tree with an estimator at each leaf node that stores counts of attribute values for each class [9]. The learning element grows the tree down from the leaf nodes by sorting a new example from the tree's root to a leaf node by traversing based on the example's values. Once the traversal reaches a leaf node, it uses the example's attribute values to update the node's estimator. If the label of the node is different than that of the example and the difference in the information gain between the top two attributes at that node is sufficiently large as measured by the Hoeffding bound [14, 29], then the algorithm splits the node on the best attribute and recursively grows the tree downward. For this study, we used WEKA [42] to implement Domingos and Hulten's algorithm for inducing Hoeffding trees [9]. The implementation requires the parameter δ, which it uses to compute the Hoeffding bound and split nodes. In our experiments, we set $\delta = 0.5$.

5.4.5 Streaming Ensemble Algorithm

The Streaming Ensemble Algorithm (SEA) maintains a fixed-capacity, unweighted collection of batch learners [39]. Let L be the collection of learners, M be the maximum number of learners, and $m \in \{1,\ldots,M\}$ be the number of learners in L. SEA processes examples in batches, so let S be a collection of training examples. Additionally, let l_i and l_{i-1} be the learners trained on the ith and $i-1$th batches of examples, respectively.

SEA accumulates p examples into S, sets l_{i-1} to l_i, and trains l_i on S. If the ensemble is not full (i.e., $m < M$), then SEA adds l_{i-1} to L. If the ensemble is full, then SEA evaluates l_{i-1} and $l_j \in L$, for $j = 1,\ldots, M$, on S. If the *quality* of some learner l_j is less than that of l_{i-1}, then SEA replaces l_j with l_{i-1}. The algorithm measures quality by examining the votes of the ensemble's learners and the new learner l_{i-1} on each example in the ith batch. The intuition behind this measure is to give preference to the new learner, but only if its addition will improve the ensemble's performance. To predict, SEA returns the majority vote of the predictions of the learners in the collection.

For this study, we used our implementation of SEA, which we implemented using WEKA [42]. Rather than using quality, we used accuracy; results from preliminary studies suggest that quality and accuracy produce comparable performance. As base learners for SEA, we used Naive Bayes and J48, and we will refer to SEA with these base learners as SEA-NB and SEA-J48, respectively. (J48 is the implementation of C4.5 [32] in WEKA. C4.5 is a batch-learning algorithm that builds decision trees from training data.) In our experiments, we varied M and p, as we describe in Sect. 5.2.

5.4.6 Accuracy Weighted Ensemble

Accuracy Weighted Ensemble (AWE) maintains a fixed-capacity, weighted collection of batch learners [41]. As before, let L be the collection of learners, M be the maximum number of learners, and $m \in \{1, \ldots, M\}$ be the number of learners in L. AWE processes examples in batches, so let S a collection of training examples. Additionally, let l_i be the learner trained on the ith batch of examples. AWE accumulates p examples into S and trains l_i on S. It then evaluates l_i on S using cross-validation and derives a weight w_i for l_i based on costs or accuracy (e.g., mean-squared error). Then, for $l_j \in L$, $j = 1, \ldots, m$, AWE applies l_j to S and computes w_j. It then sets L to the M top-weighted classifiers in $L \cup \{l_i\}$. To predict, AWE returns the weighted-majority vote of the predictions of the learners in the collection. For this study, we used our implementation of AWE, which we implemented using WEKA [42]. To compute w_i, we used the mean-squared error, so $w_i = MSE_r - MSE_i$, where MSE_r is the mean-squared error of a default rule (i.e., a classifier that predicts the majority class in S) and MSE_i is the mean-squared error for the learner l_i (see Eq. 8 in [41]). As base learners for AWE, we used Naive Bayes and J48, and we will refer to AWE with these base learners as AWE-NB and AWE-J48, respectively. In our experiments, we varied M and p, as we describe in Sect. 5.2.

5.4.7 Dynamic Weighted Majority

Dynamic Weighted Majority (DWM) [16, 19] is an extension of the weighted-majority algorithm [23] that maintains a dynamically sized, weighted collection of online-learning algorithms (see also [5]). To classify an observation, DWM obtains a prediction for the observation from each learner in the collection and predicts the weighted-majority class. Learning entails first obtaining a prediction for an example from each learner, and if a learner misclassifies the example, then DWM decreases its weight by the multiplicative factor β (e.g., .5). In addition, if the overall weighted-majority vote for the example is incorrect, then DWM adds a new, untrained learner to the collection. The algorithm also may remove a learner when its weight falls below the user-determined threshold θ. In addition to β

and θ, DWM requires the parameter p, which governs how frequently DWM updates learner weights and adds and removes learners. There is a fourth, optional parameter, M, which specifies the maximum number of learners. Although designed primarily for concept drift, recent experimental results suggest that on a number of UCI data sets [2], DWM performs no worse than a single online learner [19].

For this study, we used our implementation of DWM, which we implemented using WEKA [42]. As base learners for DWM, we used Naive Bayes and HT, and we will refer to DWM with these base learners as DWM-NB and DWM-HT, respectively. In our experiments, we set $\beta = 0.5$, $\theta = 0.01$, and varied M and p, as we describe in Sect. 5.2. We set M for DWM for two reasons: so we could draw stronger comparisons to SEA and AWE, which have fixed-capacity ensembles, and because letting DWM create an ensemble of unbounded size might lead to significant memory and processing requirements, especially for our data set, which we discuss in the next section.

5.5 From Executables to Examples

To produce a data set for the task of detecting malicious executables, we harvested executables from machines running the Windows 2000 and XP operating systems. We also retrieved popular executables from SourceForge (http://sourceforge.net). We assumed that all of these 1,971 executables were benign, and a commercial program for virus detection indicated they were benign, thereby supporting our assumption. We obtained 1,651 malicious executables from the Web site VX Heavens (http://vx.netlux.org) and from computer-forensic experts at the MITRE Corporation, the original sponsors of this work. A commercial program for virus detection indicated that most of these executables were indeed malicious, but it failed to identify 50 of them, even though they are known to be malicious. Nonetheless, we assumed these 50 executables were malicious.

To produce training data from these executables, we converted each into sequence of byte codes in hexadecimal format. We then combined these codes into n-grams, where $n = 4$, a value we determined in pilot studies [17]. This processing resulted in 255,904,403 distinct n-grams.

For simplicity, we encoded whether each n-gram was present in or absent from an executable, rather than, say, counting the frequency of occurrence therein. At this point, a vector of binary attributes represented the content of each executable. To form examples, we labeled each vector as either "malicious" (M) or "benign" (B) based on whether the vector corresponded to a malicious executable or a benign executable, respectively. Since the task is to detect malicious executables, we will refer to the malicious class as the positive class and the benign class and the negative class. Selecting the most relevant attributes – or n-grams, in our case – for prediction is a standard preprocessing step in machine learning, and we did so using information gain or average mutual information [43]. For the *j*th attribute,

$$Gain(j) = \sum_{v_j \in \{T, F\}} \sum_{y_i \in \{M, B\}} p(v_j, y_i) \log \frac{p(v_j, y_i)}{p(v_j) p(y_i)},$$

where y_i is the *i*th class, v_j is the value of the *j*th attribute, $p(v_j, y_i)$ is the proportion that the *j*th attribute has the value v_j in the class y_i, $p(v_j)$ is the proportion that the *j*th *n*-gram takes the value v_j in the training data, $p(y_i)$ is the proportion of the training data belonging to the class y_i.

After computing the information gain for each attribute, we selected the 500 attributes with the highest scores. As with the size of the *n*-grams, we determined the number of attributes to select in pilot studies. One can find the details of these pilot studies and results from batch-learning experiments elsewhere [17, 18].

5.6 Experimental Study

As discussed previously, the main motivation for our study is to evaluate online methods on the task of detecting malicious executables. We applied seven online learning methods to the examples of our data set and measured accuracy, as we discuss in the following sections. Online methods performed well, but not as well as batch methods. Nonetheless, since some online learners can update their models with as few as one new training example, it is important to understand and characterize trade-offs in performance.

5.6.1 Design

We designed an experiment with the aim of evaluating the online methods on the task of detecting malicious executables. We set the online methods to process the examples as a sequence, and for each example of the sequence, the method returned its prediction for the example before passing it to learning element. This design corresponds to how an online system would process examples and observations in a real-world environment. Researchers have used it to evaluate online learners on similar tasks (e.g., [5, 19, 35]).

Since we did not know the date when each executable appeared, we shuffled the order of the 3,622 examples and applied each online method. We presented the same sequence in the same order to each of the seven methods. We repeated this procedure 200 times, shuffling the order of the examples each time. For each method, we measured accuracy, true-positive rate (TPR), and true-negative rate (TNR). Over the 200 runs, we averaged these metrics and computed 95% confidence intervals. We also plotted these averages as learning curves, and then computed overall averages for each method.

Some of the methods required parameters, as mentioned in Sect. 5.4. Because of the time required to run each experiment, we were not able to search comprehensively for parameters. Consequently, we empirically determined parameters through smaller pilot experiments, and then used them in our study. Naturally, we make no claims that the parameters we selected are optimal; nonetheless, the parameters we used resulted in acceptable performance, and results suggest that, while parameters influenced performance, other factors also contributed, such as the base learner of an ensemble method.

For the ensemble methods, we varied two parameters: the number of learners (M) and the period (p). Let $\langle M, p \rangle$ be a tuple of these parameters. For the ensemble methods, we ran each with $\langle 3,5 \rangle$ and $\langle 10,50 \rangle$. For DWM, we also ran each instantiation with the parameters $\langle 10,3 \rangle$.[1]

5.6.2 Results and Analysis

In Table 5.1, we present the results of our empirical study of online-learning methods for detecting malicious executables. We report the method, parameters, accuracy, true-positive rate (TPR), and true-negative rate (TNR) with 95% confidence intervals.

As one can see, the best-performing methods were SEA-J48 ($M = 10$, $p = 50$), HT, and AWE-J48 ($M = 10$, $p = 50$). These three methods not only achieved the highest overall accuracies, but also performed well on the examples of the individual classes, malicious and benign. For example, SEA-J48 ($M = 10$, $p = 50$) had a true-positive rate of .951, which was the highest rate for the methods considered. Of these three methods, HT had the highest true-negative rate, which was .920, although SEA-J48 ($M = 3$, $p = 5$) with .927 achieved the highest true-negative rate overall. Generally, all of the methods performed satisfactorily in terms of the true-positive rate; all were above .82. The methods that did not perform well in terms of overall accuracy did so because of poor performance on the examples of the negative class. Stagger is a good example: It had a true-positive rate of .925, but its overall accuracy was .569 because of its true-negative rate of .269.

---

[1] We should note that the parameter p for DWM is slightly different than that for SEA and AWE. Because DWM has an ensemble of online learners, it can learn from a single new example, so it is possible for p to be 1. Since SEA and AWE use batch methods to build a new classifier from a new set of examples, we reasoned that it would not be practical to set p to 1 for these methods.

Table 5.1 Average online performance on 3,622 examples. Measures are accuracy, true-positive rate (TPR), and true-negative rate (TPN) with 95% confidence intervals

Method	Parameters	Accuracy	TPR	TNR
SEA-J48	$M = 10, p = 50$	.929±.002	.951±.002	.903±.002
HT		.928±.001	.929±.002	.920±.002
AWE-J48	$M = 10, p = 50$	.927±.002	.942±.002	.906±.002
DWM-HT	$M = 10, p = 50$	.910±.002	.917±.002	.895±.002
SEA-J48	$M = 3, p = 5$	.886±.001	.827±.001	.927±.001
AWE-J48	$M = 3, p = 5$	.881±.001	.819±.001	.934±.001
NB		.874±.001	.925±.001	.824±.001
DWM-NB	$M = 10, p = 50$	.872±.001	.926±.001	.820±.001
DWM-HT	$M = 10, p = 1$	.870±.001	.873±.002	.859±.002
SEA-NB	$M = 10, p = 50$	.860±.002	.920±.002	.803±.002
DWM-NB	$M = 3, p = 5$	.858±.001	.916±.001	.803±.001
AWE-NB	$M = 10, p = 50$	.857±.002	.919±.002	.798±.002
DWM-NB	$M = 10, p = 1$	.856±.001	.912±.001	.802±.001
DWM-HT	$M = 3, p = 5$	.854±.001	.855±.002	.846±.002
Winnow		.843±.001	.824±.002	.851±.001
SEA-NB	$M = 3, p = 5$	.842±.001	.926±.001	.766±.001
AWE-NB	$M = 3, p = 5$	.809±.001	.901±.001	.727±.002
Stagger		.569±.001	.925±.001	.269±.002

Based on these results, we concluded that methods that used decision trees generally outperformed the others. Notice that the top five methods in terms of accuracy produced a decision tree (i.e., HT) or an ensemble of decision trees.

For the ensemble methods, although parameters had an effect, the base learner seemed to more strongly influence performance. For example, SEA-J48 and AWE-J48 ($M = 10$, $p = 50$) with accuracies of .929 and .927 outperformed their respective counterparts with Naive Bayes as the base learner. SEA-NB and AWE-NB ($M = 10$, $p = 50$) achieved accuracies of .860 and .857, respectively.

In terms of the effect of parameters on performance, for the ensemble methods, fewer members of the ensemble and training over shorter periods of time translated into lower accuracy. For example, DWM-HT ($M = 10$, $p = 50$) had an accuracy of .910, whereas DWM-HT ($M = 10$, $p = 1$) achieved .870 and DWM-HT ($M = 3$, $p = 5$) yielded .854.

In Fig. 5.1, we present performance curves[2] for the four best-performing methods: SEA-J48 ($M = 10$, $p = 50$), HT, and AWE-J48 ($M = 10$, $p = 50$), DWM-HT ($M = 10$, $p = 50$). We show the individual curves for these methods in Figs. 5.2–5.5. All of the methods achieved comparable accuracy after processing

[2] As the error bars in these curves suggest, the variance across 200 runs was high, which made for curves that were difficult to interpret. Consequently, for presentation, we smoothed the curves by plotting the average of the 50 previous accuracies.

roughly 2,000 examples. SEA and AWE attained accuracies above .9 after processing about 1,000 executables and converged more quickly than did HT and DWM-HT. DWM-HT did not converge as quickly as SEA and AWE, but neither did HT. Since DWM-HT used HT as its base learner, it is possible that DWM-HT's slow convergence is due to HT, rather than to DWM itself.

In Fig. 5.6, we present performance curves for three other methods: Naive Bayes, Winnow, and Stagger. For context, we included the curve for SEA-J48 ($M = 10$, $p = 50$), one of the best-performing methods. Even though Naive Bayes, Winnow, and Stagger did not perform as well as other methods, they reached their respective asymptotes after processing 1,000 examples.

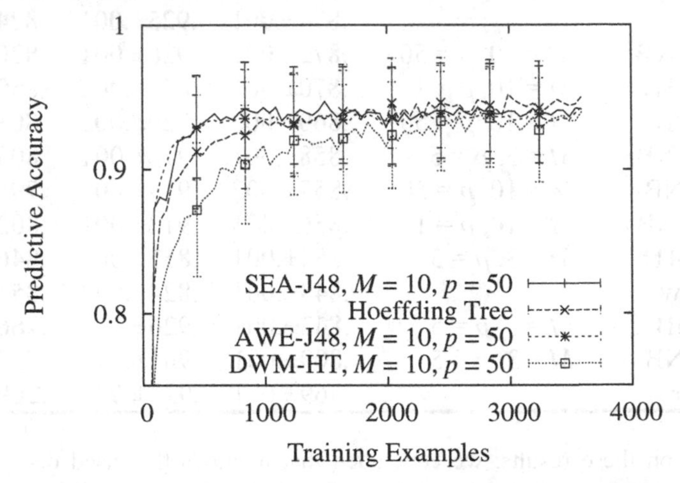

Fig. 5.1 Performance curves for the four best-performing methods: SEA-J48, HT, AWE-J48, and DWM-HT. Measures are accuracy with 95% confidence intervals

To place these results in a broader context, we compare them to results we obtained in our previous study, in which we evaluated seven batch-learning methods. Therein, we evaluated the methods using ten-fold cross-validation and plotted ROC curves. For each of the seven curves, we selected points corresponding to three desired true-negative rates and reported the associated true-positive rates. For convenience, we present this information in Table 5.2.

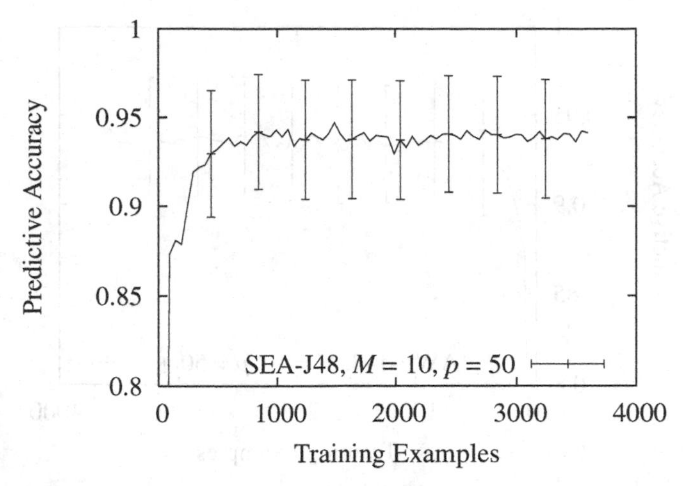

Fig. 5.2 Individual performance curve for SEA-J48. Measures are accuracy with 95% confidence intervals

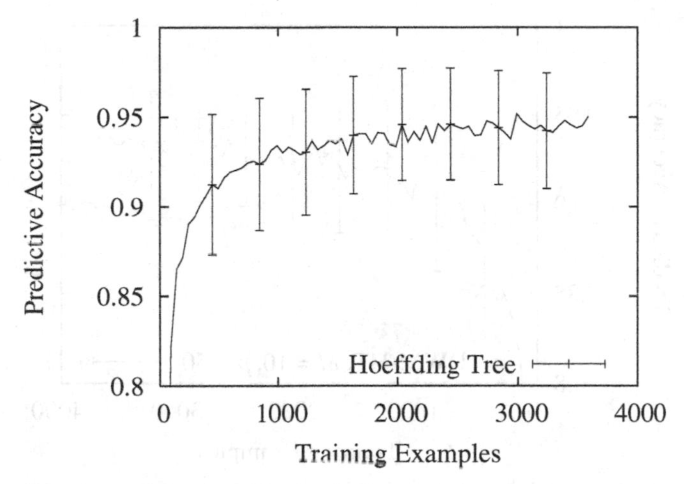

Fig. 5.3 Individual performance curve for HT. Measures are accuracy with 95% confidence intervals

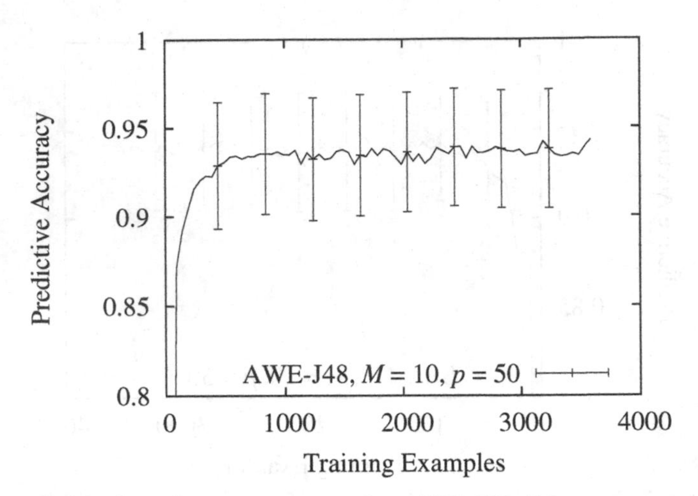

Fig. 5.4 Individual performance curve for AWE-J48. Measures are accuracy with 95% confidence intervals

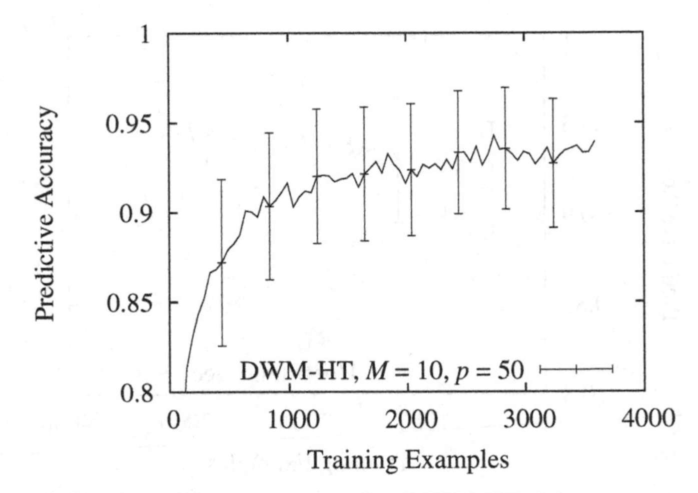

Fig. 5.5 Individual performance curve for DWM-HT. Measures are accuracy with 95% confidence intervals

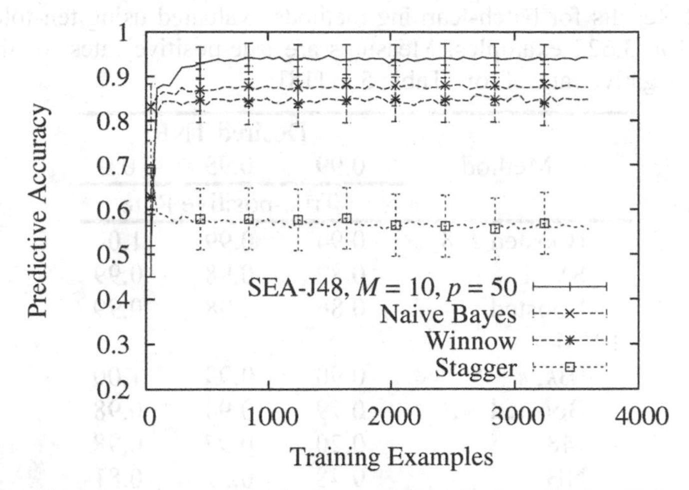

Fig. 5.6 Performance curves for the best-performing method and three others: SEA-J48, NB, Winnow, and Stagger. Measures are accuracy with 95% confidence intervals on this same data set [18].

Regarding the methods from this earlier study, we have described most in this article, but we have not mentioned SVM and IB*k*. An SVM is a support vector machine (e.g., [8]). Informally, it is similar to Winnow [22] in that it uses a linear equation to represent concepts, but an SVM sets its weights using a different algorithm and transforms training examples so they are easier to separate using a linear equation. IB*k* is an instance-based learner that classifies an observation by finding the *k*-closest matches among the training examples [1]. It then predicts the majority class.

Overall, the results in Table 5.2 suggest that batch-learning methods outperformed the online methods we considered in this study. For example, none of the online methods had a TNR of .95 and a TPR greater than .96, as did a number of the batch methods. As a specific example, boosted J48 had a TPR of .99 for a desired TNR of .95, which equals an accuracy of .97. In contrast, SEA-J48, the best performing online method, had a TPR of .951, a TNR of .903, and an accuracy of .929. While it is not entirely surprising that batch methods outperformed online methods, the important point is that the online methods achieved comparable accuracies. Additionally, we must take into account the expense of updating models using batch methods, and we may be able to adjust the decision thresholds of the online methods to obtain a better trade-off in the true-positive/true-negative rates, but these are issues we discuss further in the next section.

Table 5.2 Results for batch-learning methods evaluated using ten-fold cross-validation on 3,622 examples. Measures are true-positive rates for three desired true-negative rates. From Table 5 in [18]

Method	Desired TNR		
	0.99	0.95	0.9
	True-positive Rate		
Boosted J48	0.94	0.99	1.00
SVM	0.82	0.98	0.99
Boosted SVM	0.86	0.98	0.99
IB*k*, $k = 5$	0.90	0.99	1.00
Boosted NB	0.79	0.94	0.98
J48	0.20	0.97	0.98
NB	0.48	0.57	0.81

5.7 Discussion

In previous sections, we discussed a number of issues, such as running time and the role of online versus batch learning. In this section, we consider some of these issues and identify opportunities for future work.

Throughout the chapter, we emphasized that, while the accuracy of online methods may not surpass that of batch methods, the advantage of online methods is their ability to update their existing model with new training examples. To illustrate, we ran J48 and our implementation of HT on the full set of examples and measured the time required to build a single model.

On a Dell Optiplex GX520 with a Pentium D CPU running at 3.4 GHZ and 2 GB of RAM, J48 required 1.74 seconds to build a single decision tree from 3,622 examples. Therefore, J48 processed one example every 4×10^{-4} seconds. Our implementation of HT did so in 1.05 seconds; it processed one example every 3×10^{-4} seconds. While these implementations processed examples at nearly the same rate, if we obtained a new example, HT could use the example to update its model in 10^{-4} seconds, whereas with J48 we would have to rebuild the entire tree from scratch.

We also ran these methods on the same data set using ten-fold cross-validation. J48 required 21 seconds to complete this run and achieved an accuracy of .967. HT required 11 seconds and achieved an accuracy of .891. This illustrates the performance trade-offs between batch and online learning methods.

One technique that may let us achieve high accuracy and quick updates of a model when new examples arrive is to use batch learning (e.g., J48) to build a model from the available training data and then use online learning (e.g., HT) to update the model as new examples arrive. Domingos and Hulten call this initialization, and their results suggest that it produces higher accuracy with fewer examples as compared to an uninitialized online learner (see Fig. 5.5 in [9]). We plan to investigate initialization for our problem in future work.

One outcome of our study that we did not expect was that SEA [39] would perform comparably to AWE [41]. As we have mentioned, SEA uses an unweighted ensemble of learners, whereas AWE weights its learners. At the beginning of our study, we predicted that AWE's weighting mechanisms would let it outperform SEA, but this did not occur. We are unaware of other studies that compare SEA and AWE. Naturally, their similar performance in our study could be tied to our data set, so we hope to conduct a more focused investigation of these algorithms and evaluate the trade-offs between weighted and unweighted ensembles.

Finally, an alternative to the experimental design we used in this study is to draw randomly training and testing sets from the original set of examples, apply online learning methods to each example in the training set, and evaluate the current model on the examples in the test set. While the design we used for this study better reflects how these methods would be used in a real-world environment, this alternative design would let us conduct ROC analysis, which would give additional insight into the performance of these methods on this task. This, too, we plan to pursue in future work.

5.8 Concluding Remarks

In this chapter, we examined results from an empirical evaluation of seven online-learning methods on the task of detecting previously unseen malicious executables. Based on performance measures of accuracy, true-positive rate, and true-negative rate, these methods show promise of detecting malicious executables with accuracy sufficient for commercial application. For example, SEA-J48, HT, and AWE-J48 achieved true-positive rates above .92. While not outperforming batch methods of machine learning, online methods have the advantage of being able to update their models quickly and efficiently as new executables arrive.

Certainly, online methods have a role to play in protecting systems and in helping computer-forensic experts investigate compromised systems, as do other techniques, such as signatures, static analysis, and batch methods of machine learning. However, before we can construct systems that use online-learning methods to protect computers and their information, we must understand how these methods perform, especially with respect to batch methods. We hope that this chapter will give researchers and practitioners insight into some of these trade-offs in performance, insights that will serve as the basis for future studies on the use of methods of machine learning for cyber trust.

Acknowledgments The author would like to thank Steve Bach for the use of his implementation of Hoeffding Trees and for noting Domingos and Hulten's use of initialization. This research was conducted in the Department of Computer Science at Georgetown University.

References

[1]. Aha, D.W., Kibler, D., Albert, M.K.: Instance-based learning algorithms. Machine Learning **6**, 37–66 (1991)
[2]. Asuncion, A., Newman, D.J.: UCI machine learning repository. Web site, School of Information and Computer Sciences, University of California, Irvine, http://www.ics.uci.edu/~mlearn/MLRepository.html (2007)
[3]. Bailey, M., Oberheide, J., Andersen, J., Mao, Z.M., Jahanian, F., Nazario, J.: Automated classification and analysis of Internet malware. In: Recent Advances in Intrusion Detection, Lecture Notes in Computer Science, vol. 4637, pp. 178–197. Springer, Berlin-Heidelberg (2007). Tenth International Conference, RAID 2007, Gold Coast, Australia, September 57, 2007.Proceedings
[4]. Bishop, C.M.: Pattern Recognition and Machine Learning. Springer, Berlin-Heidelberg (2006)
[5]. Blum, A.: Empirical support for winnow and weighted-majority algorithms: Results on a calendar scheduling domain. Machine Learning **26**, 5–23 (1997)
[6]. Bruner, J.S., Goodnow, J.J., Austin, G.A.: A Study of Thinking. Wiley & Sons, New York, NY (1956). Republished in 1986 and 1990 by Transaction Publishers, New Brunswick, NJ
[7]. Cova, M., Balzarotti, D., Felmetsger, V., Vigna, G.: Swaddler: An approach for the anomaly-based detection of state violations in web applications. In: Recent Advances in Intrusion Detection, Lecture Notes in Computer Science, vol. 4637, pp. 63–86. Springer, Berlin-Heidelberg (2007). Tenth International Conference, RAID 2007, Gold Coast, Australia, September 5–7,2007. Proceedings
[8]. Cristianini, N., Shawe-Taylor, J.: An Introduction to Support Vector Machines—and Other Kernel-based Learning Methods. Cambridge University Press, Cambridge (2000)
[9]. Domingos, P., Hulten, G.: Mining high-speed data streams. In: Proceedings of the Sixth ACM SIGKDD International Conference on Knowledge Discovery and Data Mining, pp. 71–80. ACM Press, New York, NY (2000)
[10]. Domingos, P., Pazzani, M.J.: On the optimality of the simple Bayesian classifier under zero-one loss. Machine Learning **29**, 103–130 (1997)
[11]. Early, J.P., Brodley, C.E.: Behavioral features for network anomaly detection. In: M.A. Maloof (ed.) Machine learning and data mining for computer security: Methods and applications, pp. 107–124. Springer, Berlin-Heidelberg (2006)
[12]. Freund, Y., Schapire, R.E.: Experiments with a new boosting algorithm. In: Proceedings of the Thirteenth International Conference on Machine Learning, pp. 148–156. Morgan Kaufmann, San Francisco, CA (1996)
[13]. Hand, D.J., Yu, K.: Idiot's Bayes: Not so stupid after all? International Statistical Review **69**, 385–398 (2001)
[14]. Hoeffding, W.: Probability inequalities for sums of bounded random variables. Journal of the American Statistical Association **58**(301), 13–30 (1963)
[15]. Kephart, J.O., Sorkin, G.B., Arnold, W.C., Chess, D.M., Tesauro, G.J., White, S.R.: Biologically inspired defenses against computer viruses. In: Proceedings of the Fourteenth International Joint Conference on Artificial Intelligence, pp. 985–996. Morgan Kaufmann, San Francisco, CA (1995)
[16]. Kolter, J.Z., Maloof, M.A.: Dynamic weighted majority: A new ensemble method for tracking concept drift. In: Proceedings of the Third IEEE International Conference on Data Mining, pp. 123–130. IEEE Press, Los Alamitos, CA (2003)
[17]. Kolter, J.Z., Maloof, M.A.: Learning to detect malicious executables in the wild. In: Proceedings of the Tenth ACM SIGKDD International Conference on Knowledge Discovery and Data Mining, pp. 470–478. ACM Press, New York, NY (2004)

[18]. Kolter, J.Z., Maloof, M.A.: Learning to detect and classify malicious executables in the wild. Journal of Machine Learning Research **7**, 2721–2744 (2006). Special Issue on Machine Learning in Computer Security
[19]. Kolter, J.Z., Maloof, M.A.: Dynamic weighted majority: An ensemble method for drifting concepts. Journal of Machine Learning Research **8**, 2755–2790 (2007)
[20]. Langley, P., Sage, S.: Tractable average-case analysis of naive Bayesian classifiers. In: Proceedings of the Sixteenth International Conference on Machine Learning, pp. 220–228. Morgan Kaufmann, San Francisco, CA (1999)
[21]. Langley, P.W.: Elements of Machine Learning. Morgan Kaufmann, San Francisco, CA (1996)
[22]. Littlestone, N.: Learning quickly when irrelevant attributes abound: A new linear-threshold algorithm. Machine Learning **2**, 285–318 (1988)
[23]. Littlestone, N., Warmuth, M.K.: The weighted majority algorithm. Information and Computation **108**, 212–261 (1994)
[24]. Lo, R.W., Levitt, K.N., Olsson, R.A.: MCF: A malicious code filter. Computers & Security **14**(6), 541–566 (1995)
[25]. Maloof, M.A.: Concept drift. In: J. Wang (ed.) Encyclopedia of Data Warehousing and Mining, pp. 202–206. Information Science Publishing, Hershey, PA (2005)
[26]. Maloof, M.A. (ed.): Machine Learning and Data Mining for Computer Security: Methods and Applications. Springer, Berlin-Heidelberg (2006)
[27]. Maloof, M.A.: Some basics concepts of machine learning and data mining. In: M.A. Maloof (ed.) Machine learning and data mining for computer security: Methods and applications, pp. 23–43. Springer, Berlin-Heidelberg (2006)
[28]. Maloof, M.A., Stephens, G.D.: ELICIT: A system for detecting insiders who violate need-to-know. In: Recent Advances in Intrusion Detection, Lecture Notes in Computer Science, vol. 4637, pp. 146–166. Springer, Berlin-Heidelberg (2007). Tenth International Conference, RAID 2007, Gold Coast, Australia, September 5–7, 2007. Proceedings
[29]. Maron, O., Moore, A.: Hoeffding races: Accelerating model selection search for classification and function approximation. In: Advances in Neural Information Processing Systems 6, pp. 59–66. Morgan Kaufmann, San Francisco, CA (1994)
[30]. Mitchell, T.M.: Machine Learning. McGraw-Hill, New York, NY (1997)
[31]. Ng, A.Y., Jordon, M.I.: On discriminative vs. generative classifiers: A comparison of logistic regression and naive Bayes. In: Advances in Neural Information Processing Systems 14. MIT Press, Cambridge, MA (2002)
[32]. Quinlan, J.R.: C4.5: Programs for Machine Learning. Morgan Kaufmann, San Francisco, CA (1993)
[33]. Russell, S.J., Norvig, P.: Artificial Intelligence: A Modern Approach, 2nd edn. Prentice-Hall, Upper Saddle River, NJ (2003)
[34]. Schapire, R.E.: The boosting approach to machine learning: An overview. In: D.D. Denison, M.H. Hansen, C.C. Holmes, B. Mallick, B. Yu (eds.) Nonlinear Estimation and Classification, Lecture Notes in Statistics, vol. 171, pp. 149–172. Springer, Berlin-Heidelberg (2003)
[35]. Schlimmer, J.C.: Concept acquisition through representational adjustment. Ph.D. thesis, Department of Information and Computer Science, University of California, Irvine (1987)
[36]. Schlimmer, J.C., Granger, R.H.: Beyond incremental processing: Tracking concept drift. In: Proceedings of the Fifth National Conference on Artificial Intelligence, pp. 502–507. AAAI Press, Menlo Park, CA (1986)
[37]. Schultz, M.G., Eskin, E., Zadok, E., Stolfo, S.J.: Data mining methods for detection of new malicious executables. In: Proceedings of the IEEE Symposium on Security and Privacy, pp. 38–49. IEEE Press, Los Alamitos, CA (2001)
[38]. Shields, T.C.: An introduction to information assurance. In: M.A. Maloof (ed.) Machine learning and data mining for computer security: Methods and applications, pp. 7–21. Springer, Berlin-Heidelberg (2006)

[39]. Street, W.N., Kim, Y.: A streaming ensemble algorithm (SEA) for large-scale classification. In: Proceedings of the Seventh ACM SIGKDD International Conference on Knowledge Discovery and Data Mining, pp. 377–382. ACM Press, New York, NY (2001)
[40]. Tesauro, G., Kephart, J.O., Sorkin, G.B.: Neural networks for computer virus recognition. IEEE Expert **11**(4), 5–6 (1996)
[41]. Wang, H., Fan, W., Yu, P.S., Han, J.: Mining concept-drifting data streams using ensemble classifiers. In: Proceedings of the Ninth ACM SIGKDD International Conference on Knowledge Discovery and Data Mining, pp. 226–235. ACM Press, New York, NY (2003)
[42]. Witten, I.H., Frank, E.: Data Mining: Practical Machine Learning Tools and Techniques, 2nd edn. Morgan Kaufmann, San Francisco, CA (2005)
[43]. Yang, Y., Pederson, J.O.: A comparative study on feature selection in text categorization. In: Proceedings of the Fourteenth International Conference on Machine Learning, pp. 412–420. Morgan Kaufmann, San Francisco, CA (1997)
[44]. Yu, Z., Tsai, J.J.P., Weigert, T.: An automatically tuning intrusion detection system. IEEE Transactions on Systems, Man, and Cybernetics, Part B: Cybernetics **37**(2), 373–384 (2007)
[45]. Zhang, Y., Lee, W., Huang, Y.A.: Intrusion detection techniques for mobile wireless networks. Wireless Networks **9**(5), 545–556 (2003)

6 Efficient Mining and Detection of Sequential Intrusion Patterns for Network Intrusion Detection Systems

Mei-Ling Shyu, Zifang Huang, and Hongli Luo[1]

Abstract In recent years, pervasive computing infrastructures have greatly improved the interaction between human and system. As we put more reliance on these computing infrastructures, we also face threats of network intrusion and/or any new forms of undesirable IT-based activities. Hence, network security has become an extremely important issue, which is closely connected with homeland security, business transactions, and people's daily life. Accurate and efficient intrusion detection technologies are required to safeguard the network systems and the critical information transmitted in the network systems. In this chapter, a novel network intrusion detection framework for mining and detecting sequential intrusion patterns is proposed. The proposed framework consists of a Collateral Representative Subspace Projection Modeling (C-RSPM) component for supervised classification, and an inter-transactional association rule mining method based on Layer Divided Modeling (LDM) for temporal pattern analysis. Experiments on the KDD99 data set and the traffic data set generated by a private LAN testbed show promising results with high detection rates, low processing time, and low false alarm rates in mining and detecting sequential intrusion detections.

6.1 Introduction

With increasing speed and powerful function in computational and communication resources online, people have relied more and more on computer

[1] Mei-Ling Shyu and Zifang Huang

Department of Electrical and Computer Engineering, University of Miami, Coral Gables, FL 33124, USA, e-mail:shyu@miami.edu, z.huang3@umiami.edu

Hongli Luo

Department of Computer and Electrical Engineering Technology and Information System and Technology, Indiana University - Purdue University Fort Wayne, Fort Wayne, IN 46805, USA, email: luoh@ipfw.edu

J.J.P. Tsai and P.S. Yu (eds.), *Machine Learning in Cyber Trust: Security, Privacy, and Reliability*, DOI: 10.1007/978-0-387-88735-7_6,

systems to provide services. Especially, the popular web-based applications and services have enabled more intuitive and spontaneous human-computer interactions. Meanwhile, the use of the Internet is facing increasing risks regarding privacy and security as how to keep the information safe, avoid unauthorized access, prevent intrusions, etc. Nowadays, the attackers are more and more experienced to access to personal computers or to penetrate into network systems. Consequently, network security becomes an important issue that calls for the development of accurate and efficient solutions to protect the network system from various intrusions. To address this issue, powerful intrusion detection systems (IDSs) have been developed to safeguard the network systems and valuable information.

Existing intrusion detection methods can be broadly categorized into misuse detection and anomaly detection areas [3], where misuse detection is based on signature modeling of known attacks [18] and anomaly detection is based on signature modeling of normal traffic [17]. The advantage of misuse detection approaches is their higher detection accuracy for known attacks; while the advantage of anomaly detection approaches is its capability in detecting new attacks [13]. On the other hand, the disadvantage of misuse detection approaches is its incapacity of detecting previously unobserved attacks, and the disadvantage of anomaly detection approaches is its high false alarm rate. Furthermore, there have been various IDS architectures developed in the literature such as monolithic, hierarchic, agent-based, and distributed (GrIDS) systems [33]. However, to better safeguard the network systems, the researchers in IDSs are still seeking more improvements to the existing architectures.

One of the problems in the existing intrusion detection architectures is that they can only detect intrusions and respond to the intrusions passively. Antivirus software usually quickly scans the location of the weakness after the attacks were reported. However, in such a scenario, the attacks might have come into force, which leads the network or computer system to a standstill. Even though those kinds of thoughtful strategic systems can be applied independently or with the implementation of other techniques, the frequent occurrence of a series of attacks has repeatedly indicated that instead of simply detecting the occurred intrusion, it is necessary to be able to predict which kind of intrusion might occur next in the network system. In this way, the response system can act in advance to strengthen the security of the local computer and the server, so as to prevent the coming intrusion or try to minimize the damages. In addition, there might be some relations between/among different intrusive activities. That is, some of the intrusions may occur sequentially, resulting in some kind of sequential intrusion patterns. The rootkit attack is a good example in this scenario. First, an intruder would try to get the information of username and password to become a root user through a sniffer program via the Ethernet network. Second, the Trojan horse attack would occur. At the first glance, the Trojan horse attack does not seem to be harmful, which is similar to a normal background process. However, such an attack is usually followed by potential destructive processes, because it may establish back-door entry points for intruders to conduct other malevolent attacks. Finally, crucial information might be disclosed or the computer system might be crashed.

In this chapter, a novel network intrusion detection framework that is capable of mining and detecting sequential intrusion patterns is proposed. The main objective of the proposed framework is to identify the sequential intrusion patterns so that when an intrusion occurs, it can predict the next potentially harmful intrusions based on the rule set trained/derived from the attack history. The capability of detecting such sequential intrusion patterns enables proactive protection to the network system with prompt decision-making and suitable actions to be taken. The proposed framework consists of a supervised classification approach called Collateral Representative Subspace Projection Modeling (C-RSPM) [23] and an inter-transactional association rule mining method based on Layer Divided Modeling (LDM) for temporal pattern analysis. C-RSPM is a multi-class supervised classifier with high detection rate, short training and classification times, and low processing memory requirements, and provides schemes for collateral class modeling, class-ambiguity solving, and classification. Furthermore, it is capable of adaptively selecting nonconsecutive principal dimensions from the statistical information of the training data set in order to accurately model a representative subspace. LDM is a novel inter-transactional association rule mining algorithm, which fully utilizes the distributing information of the data to quicken the process of rule mining. LDM is used to find the associated attacks, such as Web server buffer overflow leading to the intruder sneaking into the networks and mapping some executable files to steal classified information. Experimental results show that the proposed LDM method can perform temporal pattern analysis and generate inter-transactional rules more efficiently (i.e., less execution time) in comparison to some exiting algorithms like EH-Aprior [31]. Using the proposed framework, it is possible to detect the intrusion in the network and get the information of potential future harmful intrusions for prompt and proactive protection. Both KDD99 data set [16] and the traffic data set generated by a private LAN testbed at the Department of Electrical and Computer Engineering at the University of Miami [26, 36] are used in the experiments. For the private LAN testbed, 10 types of traffic are generated in random time intervals. The promising experimental results demonstrate that the proposed framework can effectively identify and detect sequential intrusion patterns with high detection rates, low false alarm rates, and fast rule set extraction for IDSs.

The remainder of this chapter is organized as follows. Section 6.2 discusses existing work related to the design of IDSs and algorithms of inter-transactional rule mining. Section 6.3 presents our proposed intrusion detection framework. Experimental setup is presented in Section 6.4. Experiments and results are presented in Section 6.5. Finally, in Section 6.6, conclusions are given.

6.2 Existing Work

Due to the importance of network intrusion detection, various types of intrusion detection systems (IDSs) have been developed. Generally, existing IDSs could be broadly categorized into the following categories: network-based IDSs, protocol-based IDSs, application protocol-based IDSs, host-based IDSs, hybrid IDSs, and agent-based IDSs [2, 4, 6, 7, 12, 15, 19, 21, 24, 25, 26, 27, 30, 32, 36]. Different from the network-based IDSs, the host-based IDSs monitor and analyze the internals of a computing system rather than on its external interfaces. Agent-based IDSs become popular due to their higher level abstractions, scalability, and adaptability over the non-agent based IDSs. Based on Distributed Intrusion Detection System (DIDS) [27], many advanced DIDSs have been developed such as distributed hybrid agent-based intrusion detection and real time response system [32], MINDS [6], Distributed Intrusion Detection using Mobile Agents (DIDMA) system [15], and APHIDS++ [2].

Almost all these IDSs spontaneously use classification methods to identify each traffic received by a host or an agent. However, few of them are able to find the relationship among attacks and predict the coming attacks, so that a response system could react proactively in advance to protect the computer systems based on the prediction. Hidden Markov models or Markov chain models have been developed to predict network intrusions [39, 40]. However, these models predict the next state of the network data based solely on the current state, and cannot predict future states of the network data other than the next state. Meanwhile, this kind of method defines states as different combination of features. With an increasing number of features, the number of states will increase exponentially. In order to solve the problem and achieve the goal of prediction, temporal pattern analysis should be involved in an IDS.

Temporal pattern analysis is also an important topic in data mining. Different from association rule mining [1] which attempts to find the correlative items occurring simultaneously in one transaction, temporal pattern mining searches correlative items occurring at different time instances, i.e., asynchronously. Existing algorithms in this area can be divided into several categories: sequential patterns [5], cyclic association rules [20], frequent episodes [34], segment-wise periodic patterns [9], and inter-transaction association rule mining [10, 31]. Even though there are temporal components in all these patterns, sequential patterns, cyclic association rules, frequent episodes, and segment-wise periodic patterns mining can also be categorized as intra-transaction association rule mining, in contrast to inter-transaction association rule mining. For example, each sequence is actually taken as one transaction in sequential pattern mining, and then it finds the similar itemsets with timestamps which satisfy the requirement of minimum support.

The challenge of inter-transactional association rule mining is that it breaks the boundary of transactions, which leads to an increasing number of potential itemsets and rules. FITI [31] finds inter-transactional association rules based on intra-transactional associations and uses a special data structure for efficient mining inter-transactional frequent itemsets. PROWL [14] uses a projected window method and a depth-first enumeration approach to discover frequent patterns quickly, which has shown to outperform FITI. Algorithms for mining follow-up correlation patterns from time-related databases [37, 38] generate inter-transactional rules with the quantity constraint. In this chapter, a novel algorithm for inter-transactional association rule mining is proposed and will be presented in Section 3 which fully utilizes the distribution of the data set (or database) to cut down the number of the potential frequent itemsets. It is especially efficient for those time-related databases that the timestamps of the transactions are not consecutive.

6.3 The Proposed Network Intrusion Detection Framework

Figure 6.1 presents the architecture of the proposed network intrusion detection framework, which consists of a *Training* layer and a *Testing* layer. The main objective of the proposed framework is to facilitate the IDS with the capability of mining and detecting sequential intrusion patterns for prompt and proactive protection of the computer system.

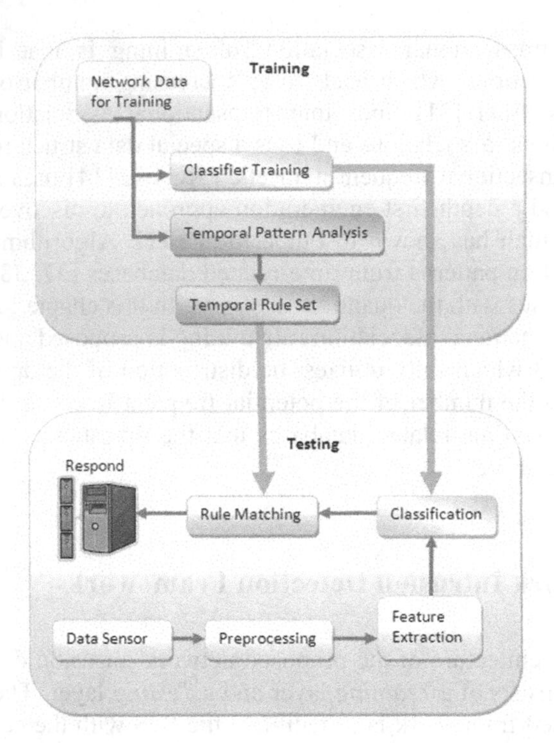

Fig. 6.1 Network intrusion detection and prevention system structure

As can be seen from this figure, in the *Training* layer, the network data for training is used by the C-RSPM supervised classification algorithm (to be briefly introduced in Section 6.3.1) as the input to attain the required parameters for the classifier in the *Testing* layer. At the same time, the network data for training is also used by the LDM *Temporal Pattern Analysis* algorithm (to be presented in Section 6.3.2) to generate the *Temporal Rule Set* for rule matching in the *Testing* layer.

In the *Testing* layer, the *Data Sensor* component attains the connection records of real network traffic from the network card or testing data. The proposed framework first preprocesses the collected network traffic connection records to get the information for the *Feature Extraction* component, and then it classifies each connection based on the features obtained from the *Feature Extraction* component. The *Classification* component uses the parameters obtained in the *Classifier Training* component in the *Training* layer. If a connection is classified as abnormal, it will be matched with the rules in the *Temporal Rule Set* to predict the coming abnormal connections so that the computer system can respond promptly and proactively.

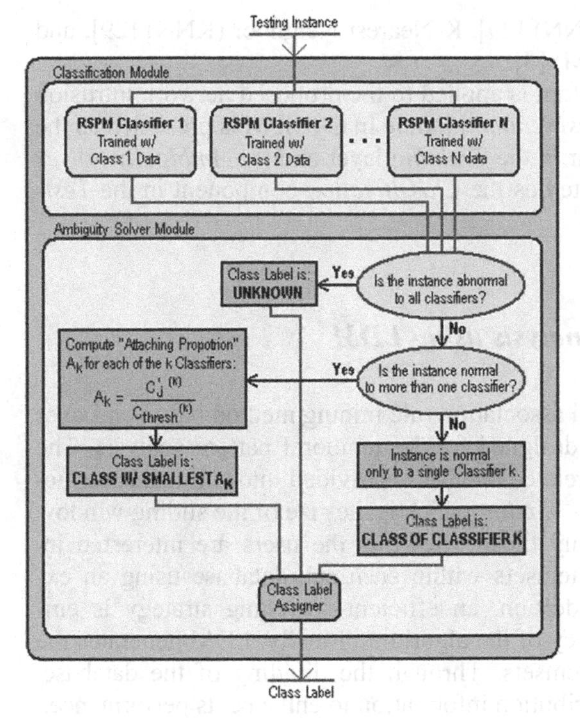

Fig. 6.2 The C-RSPM Architecture [23]

6.3.1 C-RSPM Supervised Classification

C-RSPM is a multi-class supervised classifier with promising performance. Figure 6.2 shows the architecture of the C-RSPM algorithm which includes the *Classification* module and the *Ambiguity Solver* module [23].

In C-RSPM, the number of the classifiers in the *Classification* module is set to the number of classes in the training data set. Each classifier is trained by the data instances of each individual class in the training data set, using the Representative Subspace Projection Modeling (RSPM) technique which is based on principal component analysis and adaptive representative principal component selection technique. An *unknown* class label is used to mark the testing instance which is rejected by all classifiers. The *Ambiguity Solver* module is used to coordinate and capture classification conflicts, which means more than one classifier accepts a testing instance as statistically normal to their training data. In [23], it has shown that C-RSPM outperforms other supervised classification methods such as C4.5 decision tree [22], Decision Table (DT), Nearest Neighbor (NN) [29], K-Nearest Neighbor (KNN) [29], and Support Vector Machine (SVM) [8].

When the C-RSPM algorithm is applied to the proposed network intrusion detection framework, the *Classification* module in C-RSPM is presented as the *Classifier Training* component in the *Training* layer and the *Ambiguity Solver* module in C-RSPM is presented as the *Classification* component in the *Testing* layer.

6.3.2 Temporal Pattern Analysis using LDM

A novel inter-transactional association rule mining method based on Layer Divided Modeling (LDM) is designed for the temporal pattern analysis. The core of LDM is that the time-related database is divided into sub-databases according to each interval within *MaxInt,* which is the size of the sliding window selected to concentrate on only those rules that the users are interested in. LDM searches the frequent itemsets within each sub-database using an extended FP-tree method. In addition, an efficient searching strategy is employed to improve the efficiency of the algorithm. Finally, LDM generates the inter-transactional frequent itemsets. Through the dividing of the database, LDM can fully utilize the distribution information to enhance its performance.

6.3.2.1 Problem Definition

In order to illustrate the algorithm, some definitions are given as follows.

Definition 1. Let $\Sigma=\{i_1, i_2, \ldots , i_u\}$ be a set of items, where u is an integer representing the total number of items in Σ. Let T be the timestamps within a domain -- $Dom(T)$. A time-related database is presented as (t, I), where $t \in Dom(T)$ and $I \subseteq \Sigma$.

Definition 2. A sliding window W contains $MaxInt$ + 1 continuous time intervals starting from t_0 to t_{MaxInt}. Let $s \in [0, MaxInt]$. Each transaction at t_s within W is denoted as $W[s]$.

Each sliding window corresponds to an extended-transaction which consists of every transaction within the sliding window. An extended-transaction is combined with extended-items which are items with timestamps. A set of extended-items is called an extended-itemset Σ'.

Definition 3. An extended-transaction E can be presented as

$$E = \{i_k(s) \mid i_k(s) \in W[s],\ 1 \le k \le u,\ 0 \le s \le MaxInt\}. \quad (6.1)$$

Definition 4. An extended-itemset which contains extended-items from k different timestamps is defined as a k-point extended-itemset.

Definition 5. An inter-transactional association rule is an implication of the form $X \Rightarrow Y$ [31], where

1. $X \subseteq \Sigma$ and $Y \subseteq \Sigma'$;

2. $\exists i_k(0) \in X$, $1 \le k \le u$;

3. $\exists i_k(j) \in Y$, $1 \le k \le u$, $j \ne 0$; and

4. $X \cap Y = \varphi$.

Definition 6. Let S be the number of transactions in the time-related database. Let NXY be the number of the extended-transactions which contain the set of $X \cup Y$, and NX be the number of the extended-transactions which contain X. The support and confidence of an inter-transactional rule $X \Rightarrow Y$ are defined as follows.

$$\text{Support} = \text{NXY} / \text{S}; \tag{6.2}$$

$$\text{Confidence} = \text{NXY} / \text{NX}. \tag{6.3}$$

Based on the pre-defined minimum support value (*Minsup*), minimum confidence value (*Minconf*), and maximum time interval threshold (*MaxInt*), LDM aims to generate all the rules that satisfy the following conditions: *Support* ≥ *Minsup* and *Confidence* ≥ *Minconf* within the range of *MaxInt*.

6.3.2.2 Algorithm Design

The input of LDM is a time-related database with timestamps, *Minsup*, *Minconf*, and *MaxInt*, and its output is the inter-transactional association rules. The main idea of the algorithm is to divide the time-related databasc into layers, so that we can reduce the number of potential frequent inter-transactional itemset candidates. An efficient searching strategy is adopted to reduce the search space. The algorithm consists of the following four steps.

- Step 1: Divide the time-related database into layers based on the timestamp sets with *n* timestamps.
- Step 2: For each layer, if it contains more than *Minsup* extended-transactions, mine the frequent *n-point* extended-itemsets.
- Step 3: If $n \ge \theta$, where θ is a threshold of the timestamps, derive the *n-point* extended-itemsets as candidates from connecting the frequent *(n-1)-*

point extended-itemsets, and then generate the frequent *n-point* extended-itemsets.

- Step 4: Generate the inter-transactional association rules based on the frequent extended-itemsets.

Here, we describe the LDM algorithm in details. The LDM algorithm first divides the time-related database into layers based on the 2-point timestamp set. Let each interval w ($1 \leq w \leq MaxInt$) correspond to a layer. Therefore, the number of the layers in this step is C^1_{ManInt}. Let B_w represent the number of the data instances in the wth layer, S be the total number of the data instances in the time-related database, and *Minsup* be the minimum support threshold from the user. If B_w is less than $S \times Minsup$, then empty the layer. This is due to the fact that there will be no interesting pattern derived from this layer.

Next, the LDM algorithm applies *GeneratesFreqItemset()* to generate the frequent 2-point extended-itemsets using extended FP-tree algorithm from the non-empty layers. The difference between the so-called extended FP-tree algorithm and FP-tree algorithm [11] is that the items in the extended FP-tree algorithm have time attributes with them. We use the same main idea as in the FP-tree to find the frequent itemsets. *GeneratesFreqItemset( )* is the main function of the algorithm, together with a subfunction called *GeneratenPoint-FreqItemset*(L_{n-1}) which generates all the frequent extend-itemsets based on a given input. An extended-itemset with a support value higher than *Minsup* is defined as a frequent extended-itemset. Here, L_n is the set of the frequent *n-point* extended-itemsets, C_n is the candidate set of the *n-point* extended-itemsets, θ is a threshold of the timestamps, and $SL_n(0, \ldots, i)$ is the set of the frequent *n-point* extended-itemsets at timestamp set 0, …, i (where $SL_n(0, \cdots, i) \subseteq L_n$). $\alpha(m)$ (or $\beta(m)$) means that the itemset in α (or β) at timestamp m, and 0, …, i in $SL_n(0, \ldots, i)$ represents the timestamp set in an ascending order (where $0 < i \leq MaxInt$). For example, an itemset at the timestamp set (0, 2, 5) means that the itemset contains items at time points 0, 2, and 5, respectively. The pseudo code for *GeneratesFreqItemset()* and *GeneratenPointFreqItemset*(L_{n-1}) are given below.

GeneratesFreqItemset( )
1. Read in time-related data, *Minsup*, and *MaxInt*;
2. $n = 2$;
3. While $2 \leq n \leq \theta$ do
4. Use the extended FP-tree algorithm to find the frequent extended-itemsets in each layer;
5. $n = n + 1$;
6. End While
7. While (($n \leq MaxInt$) and ($L_{n-1} \neq \varphi$)) do

```
8.    GeneratenPointFreqItemset(L_{n-1});
9.    n = n + 1;
10. End While
```

```
GeneratenPointFreqItemset(L_{n-1})
1. For each itemset α ∈ L_{n-1}
2.  For each itemset β ∈ L_{n-1}
3.   if (α ∈ SL_n(0,···,i), β ∈ SL_n(0,···,j), 0 < i < j ≤ MaxInt) then {
4.      c = α + β = {α(0) ∩ β(0),···,α(m) ∩ β(m),···,α(i), β(j)};
5.      Add c to C_n; }
6.  End for
7. End for
8. Count the support for each candidate c ∈ C_n as c.count;
9. L_k = {c ∈ C_n | c.count ≥ Minsup};
```

Table 6.1 shows the format of the frequent 2-point extended-itemsets, where w ($1 \le w \le MaxInt$), $I_w(0) = \{i_k(0) \mid 1 \le k \le u\}$, and $I_w(w) = \{i_k(w) \mid 1 \le k \le u\}$. Similarly, the layers based on the timestamp set with n timestamps, where $2 \le n \le MaxInt + 1$, can be constructed. The LDM algorithm then generates the frequent *n-point* extended-itemsets. The number of the layers should be C^n_{ManInt}. For each timestamp set, it should contain 0. That is, every extended-itemset should contain at least one item at timestamp 0.

Table 6.1 Frequent 2-point extended-itemsets

Timestamp set	*Frequent 2-point extended-itemsets*
(0, MaxInt*)*	$\bigcup\{I_{MaxInt}(0), I_{MaxInt}(MaxInt)\}$
⋮	⋮
*(0,*w*)*	$\bigcup\{I_w(0), I_w(w)\}$
⋮	⋮
(0,2)	$\bigcup\{I_2(0), I_2(2)\}$
(0,1)	$\bigcup\{I_1(0), I_1(1)\}$

Table 6.2 Time-related database [31]

T	*I*
1	*a, b, e, g, j*
2	-
3	-
4	*c, f, j*
5	-
6	*a, e, d, h*
7	-
8	-
9	*a, c, e, h*
10	*b, c, f*
11	-
12	-
13	-

Table 6.3 Divided database layers

Time stamp set	*Extended-transactions*	*Count*
(0,3)	*(1,4),(6,9)*	*2*
(0,2)	*(4,6)*	*1*
(0,1)	*(9,10)*	*1*

Take the same database given in [31] as an example to illustrate the proposed method (as shown in Table 6.2). Here, we set *MaxInt* = 4, *Minsup* = 40%, and *Minconf* = 60%. The database is first divided into layers as shown in Table 6.3. In the second column of Table 6.3, the absolute timestamp is used to present the transaction at that time. Since $S \times Minsup = 5 \times 40\% = 2$, both (0, 2) and (0, 1) layers are deleted, and only layer (0, 3) is considered. There are two extended-transactions in layer (0, 3):

1. (1, 4): {a(0), b(0), e(0), g(0), j(0), c(3), f(3), j(3)};
2. (6, 9): {a(0), e(0), d(0), h(0), a(3), c(3), e(3), h(3)}.

As can be seen from this example, the largest frequent extended-itemset is {a(0), e(0), c(3)}. Given *Minconf* = 60 %, an inter-transactional association rule is $\text{a}(0), \text{e}(0) \Rightarrow \text{c}(3)$, whose *support* is 40% and *confidence* is 66.7%.

Furthermore, when *MaxInt* increases, the size of the timestamp set will increase exponentially. Consequently, θ is introduced as a threshold. If n is larger than θ, the algorithm derives the *n-point* extended-itemsets as candidates from connecting the frequent *(n-1)-point* extended-itemsets, and then generates the frequent *n-point* extended-itemsets. Again, let L_k be the set of frequent *k-point* extended-itemsets and C_k be the candidate set of the *k-point* extended-itemsets. The algorithm utilizes the following procedure to generate L_{k+1} from L_k.

Connecting Generally, each extended-itemset in L_k contains k timestamps. All the extended-itemsets in L_k that share the same timestamp set are collected as an individual set denoted as $SL_n(0, \ldots, i)$, where $0 < i \leq MaxInt$, and $0, \ldots, i$ represents the timestamp set in an ascending order. Let $\alpha \in L_k$ and $\beta \in L_k$. α and β are connectable only when $\alpha \in SL_k(0,\cdots,i)$, $\beta \in SL_k(0,\cdots,i)$, ($0 < i < j \leq MaxInt$), and the first k-1 timestamps are the same.

If two extended-itemsets are connectable, $c = \alpha + \beta = \{\alpha(0) \cap \beta(0), \cdots, \alpha(m) \cap \beta(m), \cdots, \alpha(i), \beta(j)\}$, $\alpha(m) \cap \beta(m) \neq \varphi$, $0 \leq m < i$. That is, the two extended-itemsets are connected by getting the intersects of the itemsets at the same timestamps and the union of the itemsets whose timestamps are unique. C_{k+1} can be generated by employing the connecting process from L_k. The next step is to prune the candidate set C_{k+1} to get L_{k+1}.

Pruning C_{k+1} includes all the frequent extended-itemsets in L_{k+1}, but it may also contain some extended-itemsets that are not frequent. By simply removing every extended-itemset whose support value is smaller than the minimum support threshold (i.e., *Minsup*) from C_{k+1}, L_{k+1} is generated.

Here, an efficient searching strategy is used to facilitate the process of counting the *support* values for the extended-itemsets. Assume we have an extended-itemset (28(0), 29(0), 3(2), 11(2), 10(5)). We need to count the *support* value of this extended-itemset in the database. The timestamp set of this extended-itemset is (0, 2, 5). Go directly to the layer corresponding to the timestamp set (0, 2, 5) to match the extended-itemset with the extended-transactions in that layer, instead of searching in the whole original database. This method makes the proposed LDM algorithm much faster.

6.4 Experimental Setup

To evaluate the performance of our proposed framework, two different data sets were used in our experiments, and for each data set, the 10-fold cross-validation process was employed in both classification part and temporal pattern analysis part. The standard deviation of the classification accuracy and the rule matching accuracy are also presented. In each fold, the data set was randomly split into the training and testing data sets composed of 2/3 and 1/3 of all data instances, respectively.

The first data set is the KDD CUP 1999 [16] data set with 41 attributes for each data instance; whereas the second one is generated from the private LAN network testbed located at the Department of Electrical and Computer Engineering at the University of Miami [26, 36]. For the KDD CUP 1999 Data, 38 out of 41 attributes are used in C-RSPM. Three nominal attributes are neglected, because currently C-RSPM can only handle numerical data. In our experiments, the normal traffic and 11 types of intrusions in the KDD data are used. The total number of the data instances is 13182. The number of data instances for each type of traffic is given in Table 6.4.

Table 6.4 Dataset from KDD

Type	*Number*
normal	*10,000*
back	*250*
neptune	*400*
pod	*200*
smurf	*200*
teardrop	*100*
guess-passwd	*53*
warezclient	*499*
ipsweep	*350*
nmap	*230*
portsweep	*500*
satan	*400*

For the data set from the testbed, 10 types of traffic are generated, which include 5000 normal connections and 100 connections for each type of abnormal traffic. Typical normal connections generate a proper quantity of data and transfer them during a moderate time period in a suitable rate, frequency, and pace. Typical abnormal connections generate a large amount of data and transfer them during a quite short time period continuously. For example, connections transporting extremely huge packages are used to simulate *ping of death* attacks; connections with a lot of packages in a short time are used to simulate *mail bombing*; connections, which try to access the reserved ports, are used to simulate *Trojan infection*; connections with a lot of large packages in a short time are used to simulate *buffer-overflow*, etc. For these network traffic data, a couple of tools are used to preprocess the data, such as Windump [35] and Tcptrace [28]. Windump is used to capture the raw data (i.e., the information of each connection) from the network card directly. Tcptrace, which can analyze the output file from Windump, is used to extract 88 attributes from each TCP connection, where the extracted attributes include the elapsed time, the number of bytes and segments transferred on both ways, etc.

We have also utilized our own developed feature extraction technique [26] to extract and derive 46 features from the output file generated from Tcptrace that are useful for the proposed network intrusion detection framework. These features include some basic, time-based, connection-based, and ratio-based network features. Out of these 46 features, 43 of them are numerical ones and are currently used in the C-RSPM algorithm.

Connections are generated randomly in sequence to simulate the traffic flow in the real world. We use "second" as the time unit in the time-related database. Connections, which occur within one second, are taken as one transaction in a time-related database.

6.5 Experimental Results

Experiments are conducted to evaluate the performance of the C-RSPM classifier, the LDM inter-transactional association rule mining algorithm, and the proposed network intrusion detection framework.

6.5.1 Performance of C-RSPM

As mentioned above, C-RSPM is capable of performing high accuracy supervised classification and outperforms many other classification algorithms [23]. It also enjoys high classification accuracy when it comes to both KDD CUP 1999 Data and the data set from our testbed.

Table 6.5 shows the accuracy of C-RSPM for both data sets and their corresponding standard deviations (given in the parentheses) for the 10-fold cross-validation experiments. As can be seen from this table, C-RSPM achieves quite promising classification accuracy for each class. Some of the classes can be identified with 100% accuracy, such as back and teardrop in KDD data set and attack6 in the generated network traffic data set.

Table 6.5 Classification accuracy and standard deviation (shown in the parentheses) of C-RSPM

KDD99	*Accuracy*	*Testbed*	*Accuracy*
normal	*99.41 (±0.16)*	*normal*	*99.82 (±0.14)*
back	*100.00 (±0.00)*	*attack1*	*99.91 (±0.08)*
neptune	*99.99 (±0.01)*	*attack2*	*99.97 (±0.03)*
pod	*99.94 (±0.01)*	*attack3*	*99.96 (±0.09)*
smurf	*99.58 (±0.08)*	*attack4*	*99.97 (±0.03)*
teardrop	*100.00 (±0.00)*	*attack5*	*99.95 (±0.05)*
guess-passwd	*99.98 (±0.01)*	*attack6*	*100.00 (±0.00)*
warezclient	*99.65 (±0.06)*	*attack7*	*99.96 (±0.04)*
ipsweep	*99.11 (±0.05)*	*attack8*	*99.98 (±0.03)*
nmap	*99.11 (±0.06)*	*attack9*	*99.97 (±0.03)*
portsweep	*99.94 (±0.01)*		
satan	*99.88 (±0.00)*		

6.5.2 Performance of LDM

To evaluate the performance of the proposed LDM method, LDM is compared with another inter-transactional association rule mining algorithm called Extended Hash Apriori (EH-Apriori) [31]. EH-Apriori is an Apriori like algorithm, which takes the transactions within each sliding window as a mega-transaction and then uses the Apriori algorithm to find the frequent inter-transactional itemsets. A hashing technique is also used to improve the efficiency of the algorithm. Experiments based on changing the values of *MaxInt*, *Minsup*, and the *Size* of the database for both algorithms were conducted and the results are shown as follows.

MaxInt *MaxInt* is the range of the sliding window. It is an important parameter that is used in almost every temporal pattern analysis method. In our experiments, we used different values of *MaxInt* that range from 1 to 7 for the KDD99 data set to compare the efficiency of the proposed algorithm with EH-Apriori. The minimum support threshold is set to 1%, and there are 1147 data instances in the data set. The result is shown in Fig. 6.3.

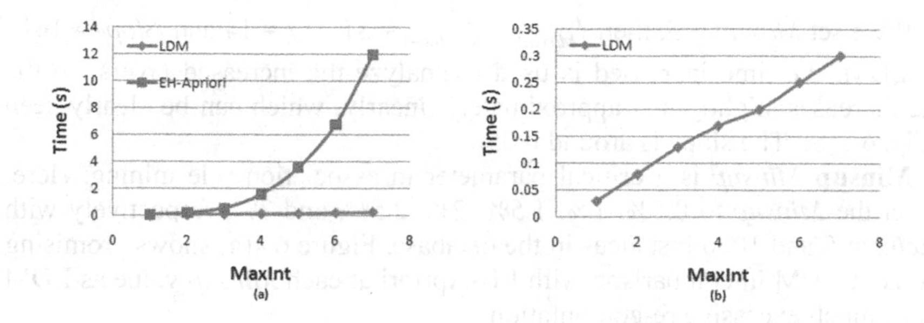

Fig. 6.3 Performance comparison on increasing *MaxInt* values

It is clear in Fig. 6.3(a) that LDM is much faster than EH-Apriori, especially when the *MaxInt* is larger. With increasing *MaxInt* values, the time required by EH-Apriori increases exponentially. This is because a larger *MaxInt* means a wider sliding window, which will lead to more transactions involved in one extended-transaction. That is, the average length of the extended-transactions will be longer. For Apriori-like algorithms, the execution time will increase exponentially when the average length increases. Therefore, *MaxInt* significantly influences the execution time of EH-Apriori, and the curve in Fig. 6.3(a) demonstrates this result. On the contrary, the execution time of LDM only mounts up slightly when the *MaxInt* becomes larger. LDM divides the data set into layers and finds the frequent inter-transactional itemsets in each layer, so the value of *MaxInt* does not affect the average length of extended-transactions directly, whereas it changes the number of the layers. The number of the layers for each *MaxInt* (i.e., l_{MaxInt}) can be calculated by Equation 6.4. The slope of the curve, which captures the relationship between l_{MaxInt} and *MaxInt*, can be calculated by Equation 6.5, where *k* represents the execution time required by each layer. The reason of using an approximate symbol in Equation 6.5 is that the time to process the data in each layer might be influenced a little by the distribution of the data and the potential frequent patterns.

$$l_{MaxInt} = \sum_{n=2}^{MaxInt+1} C_{MaxInt}^{n-1}. \tag{6.4}$$

$$Slope \approx k \times \frac{l_{MaxInt} - l_{MaxInt-1}}{MaxInt - (MaxInt-1)} = k \times \left(\sum_{n=2}^{MaxInt+1} C_{MaxInt}^{n-1} - \sum_{n=2}^{MaxInt} C_{MaxInt-1}^{n-1}\right) \tag{6.5}$$

If we set *MaxInt* = 5, then l_{MaxInt} - $l_{MaxInt-1}$ = 31 - 17 = 14 and $Slope \approx 14k$. For LDM, the time increased is used to analyze the increased layers, so the time increases slightly and approximately linearly, which can be clearly seen in Fig. 6.3(b). The slope is around 0.05.

Minsup *Minsup* is a critical parameter in association rule mining. Here, we set the *Minsup* to 0.5%, 1%, 1.5%, 2%, 2.5%, and 3%, respectively with *MaxInt* at 5 and 1000 instances in the database. Figure 6.4(a) shows promising results of LDM in comparison with EH-Apriori at each *Minsup* value as LDM avoids much excessive re-computation.

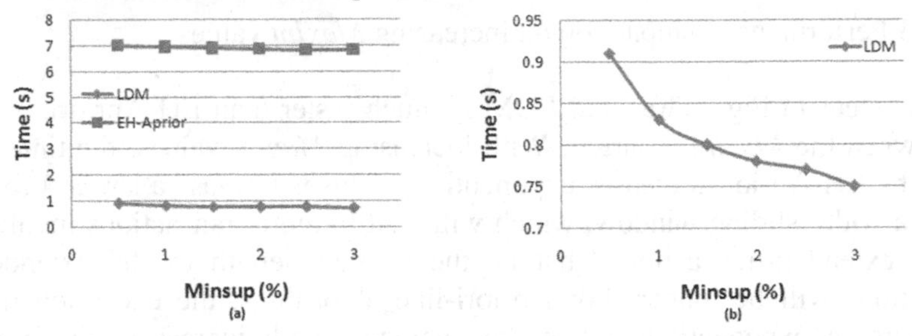

Fig. 6.4 Performance comparison on increasing *Minsup* values

To better look into the performance of LDM, Fig. 6.4(b) shows the performance of LDM on increasing *Minsup* values individually. When *Minsup* is larger, there will be fewer frequent inter-transactional itemsets. Consequently, the execution time should be reduced.

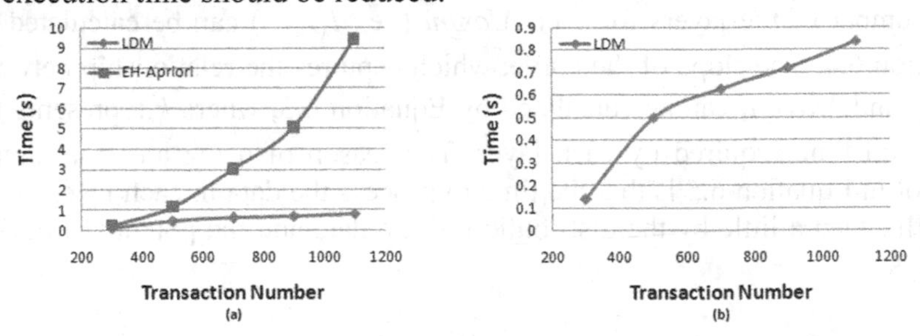

Fig. 6.5 Performance comparison on increasing sizes of the database

Size of the Database In order to investigate LDM's ability to deal with a large size of database comparing to EH-Apriori, we vary the sizes of the database from 300 to 1100, with minimum support at 2% and *MaxInt* at 5. The result presented in Fig. 6.5(a) shows that the executing time of LDM does not change much when the size of the database increases, while the time required by EH-Apriori increases significantly with the increasing number of the transactions in the database. For LDM, the database is divided into layers according to the proposed method, which, in a way, weakens the influence of the size of the database to the performance of the algorithm, though the execution time of LDM also has an ascending trend as shown in Fig. 6.5(b).

6.5.3 Performance of the Proposed Framework

In order to evaluate the performance of the proposed framework, we first generate a temporal rule set (denoted by S_1) from the training data set, and then generate a temporal rule set (denoted by S_2) from the testing data set. For convenience, we only use the 2-point inter-transactional rules for matching, because 2-point inter-transactional rules have covered all the relations among different items. The number of the 2-point inter-transactional rules in S_1 is N_1. The number of the 2-point inter-transactional rules in S_2, which can also be matched with rules in S_1, is N_2. The Rule Matching Rate is employed to evaluate the effect of the proposed framework, which is defined by Equation 6.6.

$$RuleMatchingRate = N_2 / N_1 \tag{6.6}$$

Table 6.6 Rule matching rate (%) of the KDD99 and Testbed data sets

MaxInt	*KDD99*	*Testbed*
1	*89.29(±3.14)*	*81.77(±7.51)*
2	*90.00(±2.13)*	*80.31(±5.56)*
3	*89.73(±2.64)*	*80.76(±3.74)*
4	*89.60(±3.01)*	*80.65(±4.30)*
5	*88.85(±2.38)*	*80.65(±3.88)*
6	*89.01(±3.22)*	*81.23(±4.06)*
7	*89.44(±2.67)*	*80.01(±3.67)*

Table 6.6 shows the rule matching rates for both data sets and their corresponding standard deviations (given in parentheses) with various *MaxInt* values for the 10-fold cross-validation experiments. The first column gives the various *MaxInt* values which range from 1 to 7. As can be seen from this table, our proposed framework maintains high rule matching rates for all the *MaxInt* values used in the experiments and for both data sets. This demonstrates our proposed framework can effectively identify sequential intrusion patterns from the training data set and used the trained rules to predict the coming attacks. Meanwhile, it can also be observed that the rule matching rate is quite stable when the *MaxInt* value changes, so the algorithm is robust to the *MaxInt* values. As it can also be seen that the total rule matching rate of the KDD99 data

set is higher than that of the testbed data set, which means that the rule matching rate could be influenced by the size of the data set and the distribution of the frequent inter-transactional patterns.

6.6 Conclusion

In this chapter, a novel network intrusion detection framework which consists of two main components, namely C-RSPM and LDM, is proposed to detect sequential network intrusion patterns and to predict the types of coming intrusions. The key concept in the design of the proposed framework is that it utilizes a temporal pattern analysis method to find the sequential relations among network attacks based on the detected attacks by the classifier. The KDD99 data set and a data set generated by our private LAN testbed which include both normal and abnormal network traffic are employed to evaluate the performance of the C-RSPM classifier, the LDM inter-transactional association rule mining algorithm, and the proposed framework. However, the proposed framework has not been deployed in an operational network setting, and has not been evaluated by the actual network traffic data, which could be done in the future. From the experimental results, it can be concluded that (i) C-RSPM is an effective classifier with high identification accuracy ($> 99\%$ in the data sets used); (ii) LDM is an efficient inter-transactional association rule mining algorithm, which can quickly generate all the temporal patterns under various combinations of different *MaxInt* values, *Minsup* values, and *Sizes* of the database; and (iii) the proposed framework is capable of predicting sequential intrusion patterns with high prediction rates ($> 80\%$ in the data sets used). All these promising results show that our proposed framework is efficient and effective and can enable prompt and proactive decision-making and actions to better safeguard the network system.

Reference

[1]. Agrawal R, Swami A (1993) Mining association rules between sets of items in large databases. In: Proceedings of the ACM SIGMOD conference on management of data: 207—216.

[2]. Alam M.S, Vuong S.T (2007) APHIDS++: A mobile agent based intrusion detection system. In: Proceedings of the 2nd international conference on communication systems software and middleware: 1—6. doi: 10.1109/COMSWA.2007.382483.

[3]. Anderson D, Frivold T, Anderson A (1995) Next-generation intrusion detection expert system (NIDES): A summary. In: SRI international technical report 95: 28—42. Menlo Park, CA.

[4]. Basicevic F, Popovic M, Kovacevic V (2005) The use of distributed network-based IDS systems in detection of evasion attacks. In: Proceedings of the advanced industrial confer-

ence on telecommunications/service assurance with partial and intermittent resources conference/e-learning on telecommunications workshop. AICT/SAPIR/ELETE: 78—82.
[5]. Boonjing V, Songram P (2007) Efficient algorithms for mining closed multidimensional sequential patterns. In: Proceedings of the 4th international conference on fuzzy systems and knowledge discovery 2: 749—753.
[6]. Ertoz L, Eilertson E, Lazarevic A, Tan P, Srevastava J, Kumar V, Dokas P (2004) The MINDS — Minnesota intrusion detection system. Next generation data mining. MIT Press, Cambridge, MA.
[7]. Esparza O, Soriano M, Munoz J.L, Forne J (2003) A protocol for detecting malicious hosts based on limiting the execution time of mobile agents. In: Proceedings of the 8th IEEE international symposium on computers and communication: 251—256.
[8]. Han B (2003) Support vector machines. http://www.ist.temple.edu/~vucetic/cis526fall2003/lecture8.doc.
[9]. Han J, Gong W, Yin Y (1998) Mining segment-wise periodic patterns in time-related databases. In: Proceedings of the international conference on knowledge discovery and data mining: 214—218.
[10]. Han J, Lu H, Feng L (1998) Stock movement prediction and n-dimensional inter-transaction association rules. In: Proceedings of the 1998 SIGMOD workshop research issues on data mining and knowledge discovery 12: 1—7.
[11]. Han J, Pei J, Yin Y (2000) Mining frequent patterns without candidate generation. In: Proceedings of the ACM SIGMOD international conference on management of data (SIGMOD'00): 1—12.
[12]. Helmer G, Wong J, HONAVAR V, MILLER L, WANG Y (2003) Lightweight agents for intrusion detection. J Syst Softw 67: 109—122.
[13]. Hochberg J, Jackson K, Stallings C, Mcclary J, Dubois D, Ford J (1993) NADIR: An automated system for detecting network intrusions and misuse. Comput Secur 12: 235—248.
[14]. Huang K, Chang C, Lin K (2004) Prowl: An efficient frequent continuity mining algorithm on event sequences. In: Proceedings of the 6th international conference on data warehousing and knowledge discovery (DaWak'04), Lecture Notes in Computer Science 3181: 351—360.
[15]. Kannadiga P, Zulkernine M (2005) DIDMA: A distributed intrusion detection system using mobile agents. In: Proceedings of the 6th international conference on software engineering, artificial intelligence, networking and parallel and distributed computing. 238—245.
[16]. KDD (1999) KDD cup 1999 data. http://kdd.ics.uci.edu/databases/kddcup99/.
[17]. Labib K, Vemuri V (2004) Detecting and visualizing Denial-of-Service and network probe attacks using principal component analysis. In: The 3rd conference on security and network architectures (SAR'04). La Londe, France.
[18]. Lazarevic A, Ertoz L, Kumar V, Ozgur A, Srivastava J (2003) A comparative study of anomaly detection schemes in network intrusion detection. In: Proceedings of the third SIAM conference on data mining. San Francisco, CA.
[19]. Lee W, Stolfo S (2000) A framework for constructing features and models for intrusion detection systems. ACM Trans Inform Syst Secur 3: 227—261.
[20]. Ozden B, Ramaswamy S, Silberschatz A (1998) Cyclic association rules. In: Proceedings of the 14th international conference on data engineering: 412—421.
[21]. Paek S, Oh Y, Yun J, Lee D (2006) The architecture of host-based intrusion detection model generation system for the frequency per system call. In: Proceedings of the international conference on hybrid information technology (ICHIT'06) 2: 277—283.
[22]. Quinlan J (1993) C4.5: Programs for machine learning. Morgan Kaufmann, San Fracisco, CA.
[23]. Quirino T, Xie Z, Shyu M, Chen S, Chang L (2006) Collateral representative subspace projection modeling for supervised classification. In: Proceedings of the 18th IEEE international conference on tools with artificial intelligence (ICTAI'06): 98—105.

[24]. Ramakrishnan V, Kumar R.A, John S (2007) Intrusion detection using protocol-based non-conformance to trusted behaviors. In: Proceedings of navigation and surveillance conference (ICNS '07): 1—12.
[25]. Ray P (2007) Host based intrusion detection architecture for mobile ad hoc networks. In: Proceedings of the 9th international conference on advanced communication technology 3: 1942—1946.
[26]. Shyu M, Quirino T, Xie Z, Chen S, Chang L (2007) Network intrusion detection through adaptive sub-eigenspace modeling in multiagent systems. ACM Transactions on Autonomous and Adaptive Systems 2(3): 1—37.
[27]. Snapp S, Bretano J, Dias G, Goan T, Hebrlein L, Ho C, Levitt K, Mukherjee B, Smaha S,Grance T, Teal D, Mansur D (1991) DIDS (distributed intrusion detection system)—motivation, architecture, and an early prototype. In: Proceedings of the 14th national computer science conference. Washington D.C.: 167—176.
[28]. TCPTRACE (2008) Available at http://www.tcptrace.org/.
[29]. Tou J, Gonzalez R (1974) Pattern recognition principles. Addison-Wesley, MA.
[30]. Tsai M, Lin S, Tseng S (2003) Protocol based foresight anomaly intrusion detection system. In: Proceedings of IEEE the 37th annual 2003 international carnahan conference: 493—500.
[31]. Tung A, Lu H, Han J, Feng L (2003) Efficient mining of intertransaction association rules. IEEE transactions on knowledge and data engineering 15(1): 43—56.
[32]. Vaidehi K, Ramamurthy B (2004) Distributed hybrid agent based intrusion detection and real time response system. In: Proceedings of the 1st international conference on broadband networks (BROADNETS'04): 739—741.
[33]. Verwored T, Hunt R (2002) Intrusion detection techniques and approaches. ComputComm 25: 1356—1365.
[34]. Wang Y, Hou Z, Zhou X (2006) An incremental and hash-based algorithm for mining frequent episodes. In: Proceedings of the international conference on computational intelligence and security 1: 832—835.
[35]. WinDump: tcpdump for Windows (2008) Available at http://www.winpcap.org/windump/default.htm.
[36]. Xie Z, Quirino T, Shyu M, Chen S, Chang L (2006) UNPCC: A novel unsupervised classification scheme for network intrusion detection. In: Proceedings of the 18th IEEE international conference on tools with artificial intelligence (ICTAI'06): 743—750. Washington D.C., USA.
[37]. Zhang S, Huang Z, Zhang J, Zhu X (2008) Mining follow-up correlation patterns from time-related databases. Knowl Inf Syst 14(1): 81—100.
[38]. Zhang S, Zhang J, Zhu X, Huang Z (2006) Identifying follow-correlation itemset-pairs. In: Proceedings of the 6th IEEE international conference on data mining (ICDM06): 765—774.
[39]. Gao F, Sun J, Wei Z (2003) The prediction role of hidden Markov model in intrusion detection. In: Proceedings of Canadian conference on electrical and computer engineering 2: 893—896.
[40]. Yin Q, Zhang R, Li X (2004) A new intrusion detection method based on linear prediction. In: Proceedings of the 3rd international conference on information security (InfoSecu04): 160—165.

7 A Non-Intrusive Approach to Enhance Legacy Embedded Control Systems with Cyber Protection Features

Shangping Ren*, Nianen Chen, Yue Yu, Pierre Poirot[1], Kevin Kwiat[2] and Jeffrey J.P. Tsai[3]**

Abstract Trust is cast as a continuous re-evaluation: a system's reliability and security are scrutinized, not just prior to, but during its deployment. This approach to maintaining trust is specifically applied to distributed and embedded control systems. Unlike general purpose systems, distributed and embedded control systems, such as power grid control systems and water treatment systems, etc., generally have a 24x7 availability requirement. Hence, upgrading or adding new cyber protection features into these systems in order to sustain them when faults caused by cyber attacks occur, is often difficult to achieve and inhibits the evolution of these systems into a cyber environment. In this chapter, we present a solution for extending the capabilities of existing systems while simultaneously maintaining the stability of the current systems. An externalized survivability management scheme based on the observe-reason-modify paradigm is applied, which decomposes the cyber attack protection process into three orthogonal subtasks: observation, evaluation and protection. This architecture provides greater flexibility and has a resolvability attribute- it can utilize emerging techniques; yet requires either minimal modifications or even no modifications whatsoever to the controlled infrastructures. The approach itself is general and can be applied to a broad class of observable systems.

[1] Shangping Ren, Nianen Chen Yue Yu, and Pierre Poirot, Department of Computer Science, Illinois Institute of Technology, Chicago, IL 60616, e-mail: ren@iit.edu, nchen3@iit.edu, yyu8@iit.edu, poirple@iit.edu

[2] Kevin A. Kwiat, Distributed Director, AFRL, email: kwiak@gamil.com

[3] Jeffrey J.P. Tsai, Department of Computer Science, University of Illinois at Chicago, Chicago, IL 60607, e-mail: tsai@uic.edu

* This work is supported by NSF under grant CNS 0746643.

** Approved for Public Release; distribution unlimited; AFMC-2008-0692, 22 DEC 0

J.J.P. Tsai and P.S. Yu (eds.), *Machine Learning in Cyber Trust: Security, Privacy, and Reliability*, DOI: 10.1007/978-0-387-88735-7_7,

7.1 Introduction

The following question is at the core of trust: can reliability and security be joined reliably and securely? Uncertainty over this conjoining leads to untrustworthiness, and the linking of security and reliability is non-trivial. For example, the probability of failure can be reasonably measured for natural faults, but the likelihood of failures induced by a directed attack depends on intangibles such as the skill, determination and power of the attackers. Furthermore, in the face of just natural faults, increasing system reliability follows from the application of techniques for fault tolerance; however, fault tolerance almost always implies the application of some form of redundancy. If redundancy is a typical prerequisite for fault tolerance, then replicating a secret runs contrary to normal security practices. These points go to show that coupling the treatment of failures induced by natural phenomenon with the handling of those induced by information attack goes beyond the scope of reliability.

Security of computers and communication has the following attributes: confidentiality, integrity, and availability [1]. In addition, availability is an attribute close to reliability [2].

In [3], we argued that when security is combined with reliability, distributed system features must be evaluated so that their effectiveness can be predicted – otherwise they might inadvertently magnify the attackers power. To emphasize the point, we borrowed a lesson from history: although ancient Rome was not built in a day, it did not take very long for it to fall once the barbarians took hold. We apply the same consideration to more contemporary concerns with our observe-reason-modify approach to evaluate an embedded control system and prevent the occurrence of such ”a fall”.

Recent events indicate that a modern-day assault is underway. On September 27, NN Headline News cited “Sources: U.S. power grid vulnerable to cyber attacks”. Other headlines announce: “Study: Mouse click could plunge city into darkness”; “Researchers who launched an experimental cyber attack caused a generator to self-destruct, alarming the government and electrical industry about what might happen if such an attack were carried out on a larger scale”; and “Some experts fear bigger, coordinated attacks could cause widespread damage to electric infrastructure that could take months to fix”.

Legacy distributed and embedded systems still perform important roles in many of the nation’s critical infrastructures, ranging from electric power generation, transmission, and distribution to railroads and mass transit [11]. The roles’ importance dictates stringent system reliability requirements. In general, these infrastructures often have two layers:

Physical layer. This layer consists of physical units, such as power distribution units, wirings, valves, and plumbing, etc. The physical units are used to deliver the essential services.

Supervisory and Control layer (abbreviated SC layer). This layer contains computers, networks and data gathering sensors that are used to monitor and control the physical layer.

It is understood how to protect the physical layer from being physically sabotaged by enforcing physical protection. Security locks and surveillance can effectively prevent tangible damages. Traditionally, the SC layer is located in access-controlled rooms and the communication among SC units is through a closed, stand-alone and dedicated network. However, as computer and network technologies advance, these infrastructures have gradually evolved into cyber systems that possess the flexibility and productivity that modern technology brings. Unfortunately, the side effects and risks associated with these technologies are nevertheless not fully addressed in this very special area. Some of the misconceptions that have contributed to disregard for the specialized nature of legacy systems are [12]:

The infrastructures and their control systems reside on a physically separated and stand-alone network.

Connections between the SC layer of critical infrastructures and other corporate networks are protected by strong access control schemes.

The SC layer of the embedded infrastructures requires special knowledge, making them difficult for network intruders to access and control.

In the underlying physical layer, all alarms of fault occurrences are assumed to be caused by hardware or software malfunctions and these alarms can be answered by common fault tolerance techniques.

In fact, the ability to access and control processes that were previously isolated to standalone networks has rendered them vulnerable to cyber attacks from a variety of sources including hostile governments, terrorist groups, disgruntled employees, and other malicious intruders. The 2003 incident where a disgruntled Australian engineer released tons of dirty water upon city grounds to gain revenge against his supervisor is an example [21]. Inflicting this type of damage did not require the presence of a human at either the site where the physical units are located or the control room where the SC layer is located thus making the attack much more difficult to detect and prevent.

A value of a system that has become a legacy system is its performance history. Systems that have endured over time within their environment have proven their capabilities. These capabilities include the reliability attained by the system's designers, but these designers could not have anticipated all the faults that can now be inflicted by present-day attackers. What has made the problem of protecting legacy national critical infrastructures even more challenging is that these systems have a stringent 24x7 availability goal that inhibits the "shutdown and upgrade" approach that otherwise is an effective way to handle emerging concerns. Requiring high availability while resisting system-level changes creates an impediment: any Quality of Service (QoS) enhancement must be done in a non-intrusive way. In addition, whereas the application of traditional fault tolerance measures through a legacy system's central control and administration may indeed be sufficient, survivability in a cyber environment must address highly distributed, dynamic and unbounded environments that lack central control and unified policies [23]. The proven track record of a legacy system should not be undermined by enhancements aimed at achieving attack protection; instead, reliability and security must be joined reliably and securely.

To overcome these challenges and to ensure software system dependability in cyber environments, a model that captures the characteristics of the system and the environment becomes essential. As critical information systems emerge from "closed castle" into distributed paradigms, the co-operation among distributed elements which compose the larger cyber systems inevitably become the focus of such systems.

In this chapter, we present a solution to extend existing embedded control systems. New capabilities are added; yet they are kept separate from the system's original functionality in order to maximize future system evolvability while minimizing the disturbance imposed upon the system. An externalized survivability management scheme based on the observe-reason-modify paradigm is applied, which decomposes the cyber attack protection process into three orthogonal subtasks: observation, evaluation and protection. The approach itself is general and can be applied to a broad class of observable systems. The generality of the approach also promotes assurances that changes will do no harm: the robustness introduced in the legacy system's original design is undisturbed whereas the observe-reason-modify paradigm overcomes the hitherto unforeseen challenges of attack protection.

The rest of this chapter is organized as following: Section 2 details related work with focus on cyber survivability issues in industrial distributed control systems, and software architectures. Section 3 starts with a motivation application and then presents a solution for extending a current distributed control system with new features that can protect it from cyber damage at runtime. Our architectural solution contains three modules (i.e., observation, evaluation and protection modules). Together with the controlled system, they form a control loop. From the observation module originates all the protection actions and is the most critical part of the system. Section 4 is devoted to the detailed discussion of correlating local event streams and deriving system information in the observation module. Section 5 discusses when we should make decisions under unreliable environment with real-time constraints. Section 6 briefly outlines our prototype implementation. Finally, Section 7 summarizes and concludes the chapter.

7.2 Related work

Research and continuous re-evaluation of standard practices have been conducted to study ways of improving the survivability of critical infrastructures where errant or malicious computer operations could result in a catastrophe. However, few of them demonstrate a non-intrusive approach, by integrating domain specific security knowledge into survivability solutions, to focus on cyber attacks in existing embedded systems, such as power grids, water treatment systems, or Supervisory Control and Data Acquisition (SCADA) systems.

Pollet proposes a Network Rings of Defense model to provide a layered security strategy for the SCADA system [13]. In such a structure, developing an appropriate SCADA security strategy involves analysis of multiple layers including firewalls, proxy servers, operating systems, application system layers, communications, and policy and procedures. Risk analysis are applied on all these layers and include known vulnerabilities, such as password, key stroke logging, and Denial of Service (DoS) attack protection [14]. An agent-based system is proposed to monitor a large legacy system in a distributed way to provide quick local fault recognition and response [16]. Firewalls [15] and intrusion detection techniques are also studied to help repel and localize cyber attacks [17].

Protection-Shell [19], also known as a Safety Kernel [18], is "an independent computer program that monitors the state of the system to determine when potentially unsafe system states occur or when transitions to potentially unsafe system states may occur. The Safety Kernel is designed to prevent the system from entering the unsafe state and return it to a known safe state." Leveson et al. [20] describe the term "Safety Kernel" as a technique which focus on centralizing a set of safety mechanisms. These mechanisms are used to enforce usage policies that are established in a given system to ensure system safety. Kevin G. Wika and J.C. Knight gave an evaluation of the feasibility of the safety kernel as software architecture for the enforcement of safety policies [18].

System Fault-Tree Analysis [19, 22] is a widely used safety analysis technique and also an important technology for assessment of safety-critical systems. System Fault-Tree Analysis helps to make fault dependability predictions, and identify root causes of equipment failures. Although different versions of software replications on different hardware units are used to tolerate hardware and software faults, the management of these replicas in a distributed environment is intertwined with the functional logic being protected.

Until today, most research efforts have focused on applying available general- purpose IT security technologies to large legacy and embedded infrastructures. Little effort has been put on developing large and embedded-control-specific strategies. One of the major characteristics of such systems is that it could take a decade or more to renovate the existing systems to take full advantage of general IT security technologies, but on the other side, these legacy systems still have a considerable amount of serviceable life remaining [17]. Hence, compensating and non-intrusive approaches for improving legacy systems survivability in a cyber environment must be sought-after. Furthermore, most legacy industrial systems are not designed to be self-extensible, or self-adaptive, but rather to be extremely stable in order to operate reliably in a production environment. By design, users in those systems may never customize the system or tailor its behavior at runtime.

External self-adaptive approaches have been studied recently and have shown to be a practical solution for providing self-adaptation to these systems [4, 8, 7]. These approaches adopt traditional control theory that has been used and proved to be an effective solution [9] in hardware design and industrial automation. They place general adaptation mechanisms in separate modules that can be created, modified, extended and reused across various applications. Nevertheless, while the concept of a control feedback loop is simple, i.e., sensing (observing), calculating (reasoning) and acting (adapting), what to observe, how to reason and when to act are difficult to define and the decisions regarding the what, how and when play an important role in the success of external, feedback loop based adaptation schemes.

The seminal Rainbow architecture by Garlan et. al. [4, 5] illustrates the challenges of architecture-based self-adaptation, such as latency and suitable decomposition. It provides generic self-adaptation through a gauger (monitoring), a model manager, a constraint evaluator, an adaptation engine (reasoning), an adaptation executor and effectors (adapting). To use it, it requires an extensive knowledge of the system properties, constraint rules, adaptation strategies, and adaptation operators, split among multiple components. Other works in this area include [6, 7, 8, 9].

Our non-intrusive approach is inspired by early work in the external self-adaptation domain listed above. What distinguishes our work from previous research is that in our architecture we apply both the separation of concern principle in software design and feedback loop control theory. Architectural support is thus provided to extend existing systems with new features, to reduce software system complexity, and to provide flexibility for customizing the system for different applications.

7.3 Event-Based Non-intrusive Approach for Enhancing Legacy Systems with Self-Protection Features

Our approach is different from those mentioned in the related work section because our solution aims to streamline the extension process through creation of a modular control loop that encapsulates attack protection behaviors into separate modules that can be created, modified and reused across multiple systems. Our solution has two major thrusts: (1) decomposition of a complex attack-protection system into three orthogonal modules, i.e., observation, evaluation and protection modules; and (2) non-intrusiveness. Extending features is non-intrusive and external to existing controlled systems. Also supported are independent specifications for observation focus, reasoning logic, and protection procedures.

7.3.1 Motivating Example

Before we propose an architecture for the purpose of extending legacy systems with new features, we first take an example from a real life application and study the properties that are intrinsic to this set of applications.
Consider a simplified version of water treatment systems. In the water treatment system, there are six valves (V1 ~ V6) which control the fluid velocity and four pumps (P1 ~ P4) which are used to pump raw water into the process system and distribute the purified water to consumers. In the normal condition, only P3 and P4, called primary pumps, will operate. P1 and P2 are backups and are activated only when the primary pumps are out of order.
The supervisory and control (SC) layer monitors the pumps through two sensors (S1 and S2) attached to the primary pumps. As soon as the status of the primary pump becomes abnormal, the backup pumps are activated. We also have a Pressure Vessel in which raw water is buffered and elementary filtering is applied. Normally, the valve opening and the pump speed are tuned to ensure that all the containers are not over pressured.

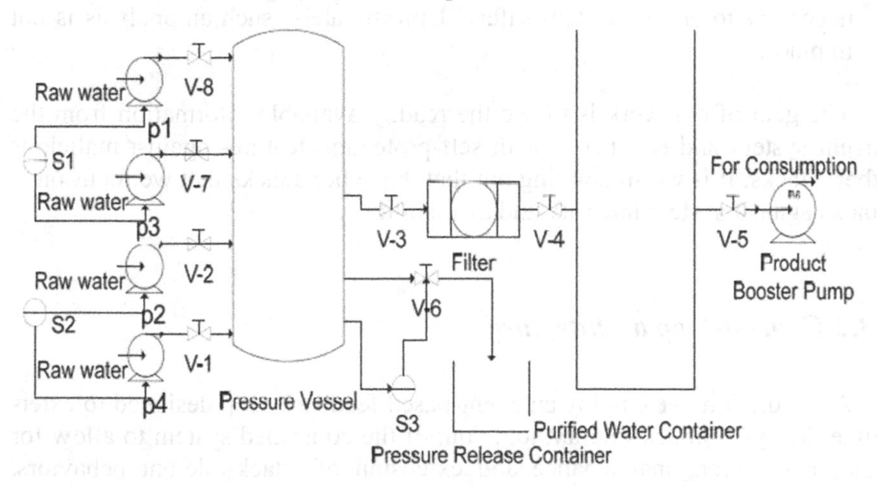

Fig. 7.1 A Simplified Water Treatment System

In case of over pressurization, a pressure release container is attached to the pressure vessel as a safety mechanism. A sensor (S3) is used to detect the pressure level in the pressure vessel. When its value exceeds a threshold, the valve (V6) for the pressure release container is activated to release water from the pressure vessel. We assume that both sensors are highly reliable. Figure 7.1 depicts the simplified water treatment system.

A few key facts about the system are as following:

1. The physical units are monitored and controlled by the SC units. The correct relation among the physical units (such as the status of each valve, the pump and the container) is predefined and the control logic, in responding to known failures, is also programmed in the SC layer's control processes.
2. The critical status of the system, such as the velocity of the pump and the velocity of the water flow, is externally observable and can be used to analyze the health of the system.
3. Only hardware failures, such as a pump or valve not working, are considered in the current systems and hardware redundancy are in place for fault tolerance purposes. However, situations in which all individual units are working properly, but the system, as a whole, is malfunctioning, are not considered.
4. A certain action may not be harmful; however, a sequence of non-harmful actions can be dangerous. For instance, increasing the velocity of raw water being pumped is a normal action, but if it is followed by turning off both the filter valve and pressure release valve, these actions can cause catastrophic damage to the system. Therefore, before a command can be issued on the physical unit, a state, time and history dependent analysis is necessary to ensure system safety. Unfortunately, such an analysis is not in place.

The goal of our work is to use the readily-available information from the current system and enhance it with self-protection features against malicious cyber attacks. It is worth pointing out that the cyber attacks that we focus on is attacks against system integrity and availability.

7.3.2 Control-loop architecture

As a solution, we employ an event-based feedback loop designed to externalize the cyber attack-tolerant logic out of the controlled system to allow for easier conception, maintenance and extension of attack-tolerant behaviors. Under this architecture, a controlled system is monitored and compared with a system model that represents the essential components and their relationship with the controlled system to determine the health of the system. Figure 7.2 depicts the high-level view of our proposed architecture.

As shown in the figure, the newly added protection logic is separated from the existing controlled system and its activation is only through even observations. Furthermore, the software extension is deployed on platforms that are independent from the controlled systems. In addition, the observation, reasoning and action schemes are separated into independent modules. Such architecture allows us to change and incorporate different observation interests, reasoning schemes and action strategies without much modification to the controlled systems, or the other modules.

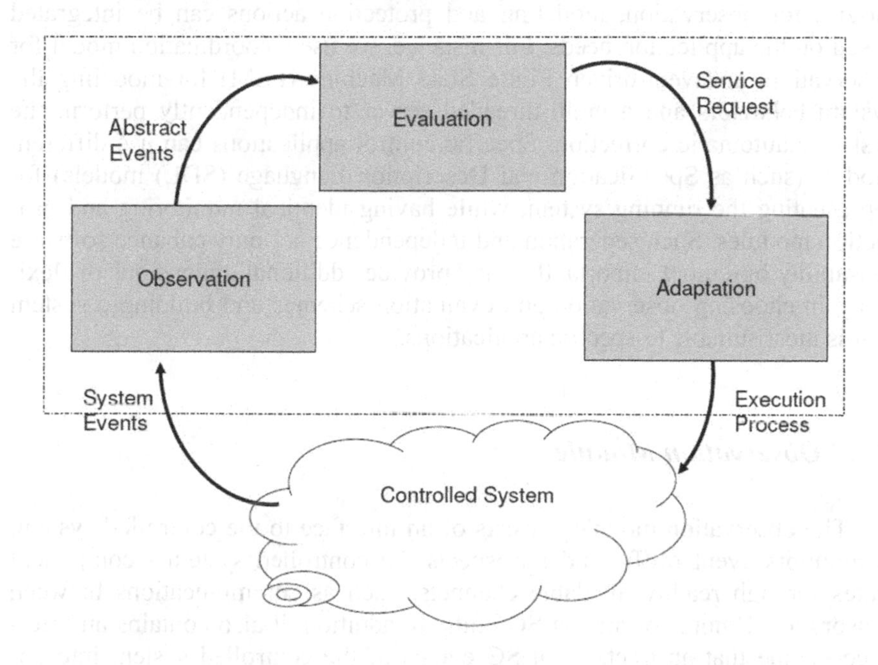

Fig. 7.2 External Control Loop.

More specifically, the external layer contains three modules, i.e., Observation, Evaluation, and Protection modules. These three modules communicate with each other through standard interfaces. Various technologies such as direct calls, JNI calls, or interprocess communications can be applied for such communications. The Observation module observes events generated by the controlled system and maps them into a high level abstraction so that the Evaluation module does not have to be tied with a specific system or system specific events; instead, the information will be provided to the Evaluation with high level abstractions to promote the separation of reasoning logic from individual systems. The Evaluation module is responsible for reasoning about the controlled system from the information provided by the Observation module and decides how the system should act. The Protection module interfaces with the controlled system and decides if an action should be issued on the physical units, or if a system reconfiguration should take place, such as activating backups, excluding cyber-infected SC units from the system, and to prevent potential catastrophe.

This architecture employs minimum dimensions required for automatic controlling the behavior of an embedded software system: observation, reasoning and adjustment. These modules are separated, so that the extension loop can be customized easily. In other words, different strategies or methodologies for observation, modeling and protection actions can be integrated based on the application needs. For instance, we use a coordination model for observation, an event-driven Finite State Machine (FSM) for modeling the system behaviors and a multi-threaded server to independently perform the system's automatic correction. Specific control applications can use different models (such as Specification and Description Language (SDL) models) for representing the running system, while having identical monitoring and protection modules. Such separation and independence not only enhance software reusability but, more importantly, they provide additional dimensions of flexibility in choosing observation and evaluation schemes and building a system that is most suitable to specific applications.

7.3.3 Observation Module

The observation module consists of an interface to the controlled system. It monitors event traffic and introspects the controlled system's component states through readily available channels, such as communications between sensors to SC units, or among SC units. In addition, it also contains an inference engine that abstracts local SC events of the controlled system into abstract events. The purpose of the inference engine is to transform a stream of SC monitored events into abstract events. For instance, observing water flow decrease and pump velocity increase can be interpreted as a speed mismatch and the observation module's inference engine will generate a mismatch event. The evaluation module works exclusively on the model related abstract events. One of the advantages of mapping process-level events to model-level events is that it shields the evaluation module from numerous low-level events coming from the controlled system and decouples the system reasoning from platform specific and platform dependent controlled systems and also from possible noises.

In addition, how the mapping is performed is encapsulated within the observation module and hence its modification will not disrupt the rest of the protection system.

Observing controlled system event streams is not difficult, but correlating the streams and deriving model level information from these local streams are challenging. Section 4 presents a detailed solution to face the challenge.

7.3.4 Evaluation Module

The evaluation module contains an abstract model representing the essential functionalities of the controlled systems and an inference engine. The inference engine examines how the controlled system should reconfigure or react based on the system abstraction and the events provided from the observation module. Encapsulated within the evaluation module is the logic that bridges observation and reaction.

In the current version, the evaluation model of the system uses event-driven Finite State Machine (FSM) that defines the desired safety property of the system. The event-driven FSM makes the sophisticated cyber attack-tolerant logic much more comprehensible and much easier to maintain and update. The states used in the FSM may differ from states in the controlled system as not all the various states of the controlled system need to be considered during the protection process. In our current design, the result of the evaluation system is a service request to the protection module when it transitions from one state to another in response to the events generated by the observation module.

For instance, consider a scenario that the observation module has observed an event from the water treatment system that the water level in pressure vessel has exceeded a predefined threshold. Through the observation module's inference engine, this event is mapped to a abstract model event — raw water pump velocity and filter velocity do not match. Upon receiving a mismatch event, depending on the current state of the controlled system (which is reflected as a node in the FSM), the evaluation model may conclude that one of the raw water pump's SC units has a suspicious behavior and further control command from the SC unit to increase the pump's velocity should exclude the pump and its SC unit from the system. It will send the action module a service request to prevent certain actions on the suspicious SC. However, how such actions are to be prevented on specific physical units through the supervisory and control (SC) units is fully encapsulated in Protection module.

Figure 7.3 is a section of the water treatment control model. In particular, if the water level in the vessel is low, a mismatch between water in and water out for a certain time duration (t) will only move the system into state of water level medium; however, in any other states, such a mismatch will trigger issuance of a certain prevention service request to the protection modules.

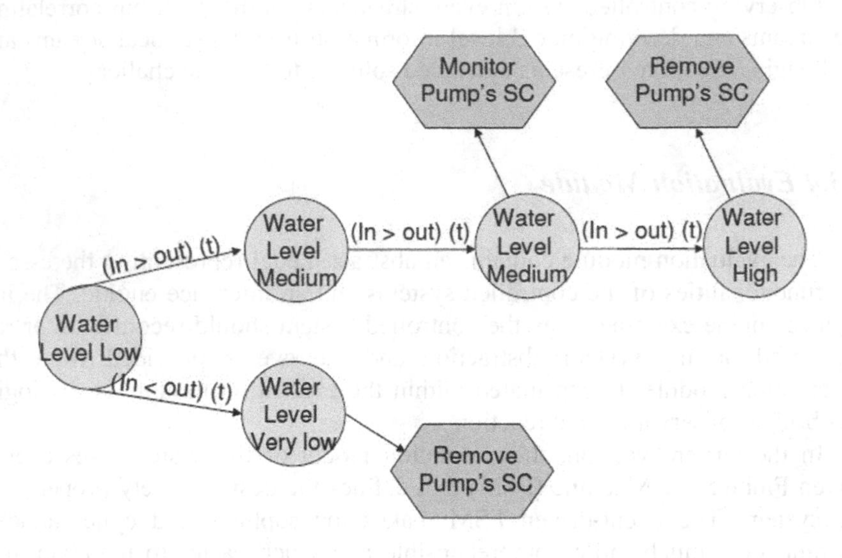

Fig. 7.3 System Abstract FSM

7.3.5 Protection Module

The Protection module executes the service requests determined by the Evaluation module. The execution of a service request is hence decoupled from its selection.

The protection module is designed to execute service requests while maintaining the system's consistency. Inspired by [4, 5], service requests are currently implemented as scripts operating on the controlled system. These scripts implement the activation procedures used to perform functional or structural reconfiguration (such as cancel a suspicious SC unit). Figure 7.4 depicts the concept.

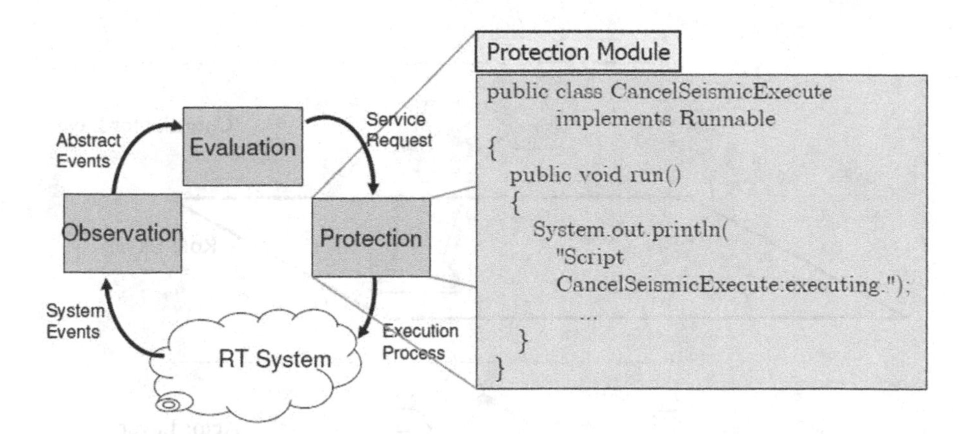

Fig. 7.4 Protection Module

In addition, the protection module is designed to hide the details of the action scripts (for instance, the steps for deactivate a primary device and activate a backup) from the observation and evaluation modules and the controlled systems. This significantly simplifies the survivability model of the system, as the details of implementing these requests (e.g. move SC unit to a new node) can be redefined for different applications without the need to make other changes on the architecture.

7.4 Observation and Inference

As we discussed in the previous section, one of the responsibilities of the observation module is to infer from the observed distributed system events and correlate them to generate abstract events for the evaluation module. In order to infer that two distributed and independent system events indicate a potential malfunction, the observation module must have the knowledge on how to coordinate a controlled system's distributed SCs.

If we use asynchronous objects, called actors, to represent each supervisory and control (SC) unit, we can group these actors, i.e., SC units, based on the roles they play in the system. For instance, in the water treatment systems, we have two types of SC units — controlling incoming water and outgoing water. In order for the whole system to work properly, these different roles must be appropriately coordinated. A violation of coordination requirements is an indication of a possible malfunctioning among SC units. Hence, we developed a coordination model, called Actor-Role-Coordinator [24], as a basis for the observation module. Figure 7.5 depicts the relation among the composing elements of the coordination model.

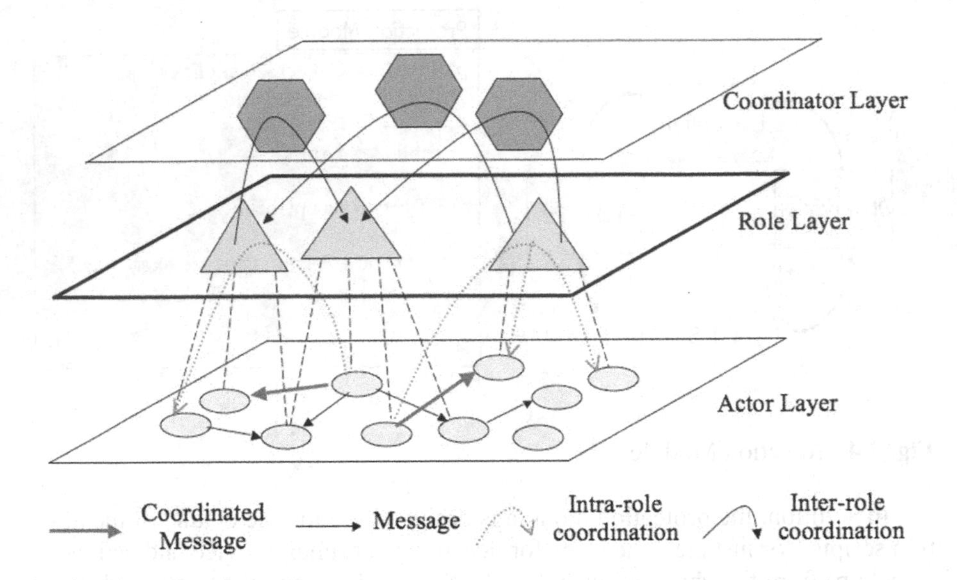

Fig. 7.5 The Actor, Role and Coordinator Model

More specifically, the ARC model has the following characteristics:

- The SC units are modeled as asynchronous active objects (actors) and they communicate with each other through messages.
- Role is an abstraction for coordinated behaviors that may be shared by multiple actors and also provides localized coordination among its players.
- Coordinators are responsible for different roles in the system. This setting further ensures that both the coordination constraints and coordination activities are decentralized and distributed among the coordinators and the roles.
- The valid relations between observed events are mapped as coordination constraints and are transparently and concurrently evaluated by roles and coordinators.

The ARC model may be conceptualized as the composition of three layers, with each of the three components of the model associated with a dedicated layer, as illustrated in Figure 7.5. The separation of concerns is apparent in the relationships involving the layers. From the perspective of a coordinator, a role enables the coordination of a set of actors that share the static description of abstract behavior associated with the role without requiring the coordinator to have fine-grained knowledge of the individual actors that play the role.

As active objects, roles have states, and based on its state and a set of intra-role constraint policies, a role actively coordinates the actors sharing the role and maintains the integrity of itself. In particular, the time dependent history information about the underlying layer (the actor layer) can be stored the role's state variables. This information can help the role to review history and identify abnormal behaviors of its member actors and further suggest to the evaluation module the possibility of excluding an attack-infected actor (processes, or control units) from the system.

The declarative criteria in the roles not only abstract the behaviors of actors, but also present a static interface to coordinators. Coordinators, therefore, do not have to directly coordinate actors, but implicit groups of actor, i.e., roles. Although in an open environment actors are very dynamic, they join or leave the system frequently; with the roles abstraction, coordinators are refrained from such dynamics.

In the ARC model, coordination and constraint policies are distributed among coordinators and roles. One of the obvious advantages of such a distribution, other than avoiding single points of failure and permitting control scalability, is localization of a fault's impact on the observers — if the role becomes faulty, it only affects those actors that play the role.

Similar to the roles and actors, coordinators also have states and are active. They are able to observe events and make corresponding state changes. The declarative constraint policies are state-based and apply to roles only. Again, the state-based policies provide a mechanism to use the history information when making judgments concerning abnormalities. The actors and coordinators are mutually transparent: though changes on actors or coordinators may have an impact on each other, such impacts are only passed through the roles.

Generally speaking, in an open environment, the number of SC entities (i.e., actors), can be very large and quite dynamic. However, the functional categories of the actors in a system (i.e., the roles that actors play), are relatively small in number and are stable. Hence, from a coordination perspective, role-based coordination is lightweight and more scalable compared with individual functional entity-based coordination, especially in open systems. Furthermore, as the roles shield the dynamics of the actors, the logic of role-based coordination becomes simpler and more reusable.

The observation model also has a rule-based inference engine. The coordination model helps to decide if a coordination constraint is violated; based on the result, the inference engine will generate abstract events for evaluation module to decide what necessary action shall take place.

The integration of the ARC model and the external control loop architecture is shown in Fig. 7.6.

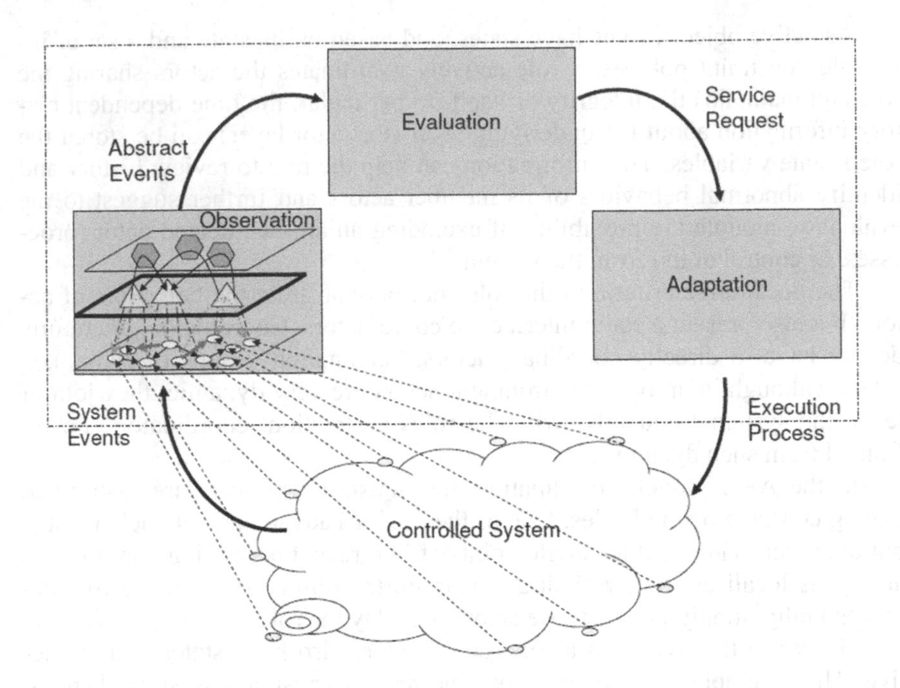

Fig. 7.6 Integration of the ARC Model and the External Control Loop

7.5 Making Decisions in Real-Time

As the whole protection process is based on the observation on the controlled systems. Therefore, the accuracy and reliability of the observation is critical. To avoid the observer themselves being blinded by attackers, or fail naturally, different observers may be placed to monitor for the same events and a voting scheme is taken to get a consensus. However, in real-time environment, data obtained through observation usually has a lifespan associated with it. Hence, the process of getting a consensus data from a group of replicated observation units must not take longer time than the lifespan of the data.

In this section, our discussion is based on the assumption that all the n sensor units provide datum *Di* to the decision unit(s) and the inherently correct data value is *D*. The information credibility may not be at the fixed 100% level, that is, *Di* may not always be the same as *D*. Instead, it may be time dependent. We use a credibility function *Ci(t)* to describe the probability that *Di* is the same as *D* at time t.

The following voting schemes are discussed here:

1. **out-of-n scheme.** Under truthful assumption, we have that *Di* = *D*, that is, every sensor unit provides correct data and *Ci(t)* = 1. In this case, once the decision unit gets a datum *Di* from any sensor, it can deliver *Di* to the user without waiting for data from other sensors.
2. **k-out-of-n scheme.** In the presence of faulty voters, a datum *Di* given by a faulty voter may not be in agreement with the data of non-faulty voters. However, a datum *Di* given by a non-faulty voter will be in close agreement with (or simply the same as) the data *D* of all the other non-faulty voters. We assume that the inherently correct data *D* is in the majority so that *D* can be determined by majority voting protocols. The credibility function *Ci(t)* is given to be monotonic with bound of [0, 1]. The monotonicity indicates that with more time, we would get more trustworthy data.

We further assume that the probability distribution function for the time a sensor i takes to obtain and transmit data is given as Vi(t). In other words, the probability that the decision unit get a datum from a sensor i by time t is given by Vi(t).

To formulate the problem, let Xi be the random variable representing if the decision unit get a vote from the ith sensor

$$X_i = \begin{cases} 1, & \textit{if } \textbf{the vote of the i'th sensor is given} \\ 0, & \textit{otherwise} \end{cases} \quad (1)$$

Thus, $P\{Xi = 1\} = Vi(t)$, $P\{Xi = 0\} = 1 - Vi(t)$. Moreover, we interpret data credibility as the probability that a given data *Di* agrees with the inherent correct data *D*. Let *Yi* be the random variable representing whether the data *Di* agrees with *D*, that is

$$Y_i = \begin{cases} 1, & \textit{if } \textbf{the vote given by the i'th sensor is } D \\ 0, & \textit{otherwise} \end{cases} \quad (2)$$

Thus, $P\{Yi = 1|Xi = 1\} = Ci(t)$, $P\{Yi = 0|Xi = 1\} = 1 - Ci(t)$. Therefore, the probability that the decision unit get a correct vote from the ith sensor is

$$\begin{aligned} pi &= P\{Yi = 1 \cap Xi = 1\} \\ &= P\{Yi = 1 \mid Xi = 1\} \times P\{Xi = 1\} = Ci(t)Vi(t) \end{aligned} \quad (3)$$

and the probability that the decision unit cannot get a correct vote (either the vote is not given, or the given vote is incorrect) from the ith voter is

$$\begin{aligned} qi &= P\{Yi = 0 \cap Xi = 0\} \\ &= P\{Yi = 1 \mid Xi = 1\} = 1 - pi = 1 - Ci(t)Vi(t) \end{aligned} \quad (4)$$

When all sensors are homogeneous, i.e., their $Ci(t)$ and $Vi(t)$ are identical, the probability that at least k similar (or the same as D) votes are collected is the summation of binomial distributions:

$$P\left\{\sum_{i=1}^{n}(X_i \wedge Y_i) \geq k\right\} = \sum_{i=k}^{n}\binom{n}{i} p^i (1-p)^{n-i} \tag{5}$$
$$\text{where}$$
$$p = p_1 = \cdots = p_n = C(t)V(t)$$

Note that p is a function of t, it follows that equation (5) is the probability that at least k similar votes are collected before time t. Let random variable T represent the time point at which enough similar votes (at least k) are collected, i.e., the decision time, we have,

$$P\{T \leq t\} = \sum_{i=k}^{n}\binom{n}{i} p^i (1-p)^{n-i} \tag{6}$$
$$\text{and}$$
$$P\{T > t\} = 1 - \sum_{i=k}^{n}\binom{n}{i} p^i (1-p)^{n-i} = \sum_{i=0}^{k-1}\binom{n}{i} p^i (1-p)^{n-i}$$

Therefore, the expected time that at least k same/similar votes are collected by the decision unit is

$$E[T] = \int_0^{\infty} P\{T > t\} dt \tag{7}$$
$$= \int_0^{\infty} \sum_{i=0}^{k-1}\binom{n}{i} (C(t)V(t))^i (1 - C(t)V(t))^{n-i} dt$$

Note that in (7), different k's are used in distinct voting schemes. In *1-out-of-n* scheme where all sensors are truthful, we have $k = 1$. Whereas in *k-out-of-n* scheme, we have $k = \lceil (n+1)/2 \rceil$ in majority voting protocols and $k = \lceil 2n/3 \rceil$ in the more stringent Byzantine voting protocols. In the following subsections, we discuss these schemes separately, assuming $C(t)$ and $V(t)$ are given.

7.5.1 Truthful Voters

Under this scheme, we have $k = 1$ and $C(t) = 1$ in (7). We further assume that $V(t)$ is uniformly distributed over the interval $[0, T1]$, i.e.,

$$V(t) = \begin{cases} \frac{t}{T_1}, & if\ t \in (0, T_1) \\ 1, & otherwise \end{cases} \tag{8}$$

Substitute k, C(t) and V(t) in (7), we have

$$E[T] = \int_0^{T_1} \left(1 - \frac{t}{T_1}\right)^n dt + \int_{T_1}^{\infty} (1-1)^n dt = \frac{1}{n+1} T_1 \tag{9}$$

Equation (9) indicates that as n increases, $E[T]$ decreases. In other words, under truthful assumption, resource availability positively impact data availability and system dependability. More careful observation reveals that the voting subsystem under truthful assumption is in fact a parallel system where the probability that the decision unit get at least one correct data from n sensors is

$$\begin{aligned} P\{\sum_{i=1}^{n} (X_i \wedge Y_i) \geq 1\} &= 1 - P\{\sum_{i=1}^{n} (X_i \wedge Y_i) = 0\} \\ &= 1 - \prod_{i=1}^{n} q_i = 1 - \prod_{i=1}^{n} (1 - C_i(t) V_i(t)) \end{aligned} \tag{10}$$

in which $\Pi q i$ characterizes a parallel system. In such a system, sensor units work in a "cooperative" way. Therefore, adding resources (more homogenous sensor units) to the subsystem improves its performance and thus reduces the expected decision time.

Similarly, consider a situation in which the data coming from the sensors are at constant rate (λ) for any unit interval, i.e., the number of data within a unit time is constant over time. Based on probability theory, we know that such event probability distribution can be modeled as exponential distribution, with probability distribution function given below:

$$V(t) = 1 - e^{-\lambda t}, t \geq 0 \tag{11}$$

Substitute k, $C(t)$, and $V(t)$ in (7), we have

$$E[T] = \int_0^{\infty} e^{-n\lambda t} dt = \frac{1}{\lambda} \cdot \frac{1}{n} \tag{12}$$

Therefore, though the probability distribution functions for voting time are different, if all the sensors are truthful, increasing *n*, i.e., the number of resources, reduces the expected time to obtain assured votes.

7.5.2 Untruthful Voters

Under untruthful voter scenario, we have *k* determined by the specific majority voting protocol (where $k = \lceil (n+1)/2 \rceil$ in majority voting protocols and $k = \lceil 2n/3 \rceil$ in the more stringent Byzantine voting protocols). We further assume that $C(t)$ is uniformly distributed over the interval $[0, T_2]$ and $V(t) = 1$[4]. From (5), we can derive the probability of getting a valid data before time *t*:

$$P(t) = \sum_{i=\lceil (n+1)/2 \rceil}^{n} \binom{n}{i} \left(\frac{t}{T_2}\right)^{i} \left(1 - \frac{t}{T_2}\right)^{n-i} \quad (t \in [0, T_2]) \tag{13}$$

As can be seen that when $t = 0.4T_2$, which means that the probability of getting a valid vote from an individual voter by time t is less than 50%, adding more homogeneously untruthful resources only makes it harder to get a consensus within given time. Intuitively, if over 50% chance a voter is to lie, adding more such voters only reduce the probability of getting valid votes within a given time.

However, when $t = 0.6T_2$, which means that the probability of getting a valid vote from an individual voter by time t is greater than 50%, adding more homogeneous resources facilitates the decision process, thus resulting in an increasing probability of obtaining a valid vote. The question now is: how does the resource availability influence the average decision time and thus the data availability?

Substitute $C(t)$, and $V(t)$ in (7), we have

$$\begin{aligned} E[T] &= \int_0^{T_2} \sum_{i=0}^{k-1} \binom{n}{i} \left(\frac{t}{T_2}\right)^{i} \left(1 - \frac{t}{T_2}\right)^{n-i} dt + \int_{T_2}^{\infty} \sum_{i=0}^{k-1} \binom{n}{i} (1)^{i} (1-1)^{n-i} dt \\ &= \sum_{i=0}^{k-1} \binom{n}{i} \int_0^{T_2} \left(\frac{t}{T_2}\right)^{i} \left(1 - \frac{t}{T_2}\right)^{n-i} dt \end{aligned} \tag{14}$$

---

[4] Although it is unreasonable to assume $V(t) = 1$, i.e. a sensor is constantly giving out vote to the decision unit, we do this to simplify calculations and because not $V(t)$ alone but $C(t) \times V(t)$ characterizes the possibility that the decision unit gets a vote valued *D*, which is the inherently correct data.

Make the substitution $x = t/T2 \Rightarrow dx = (1/T_2)dt$ in (14), we have

$$E[T] = \sum_{i=0}^{k-1} \binom{n}{i} \int_0^1 x^i (1-x)^{n-i} T_2 dx \tag{15}$$

Integrate by parts, we have

$$\begin{aligned} \int_0^1 x^i (1-x)^{n-i} dx &= \frac{1}{i+1}\left[x^{i+1}(1-x)^{n-i}\Big|_{x=0}^{1} - \int_0^1 x^{i+1} d(1-x)^{n-i} \right] \\ &= \frac{n-i}{i+1} \int_0^1 x^{i+1}(1-x)^{n-i-1} dx \end{aligned} \tag{16}$$

Use mathematical induction on, we can prove that

$$\int_0^1 x^i (1-x)^{n-i} dx = \frac{i!(n-i)!}{(n+1)!} \tag{17}$$

Therefore, from (15) and (17), we have

$$E[T] = T_2 \sum_{i=0}^{k-1} \binom{n}{i} \frac{i!(n-i)!}{(n+1)!} = T_2 \sum_{i=0}^{k-1} \frac{1}{n+1} = \frac{k}{n+1} T_2 \tag{18}$$

Given that $k = \lceil (n+1)/2 \rceil$ and n is large, we have

$$E[T] = T_2/2 \tag{19}$$

Therefore, in an open hostile environment where not all voters are truthful, adding homogeneous resource does not impact the expected time of getting a valid vote.

Similarly, when the credibility function $C(t)$ is exponentially distributed over the interval $[0,\infty)$ with average rate λ, that is,

$$C(t) = 1 - e^{-\lambda t}, t \in [0,\infty) \tag{20}$$

Using equation (7), we have

$$E[T]=\sum_{i=0}^{k-1}\binom{n}{i}\int_{0}^{\infty}\left(1-e^{-\lambda t}\right)^{i}\left(e^{-\lambda t}\right)^{n-i}dt \tag{21}$$

Make the substitution where $x=e^{-\lambda t}\Rightarrow dx=-\lambda e^{-\lambda t}dt=-\lambda x dt$, we have

$$\begin{aligned}E[T]&=\sum_{i=0}^{k-1}\binom{n}{i}\int_{1}^{0}(1-x)^{i}x^{n-i}\frac{1}{-\lambda x}dx\\&=\frac{1}{\lambda}\sum_{i=0}^{k-1}\binom{n}{i}\int_{0}^{1}(1-x)^{i}x^{n-i-1}dx\end{aligned} \tag{22}$$

Integrate by parts and use mathematical induction, we can prove that

$$\int_{0}^{1}(1-x)^{i}x^{n-i-1}dx=\frac{i!(n-i-1)!}{n!} \tag{23}$$

Therefore, from (22) and (23), we have

$$\begin{aligned}E[T]&=\frac{1}{\lambda}\sum_{i=0}^{k-1}\binom{n}{i}\frac{i!(n-i-1)!}{n!}\\&=\frac{1}{\lambda}\sum_{i=0}^{k-1}\frac{n!}{i!(n-i)!}\frac{i!(n-i-1)!}{n!}=\frac{1}{\lambda}\sum_{i=0}^{k-1}\frac{1}{n-i}\end{aligned} \tag{24}$$

where $k=\lceil (n+1)/2\rceil$. The relationship between $E[T]$ and n in case of exponential distribution is illustrated in Figure 7.2. As can be seen, when the number of working sensors are small, increasing the number of sensors generally decreases expected decision time. However, since

$$\lim_{n\to\infty}\sum_{i=0}^{\lceil (n+1)/2\rceil-1}\frac{1}{n-i}=\ln n-\ln\frac{n}{2}=\ln 2\approx 0.6931 \tag{25}$$

The expected decision time converges at $ln2/\lambda$ and no further decrease can be achieved by adding more resources. For example, with 11 sensors, the expected decision time is $0.7365/\lambda$, while with 23 sensors, the expected decision time is $0.7144/\lambda$ — a 3.0% time gain is at the cost of more than twice the resources.

7.6 Current Implementation

The current framework is centered around the concept of modules, which are components realizing a portion of the adaptation logic (such as event transformation, model evaluation, etc.), events which are discrete tuples of information transiting between modules, and channels that transfer abstract events between modules. Modules are combined by associating them with input and outputs channels, following the configuration given by the user. Input and output channels abstract the transfer of events, so that different technologies such as JNI or CORBA can be transparently used. Channels are used to generalize the applicability of the architecture.

7.6.1 Event and channels

Events define abstract types of information that modules can transmit, and realize a common ancestor to all the specialized information that the modules can communicate. Events and channels form the basic interfaces by which components in the framework may receive or send information. They define the service capabilities for any components in the framework to receive or send events, and handle system information without regards to the actual technologies and implementations that provide them (such as JNI calls or TCP sockets). For example, the current type of channels provided within the framework create shared queues that permit transmission of objects representing the event interface in a low-overhead manner. The use of a channel interface allows them to be substituted later on by future types of channels, without exposing their implementation to the modules.

7.6.2 Modules

In the current prototype, the Observation, Evaluation, Protection modules are implemented through a Jess-based inference engine [10], a Finite State Machine model evaluator, and a multi-threaded scripting engine, respectively. Figure 7.7 is a snapshot view of a running system. In the figure, on the left is a GUI interface to permit the generation of system events; in the middle is the console for the protection framework, showing debug messages from the Evaluation module; and the right shows the response from the protection module.

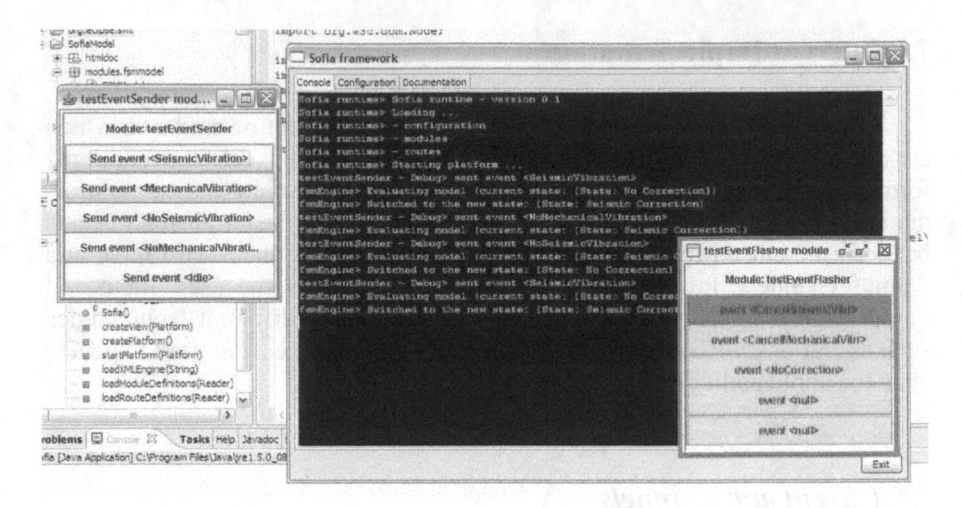

Fig. 7.7 Prototype Implementation – GUI

7.6.2.1 Observation module

A sophisticated, Jess-based observation module is provided with the prototype to perform event transformations. The sophistication of Jess provides the user with the capacities to do extensive manipulation and verification on multiple events. The user can provide inference rules either in a Jess-based language or in XML. The module automatically encapsulates and forwards the events generated by the Jess rules. The current transformation possibilities are limited by Jess's functionality. Although the sophistication of the Jess engine still posts a performance challenge, its well-declared inference rules and well-arranged rule orders can, on the other hand, offset the performance latencies. In addition, the electrical (or computer) speed and mechanical speed are at different time scales; therefore, in most cases, the performance cost of running the Jess inference engine is acceptable for real-time control systems, especially for legacy control systems.

7.6.2.2 Evaluation module

In our prototype evaluation module, the protection model of the system is abstracted through a Finite State Machine (FSM) that defines the desired protective behavior of the system. The states used in the FSM may differ from the functional model of the system, as various states of the system may not be considered needing protection. In our current design, the FSMs are specifically modeled after the Mealy machine model, which triggers protection requests during transitions between its states, following events generated by the observation module, which interfaces with the system.

7.6.2.3 Protection module

The current protection module provided with the framework is a multi-threaded protection engine relying on a configurable pool of threads to service protection requests. The configuration provided by the user associates event names with Python scripts which are fed with the event's name/value pairs at launch.

7.7 Conclusion

We have presented a framework and methodology for adding new features to existing systems. The solution emphasizes externalization, separation of concerns, and modularity when adding new features into controlled systems. Similar to other efforts, its purpose is to design general mechanisms for developing systems that can adapt to their new operating conditions. Our approach, however, is unlike other efforts because it is aimed at joining reliability and security both reliably and securely.

Our solution relies on previous work in external, control-based and model-based architectures, but is distinguished by its decomposition of the control process. The framework is decomposed into the observation, evaluation and protection modules that are virtually independent from each other. Hence, it supports modularization and allows for the cyber protection process to be customized for specific needs or application domains. The observation, evaluation and protection modules can be substituted to fulfill a particular need. The framework relies on the first layer, the observation module, to infer information about the controlled system and allow protection decisions to be based on high level information rather than many low-level events. Then, the evaluation module maps these observations onto an actual model of the system, which in turn permits situational protection. Finally, the protection module independently executes appropriate tasks for fault tolerance and abstracts away the concerns necessary to tolerate faults caused by cyber attacks. With our solution we achieve a dual assurance: by leaving the original system design undisturbed, its established reliability remains unchanged; yet the external enhancements of our approach secure the system from unanticipated attacker-induced faults.

References

[1]. R. Panko, Corporate Computer and Network Security, Prentice Hall, Upper Saddle River, NJ, 2004.

[2]. V.P. Nelson, Fault-Tolerant Computing: Fundamental Concepts, Computer, IEEE, July 1990, pp. 19-25.

[3]. K. Kwiat, Can Reliability and Security be Joined Reliably and Securely?, Proceeding of the Symposium on Reliable Distributed Systems (SRDS), IEEE, October 2001.

[4]. D. Garlan, S.-W. Cheng, A.-C. Huang, B. Schmerl, and P. Steenkiste, "Rainbow: Architecture-based self-adaptation with reusable infrastructure" IEEE Computer, vol. Volume 37, Issue 10, pp. 46–54, October 2004.

[5]. D. Garlan and B. Schmerl, "Model-based adaptation for self-healing systems" in WOSS '02: Proceedings of the first workshop on Self-healing systems, New York, NY, USA, 2002, pp. 27–32.

[6]. G. Karsai, A. Ledeczi, J. Sztipanovits, G. Peceli, G. Simon, and T. Kovacshazy, "An approach to self-adaptive software based on supervisory control" in IWSAS, 2001, pp. 24–38.

[7]. G. Kaiser, J. Parekh, P. Gross, and G. Valetto, "Kinesthetics eXtreme: An external infrastructure for monitoring distributed legacy systems" in Autonomic Computing Workshop Fifth Annual International Workshop on Active Middleware Services (AMS'03), 2003.

[8]. Y. Qun, Y. Xian-Chun, and X. Man-Wu, "A framework for dynamic software architecture-based elf-healing" SIGSOFT Softw. Eng. Notes, vol. 30, no. 4, pp. 1–4, 2005.

[9]. Y. Diao, J. Hellerstein, S. Parekh, R. Griffith, G. Kaiser, and D. Phung, "Self-managing systems: A control theory foundation" in 12th IEEE International Conference and Workshops on the Engineering of Computer-Based Systems (ECBS '05), 2005.

[10]. S. N. L. Ernest Friedman-Hill, "Jess" http://herzberg.ca.sandia.gov/jess/.

[11]. United States General Accounting Office. Critical Infrastructure Protection Challenges and Efforts to Secure Control Systems. Report to Congressional Requesters. March 2004.

[12]. Understanding SCADA Security Vulnerabilities. Technical Report. Riptech, Inc. 2001.

[13]. J. Pollet. Developing a Solid SCADA Security Strategy. SICON. Houston. TX. 2002.

[14]. F. Haji. L. Lindsay. S. Song. Practical Security Strategy for SCADA Automation Systems and Networks. CCECE/CCGEI, Saskatoon. May 2005.

[15]. C. L. Bowen. T. K. Buennemeyer. R. W. Thomas. Next Generation SCADA Security: Best Practices and Client Puzzles. In Proceedings of the IEEE Workshop on Information Assurance and Security. West Point, NY. 2005.

[16]. D. Gamez. S. N. Tehrani. J. Bigham. C. Balducelli. K. Burbeck. T. Chyssler. Dependable Computing Systems: Paradigms, Performance Issues, and Applications. Wiley, Inc. 2000.

[17]. InTech Inc. Intrusion Detection and Cybersecurity. Technical Report. May 2004.

[18]. Wika, K.J., Safety Kernel Enforcement of Software Safety Policies, Ph.D. dissertation, Department of Computer Science, University of Virginia, Charlottesville, VA, 1995.

[19]. Knight J. C. Nakano L. G. Software test techniques for system fault-tree analysis. In Proc. SAFECOMP 97, 1997, pp. 369-380

[20]. Leveson, N. G., T. J. Shimeall, J. L. Stolzy, and J. C. Thomas, "Design for Safe Software" in Proceedings AIAA Space Sciences Meeting, Reno, Nevada, 1983.

[21]. Wayne Labs. Technology Brief (Issue 2, 2004). How secure is your control system? http://www.automationnotebook.com/2004 Issue 2/ technologybrief September2004.html

[22]. NUREG-0492, Fault Tree Handbook, U.S. Nuclear Regulatory Commission, January, 1981.

[23]. Shangping Ren, Limin Shen, Jeffrey J.P. Tsai: Reconfigurable Coordination Model for Dynamic Autonomous Real-Time Systems. SUTC (1) 2006: 60-67

[24]. Shangping Ren, Yue Yu, Nianen Chen, Kevin Marth, Pierre-Etienne Poirot, Limin Shen: Actors, Roles and Coordinators - A Coordination Model for Open Distributed and Embedded Systems. COORDINATION 2006: 247-265

8 Image Encryption and Chaotic Cellular Neural Network

Jun Peng[1, 2] **and Du Zhang**[2]

Abstract Machine learning has been playing an increasingly important role in information security and assurance. One of the areas of new applications is to design cryptographic systems by using chaotic neural network due to the fact that chaotic systems have several appealing features for information security applications. In this chapter, we describe a novel image encryption algorithm that is based on a chaotic cellular neural network. We start by giving an introduction to the concept of image encryption and its main technologies, and an overview of the chaotic cellular neural network. We then discuss the proposed image encryption algorithm in details, which is followed by a number of security analyses (key space analysis, sensitivity analysis, information entropy analysis and statistical analysis). The comparison with the most recently reported chaos-based image encryption algorithms indicates that the algorithm proposed in this chapter has a better security performance. Finally, we conclude the chapter with possible future work and application prospects of the chaotic cellular neural network in other information assurance and security areas.

8.1 Introduction to image encryption

With the rapid development of the Internet and multimedia technologies, security is becoming an important issue in the storage and communication of digital images. In many secure applications fields, images such as military satellite images, radar tracking object images, fingerprint images for identity authorization are required to be encrypted. Even in a remote medical system, according to the law, the electronic forms of medical records that include patient's medical images should be encrypted before they are sent over networks (Dang and Chau 2000).

---

[1] College of Electronic Information Engineering, Chongqing University of Science and Technology, Chongqing 400050, China, e-mail: pengjun70@126.com

[2] Department of Computer Science, California State University, Sacramento, CA 95819, USA, e-mail: zhangd@ecs.csus.edu

J.J.P. Tsai and P.S. Yu (eds.), *Machine Learning in Cyber Trust: Security, Privacy, and Reliability*, DOI: 10.1007/978-0-387-88735-7_8,

Basically, there are two fundamental technologies for protecting digital images, namely image encryption and digital watermark. The thought of digital image encryption is originated from early classical encryption theory, and its aim is to transform a given image into a disorderly and unsystematic one by applying a kind of transformation regulation with a secret key in the spatial domain or frequency domain so as to hide the true information of the source image. The encrypted image cannot be recognized by the attacker without a correct secret key. On the other hand, digital watermark adopts a different technique, which is intended to complement cryptographic processes. It is a visible, or preferably invisible, identification code that is permanently embedded in the image and remains present within the image after any decryption process (Cox et al. 1997). A simple example of a digital watermark would be a visible "seal" placed over an image to identify the copyright owner.

In this section, we mainly focus on the image encryption techniques. Compared with the original image, the encrypted image can have the following two kinds of changes: the first is that the location relationships between the pixels are rearranged due to the image scrambling transformation that can effectively reduce the correlation of adjacent pixels; the second is that the gray-scale value of pixels are changed so as to make the information entropy of the encrypted image very close to the maximum value and the histogram of the encrypted image more smooth. The purpose of these changes is to make the encrypted image looks like a random one meanwhile keeping a capability of withstanding most common cryptographic attacks such as brute-force attack, statistical attack, known-plaintext attack, etc. It's well known that cryptographic attacks are designed to subvert the security of cryptographic algorithms such as image encryption algorithm, and they are used to attempt to decrypt data without prior access to a secret key. Therefore, these two kinds of changes are very important to the information security of the image.

In the following part, we briefly introduce the main techniques for image encryption.

8.1.1 Based on image scrambling technique

Image scrambling is one of the most prevailing encryption algorithms, which exploits a matrix transformation or pixel permutation to scramble the source image. Arnold's cat map is a mostly used transformation for this purpose. Qiu (Qiu and Ma 2003) and Chen (Chen et al. 2004) obtained the two-dimensional and three-dimensional discrete form for Arnold's cat map, which is suitable to the image encryption application. If a source image is transformed several times by this map then a desirable scrambled image would be obtained. Other techniques are also investigated including affine transformation (Chang 2004), magic square transformation (Arthur and Kan 2001) and knight-tour transformation (Parberry 1997, Charilaos et al. 2000). From the cryptography point of view, these scrambling techniques can efficiently hide the statistical property of the plaintext (source image) to withstand a statistical attack. Besides, they also can serve as a pre-processing of a cryptographic system, or a permutation part of a substitution-permutation network (SPN).

8.1.2 Based on SCAN pattern

Bourbakis and Maniccam proposed an image encryption algorithm based on SCAN pattern (Bourbakis 1997; Bourbakis and Alexopoulos 1992, 1999; Maniccam and Bourbakis 2004). The SCAN is a formal language based two dimensional spatial accessing methodologies which can represent and generate a large number of wide variety of scanning paths or space filling curves easily. The SCAN is a family of formal languages such as Simple SCAN, Extended SCAN, and Generalized SCAN, each of which can represent and generate a specific set of scanning paths. For an image, a scanning path is an order in which each pixel of the image is accessed exactly once. The basic steps of image encryption using SCAN pattern are described as follows: Firstly, a set of scanning paths for each sub-block image are chosen by the user. The scanning paths which are used as encryption keys are defined by a specific SCAN word. Secondly, the pixels of each sub-block image are rearranged according to the scanning paths. Finally, as the combination of the different SCAN words will result in a different image ciphertext, the SCAN words are furthermore encrypted through the standard commercial encryption ciphers such as DES (Schneier 1996) and IDEA (Lai and Massey 1990).

8.1.3 Based on tree data structures

Since many tree data structures such as binarytree, quadtree, wavelet zerotree have already been widely used in image and video compression coding fields (Radha et al. 1996, Clarke 1995, Jessl et al. 2005), some encryption schemes combined with compression algorithm that was based on tree structures were proposed. To protect sensitive data in wireless image communication, Li (Li et al. 1997) proposed a partial encryption scheme, in which a quadtree was used to decompose the image into two parts in the spatial domain. Then the small but crucial part was encrypted directly by a public key algorithm such as RSA (Rivest et al. 1978) while the remaining part will be sent without encryption. Martin (Martin et al. 2005) investigated an efficient secure color image coder, in which a color set partitioning in hierarchical trees (C-SPIHT) compression algorithm was used to direct the stream cipher to encrypt only the significance bits of individual wavelet coefficients encountered during K iterations of the C-SPIHT algorithm. The use of a stream cipher and the encryption of a small number of bits keep computational demands at a minimum and make the technique suitable for hardware implementation. Other similar proposed schemes can be found in (Cheng and Li 2000, Lian et al 2004). The hybrid of the compression and encryption techniques may be one of ways to balance the purpose of both encryption and real-time requirements for multimedia data communication.

8.1.4 Based on chaotic systems

It has been established that chaotic systems have several appealing features for information security applications, especially on the secure communications and cryptography using chaos. In early 90's, Pecora and Carroll were the first to propose a method to synchronize two identical chaotic systems to implement a secure communication (Pecora and Carroll 1990,1991), and Matthews presented an earliest chaotic stream cipher (Matthews 1989). The close relationship has been observed between chaotic maps and cryptographic algorithms (Fridrich 1997, 1998; Kocarev 2001; Alvarez and Li 2006). In particular, the following connections would be found between them: (1) Ergodicity in chaos vs. confusion in cryptography; (2) Sensitive dependence on initial conditions and control parameters of chaotic maps vs. diffusion property of a good cryptosystem for a slight change in the plaintext and in the secret key; (3) Random-like behavior of deterministic chaotic-dynamics which can be used for generating pseudorandom sequences as key sequences in cryptography. However, an important difference between chaos and cryptography is that encryption transformations are defined on finite sets, while chaos has meaning only on real numbers. Moreover, for the time being, the notions of cryptographic security and performance of cryptographic algorithms have no counterpart in chaos theory (Kocarev 2001). Hence, the in-depth relationship between chaos and cryptography is still worthy of further research.

Despite of this, the chaotic properties have been widely employed to design cryptographic systems in recent years (Yang et al. 1997, Baptista 1998, Kocarev and Jakimoski 2001, Masuda and Aihara 2002, Li et al. 2003, Wong 2003, Amigo et al. 2007). Here, we mainly offer a brief overview to some of chaos-based image encryption schemes. Scharinger (Scharinger 1998) introduced an efficient image and video data encryption approach, which involves keyed permutations on large data blocks induced by specific chaotic Kolmogorov flows. At the same time, Fridrich (Fridrich 1998) represented a method to adapt an invertible two-dimensional standard baker map on a torus or on a square to create symmetric block encryption scheme. In Fridrich's scheme, firstly, the standard baker map is generalized by introducing parameters, and then discretized to a finite rectangular lattice of points. After that the map is extended to three dimensions to obtain a more complicated substitution cipher that can be used for the purpose of image encryption. Yen and Guo (Yen and Guo 2000) proposed a chaotic key-based algorithm (CKBA) in which a binary sequence as a key is generated using a chaotic system, and the image pixels are arranged according to the generated binary sequence, and then the scale-gray values of pixels are XORed or XNORed bit-by-bit to one of the two pre-determined keys. However, the analysis conducted in (Li and Zheng 2002) showed that Yen's algorithm has some drawbacks: it is vulnerable to the chosen or known-plain-text attack using only one plain-image, and its security to brute-force attack is also questionable. Recently, Chen (Chen et al. 2004) proposed a symmetric image encryption in which a 3D cat map was exploited to shuffle the positions of image pixels and another chaotic map was also used to confuse the relationship between the original image and its encrypted image. In (Pareek et al. 2006), Pareek employed an external secret key of 80-bit and two chaotic logistic maps to construct an encryption scheme, in which the initial conditions of the both logistic maps were derived from the external key and eight different types of operators were used to encrypt the pixels of the original image. Furthermore, Pisarchik (Pisarchik et al. 2006) suggested a new practical algorithm based on a chaotic map lattice (CML). The main idea is to convert, pixel by pixel, the image color to lattices of chaotic maps one-way coupled by initial conditions. After small numbers of iteration and cycles, the source image will become an indistinguishable one due to the intrinsic properties of the chaotic system. Gao (Gao et al. 2006) presented a new nonlinear chaotic algorithm (NCA) which uses power function and tangent function instead of linear function to overcome the drawbacks of small key space and weak security in the widely used one-dimensional Logistic system, while maintaining acceptable efficiency. Very recently, similar to (Chen et al. 2004), Gao (Gao and Chen 2008) exploited an image total shuffling matrix to shuffle the position of image pixels and then used a hyper-chaotic Chen system to confuse the relationship between the original image and encrypted image.

Machine learning has been playing an increasingly important role in information security and assurance. Machine learning falls into the following broad categories: *supervised* learning, *unsupervised* learning, *semi-supervised* learning, *analytical* learning, *reinforcement* learning, and *multi-agent* learning. Each of the categories in turn includes various learning methods. Supervised learning deals with learning a target function from labeled examples. Unsupervised learning attempts to learn patterns and associations from a set of objects that do not have attached class labels. Semi-supervised learning is learning from a combination of labeled and unlabeled examples. Analytical learning relies on domain theory or background knowledge to learn a target function. Reinforcement learning is concerned with learning a control policy through reinforcement from an environment. Multi-agent learning is an extension to single-agent leaning. There are many emerging learning methods such as argument based machine learning (Mozina 2007), interactive learning (Guzman 2008, Shiang 2008), transfer learning (Lee and Giraud-Carrier 2007, Quattoni 2008), and so forth. In this chapter, our focus is on the use of cellular neural networks (CNN) for image encryption.

Obviously, chaos-based algorithms have suggested a new and efficient way to deal with the intractable problem of fast and highly secure image encryption, and it is true that the research achievements mentioned above are very useful to the latter studies on the chaos-based image encryption algorithm. In the remainder of this chapter, we will describe a novel image encryption scheme based on a hyper-chaotic CNN and analyze its security features.

8.2 Image encryption scheme based on chaotic CNN

As we know, low-dimensional chaotic maps usually have only one positive Lyapunov exponent, the cryptography systems based on this kind of maps face a potential risk for security (Perez and Cerdeira 1995, Yang et al. 1998, Parker and Short 2001). For example, the intruder can exploit nonlinear dynamics forecasting or return map technique (Perez and Cerdeira 1995), or neural network method (Yang et al. 1998) to reconstruct the keystream. Parker et al. (Parker and Short 2001) use the sufficient geometric information in the transmission to extract an estimate of the keystream and exploit the characteristics of the encrypting function to recover the encryption function.

However, since hyper-chaos has more than one positive Lyapunov exponent and contains more complex dynamic features, many secure communication schemes and cryptosystems based on hyper-chaotic systems have been developed (Mascolo and Grassi 1998, He and Li 2000, Vladimir et al. 2001, Goedgebuer et al. 2002, Li et al. 2007, Gao and Chen 2008). In this section, we will focus on the research on the digital image encryption algorithm based on the high-dimensional chaotic system such as a hyper-chaotic CNN.

8.2.1 Introduction to Cellular Neural Network

Cellular neural networks were invented by Leno O Chua and Lin Yang in UC Berkeley in 1988 (Chua and Yang 1988a). CNN is a kind of circuit architecture, which possesses some of the key features of neural network while having many potential applications in the fields related to massive parallel computation, image processing, pattern recognition and information security.

8.2.1.1 Basic structure of CNN

The basic circuit unit of a cellular neural network is called a cell or artificial neuron, which typically consists of linear and nonlinear circuit elements such as linear resistors, linear capacitors, and some voltage-controlled current sources. Like a cellular automata, CNN consists of lots of cells, which are arranged regularly to form a net architecture. An example of a two-dimensional CNN with circuit size 4×4 is shown in Fig. 8.1. Any cell in a CNN is connected only to its neighbor cells, which resulting this kind of structure is very easily implemented using VLSI technique. Adjacent cells can interact directly with each other and cells not directly connected together may affect each other indirectly because of the propagation effects of the continuous-time dynamics of the network (Chua and Yang 1988b).

Firstly, to describe the structure of a CNN, let us define a neighborhood of a cell $C(i,j)$.

Definition 1. r-neighborhood. The r-neighborhood of a cell $C(i,j)$ in a CNN is defined to be the set of all the neighborhood cells meeting the following formula,

$$N_r(i,j) = \{C(k,l) \mid \max\{|k-i|,|l-j|\} \leq r, 1 \leq k \leq M; 1 \leq l \leq N\}$$

where r is a positive number. Sometimes, we also can refer to $N_r(i,j)$ as a $(2r+1)\times(2r+1)$ neighborhood.

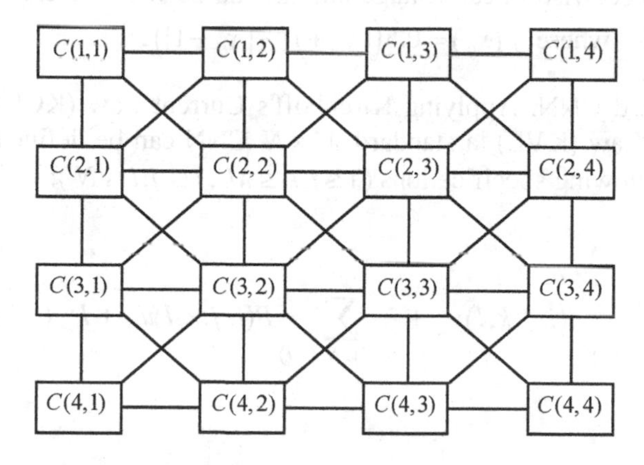

Fig. 8.1 A two-dimensional CNN with 4×4 size.

For a $C(i,j)$, it can be implemented by an equivalent circuit, which is shown in Fig.8.2.

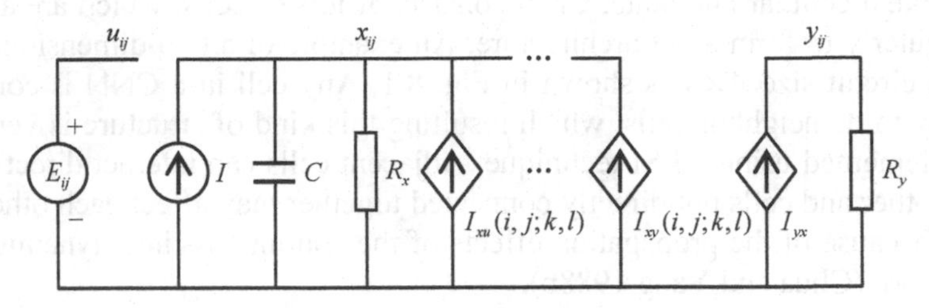

Fig. 8.2 Equivalent circuit of a cell.

In Fig.8.2, u_{ij}, x_{ij} and y_{ij} are called the input, state and output of the cell, respectively; E_{ij} is an independent voltage source; I is an independent current source; C is a linear capacitor; R_x and R_y are linear resistors; $I_{xy}(i,j;k,l)$ and $I_{xu}(i,j;k,l)$ are linear voltage controlled current sources with the characteristics $I_{xy}(i,j;k,l) = A(i,j;k,l)y_{kl}$ and $I_{xu}(i,j;k,l) = B(i,j;k,l)u_{kl}$ for all $C(k,l) \in N_r(i,j)$, $A(i,j;k,l)$ and $B(i,j;k,l)$ are feedback and control cloning templates, also called feedback and control operator, which gives the dynamic rules in CNN; I_{yx}, the only nonlinear element, is a piecewise-linear voltage current source with the characteristics $I_{yx} = f(x_{ij}) / R_y$ where $f(x_{ij}) = 0.5\left(|x_{ij}+1| - |x_{ij}-1|\right)$.

Definition 2. Standard CNN. Applying Kirchhoff's Current Law (KCL) and Kirchhoff's Voltage Law (KVL), a standard $M \times N$ CNN can be defined mathematically by the following specifications ($1 \le i,k \le M$; $1 \le j,l \le N$):

1) State equation:

$$C\frac{dx_{ij}}{dt} = -\frac{1}{R_x}x_{ij} + \sum_{c(k,l)\in N_r(i,j)} A(i,j;k,l)y_{kl} + \sum_{c(k,l)\in N_r(i,j)} B(i,j;k,l)u_{kl} + I \quad (8.1)$$

2) Output equation:

$$y_{ij}(t) = \frac{1}{2}\left(|x_{ij}(t)+1| - |x_{ij}(t)-1|\right) \quad (8.2)$$

3) Input equation:

$u_{ij} = E_{ij}$, where E_{ij} is a time-invariant independent voltage source.

4) Constraint conditions:

$$|x_{ij}(0)| \le 1,\ |u_{ij}| \le 1.$$

5) Parameter consumptions:

$$A(i,j;k,l) = A(k,l;i,j),\ C > 0,\ R_x > 0.$$

8.2.1.2 Dynamica behavior of the CNN

Chua and Yang have investigated the complex dynamic characteristics of the CNN (Chua and Yang 1988a, 1988b). The bifurcation phenomenon and chaotic behaviors are also observed in two-order CNN (Fan and Josef 1991, Chen et al. 1994, Slavova 1998). By considering simple chaotic units as neural cells, the interconnections of a sufficiently large number of neurons can exhibit extremely complex behaviors, such as spiral and scroll waves, turing patterns, and high-dimensional chaotic attractors (Chua et al. 1995). The applications of the CNN are mostly depended on its dynamic behaviors. For example, when applied in such as image processing, pattern recognition and control fields a CNN is needed to converge to a stable equilibrium state. While being applied in secure communication, image encryption and physics, it must have chaotic attractors or limit cycles.

In this section, let us consider the following three-order CNNdynamic system with full connection which exhibits chaotic phenomena (He et al. 1999):

$$\frac{dx_j}{dt} = -x_j + a_j y_j + \sum_{k=1,k\neq j}^{3} a_{jk} y_k + \sum_{k=1}^{3} S_{jk} x_k + i_j,\ j = 1,2,3 \tag{8.3}$$

where x_j and y_j denote the state variables and cell outputs of the CNN respectively. y_j is related to x_j, and their relationship (output equation) is defined as follows:

$$y_j(x_j) = 0.5(|x_j + 1| - |x_j - 1|),\ j = 1,2,3 \tag{8.4}$$

If the parameters of three cellulars are given by (He et al. 1999):

$$a_{12} = a_{13} = a_2 = a_{23} = a_{32} = a_3 = a_{21} = a_{31} = 0,$$

$$S_{13} = S_{31} = S_{22} = 0; \; i_1 = i_2 = i_3 = 0; \; S_{21} = S_{23} = 1.$$

then the dynamic model in (8.3) can be abbreviated as:

$$\frac{dx_1}{dt} = -x_1 + a_1 y_1 + S_{11}x_1 + S_{12}x_2, \tag{8.5a}$$

$$\frac{dx_2}{dt} = -x_2 + x_1 + x_3, \tag{8.5b}$$

$$\frac{dx_3}{dt} = -x_3 + S_{32}x_2 + S_{33}x_3. \tag{8.5c}$$

where $y_1(x_1) = 0.5(|x_1 + 1| - |x_1 - 1|)$.

The coefficient matrix of the linear sub-system of model (8.5) is

$$A = \begin{pmatrix} S_{11} - 1 & S_{12} & 0 \\ 1 & -1 & 1 \\ 0 & S_{32} & S_{33} - 1 \end{pmatrix}$$

If let $S_{11} = -0.367$, $S_{12} = 7.278$, $S_{32} = -12.025$ and $S_{33} = 0$, then the three eigenvalues of A are -1.8546, $-0.7562 + 2.2593i$ and $-0.7562 - 2.2593i$. All the real part of these eigenvalues are negative, hence the evolvement trajectory of model is tending to a stable state. However, if let $S_{33} = 1$ while keeping the identical value of S_{11}, S_{12} and S_{32}, then the three eigenvalues of A are -2.5245, $0.0787 + 2.5506i$ and $0.0787 - 2.5506i$. As there is at least one eignvalue with a positive real part, the evolvement trajectory of model is becoming unstable even chaotic.

As we mainly focus on the application in the image encryption using its chaotic property, we investigate the dynamical behavior of the model (8.5) through the adjustment of value of parameter a_1 when $S_{33} = 1$ and S_{11}, S_{12} and S_{32} are still assigned to the above same values. The initial condition $x_1(0)$, $x_2(0)$ and $x_3(0)$ is 0.1, 0.1 and 0.1, respectively. By exploiting the four-order Runge-Kutta method, the computer numerical simulation results are shown in Fig.8.3.

Generally, there are four behaviors with a nonlinear dynamic system: equilibrium point, periodic trajectory, quasi-periodic trajectory and chaotic trajectory. From Fig.8.3 (a)-(j), we found that the CNN model (8.5) has different dynamica behavior changing from the equilibrium point to double scroll attractor with the increasing of value of parameter a_1. It is well known that the chaotic attractor can be characteristic measured by Lyapunov exponents, which describe the extent of adjacent trajectories departing. By using Wolf algorithm (Wolf et al. 1985), the maximum Lyapunov exponent when parameter $a_1 = 3.581$ and $a_1 = 8.861$, is 0.2858 and 2.7022, respectively. Due to the fact that the maximum Lyapunov exponent is greater than zero, the system in Fig.8.3 (i) and (j) are in chaotic states.

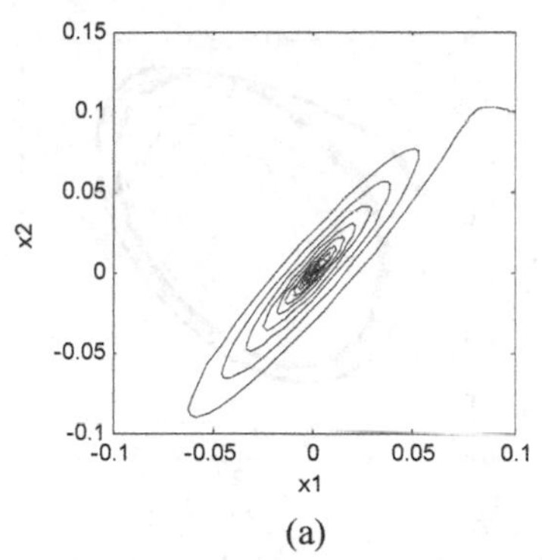

(a)

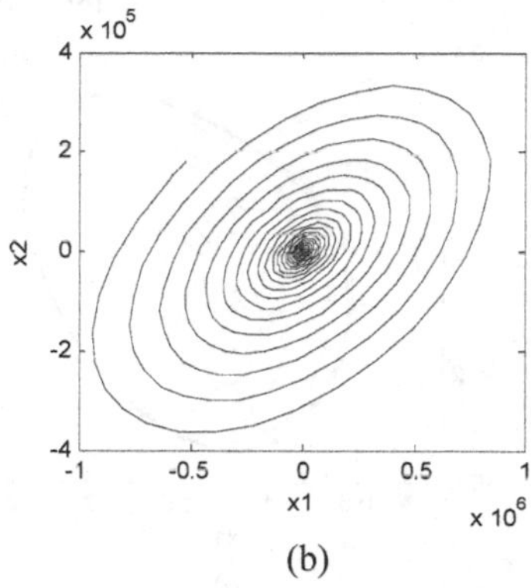

(b)

(a) $a_1 = -8.68$, stable equilibrium point (0,0,0), (b) $a_1 = 0.158$, unstable trajectory.

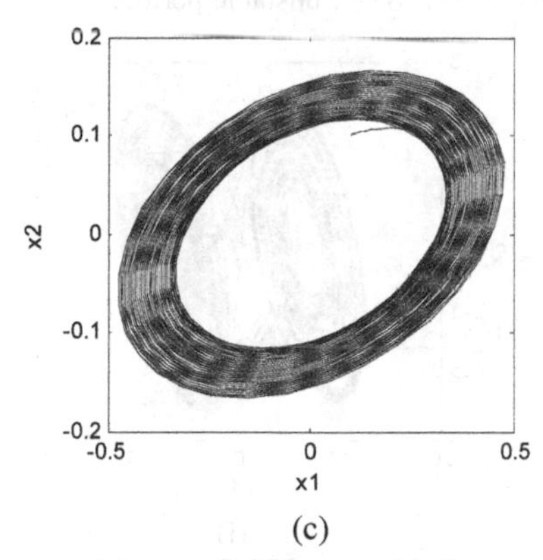

(c)

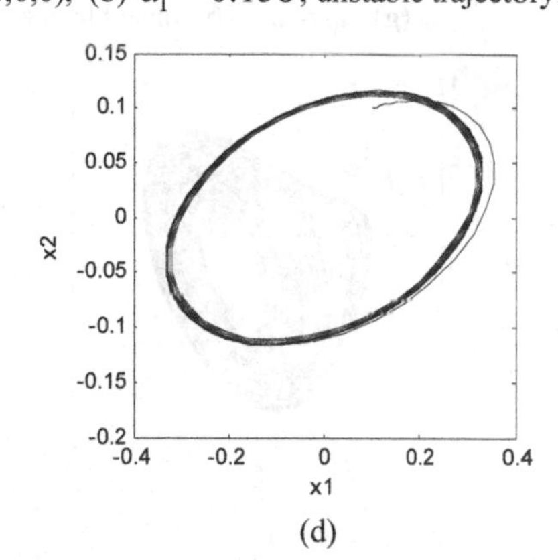

(d)

(c) $a_1 = 0.480$, unstable long-period, (d) $a_1 = 0.488$, stable long-period.

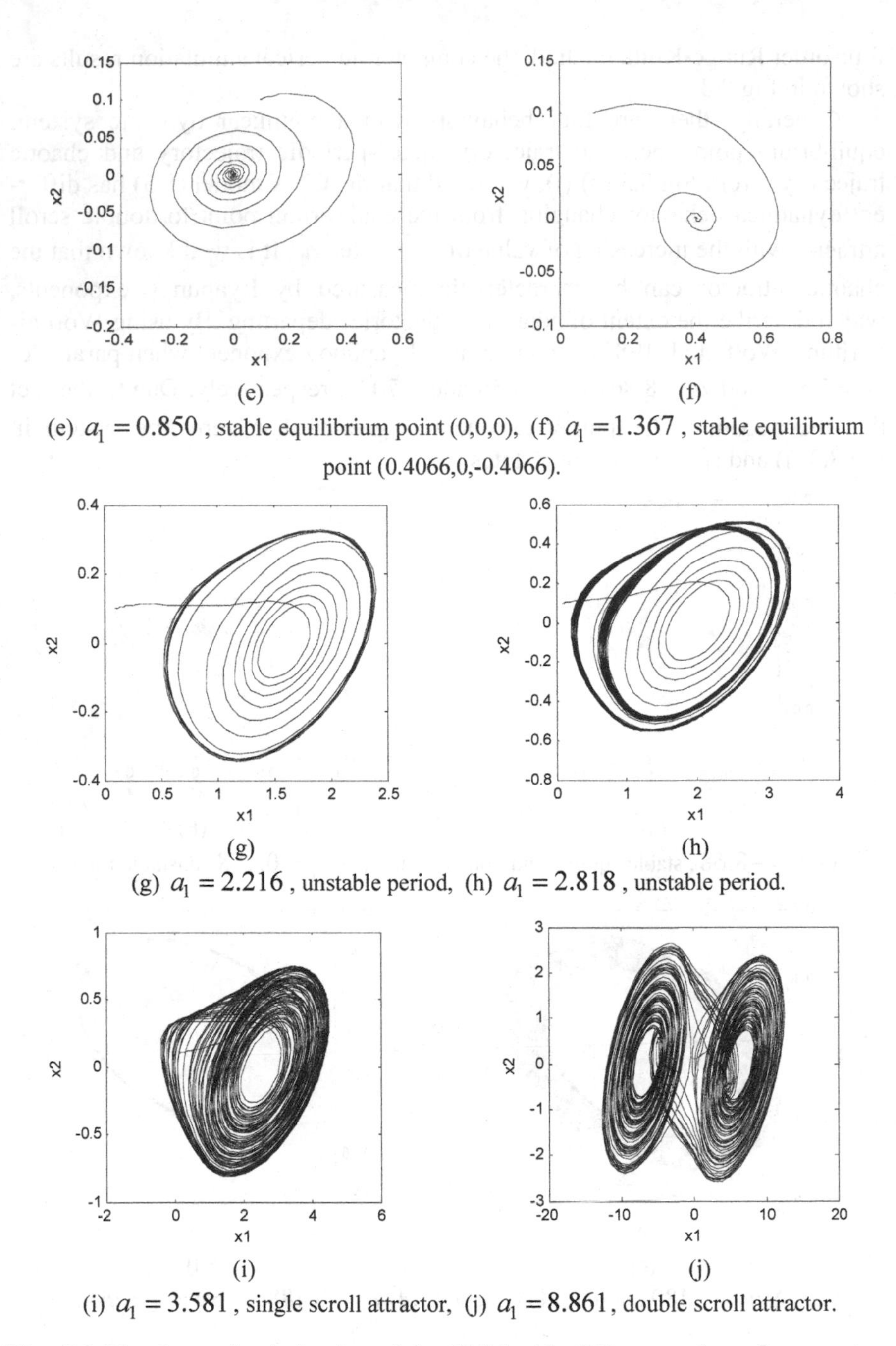

(e) $a_1 = 0.850$, stable equilibrium point (0,0,0), (f) $a_1 = 1.367$, stable equilibrium point (0.4066,0,-0.4066).

(g) $a_1 = 2.216$, unstable period, (h) $a_1 = 2.818$, unstable period.

(i) $a_1 = 3.581$, single scroll attractor, (j) $a_1 = 8.861$, double scroll attractor.

Fig. 8.3 The dynamica behavior of the CNN with different value of parameter a_1.

8.3 Description of image encryption algorithm

Basically, the chaos-based image encryption algorithm proposed in this chapter is a symmetrical algorithm, which means an identical secret key will be used to encrypt and decrypt image. We use a 128-bit key to drive a CNN hyper-chaos system, i.e., a 128-bit key is mapped to the initial conditions of the CNN and used to calculate system parameters, which will then be deployed in the algorithm to generate binary chaotic sequences for the encryption process. The diagram for the algorithm is given in Fig.8.4.

Before delving into the details of the algorithm, we give some explanations on several symbols used in the algorithm: S denotes the original image with its size $ih \times iw$, where ih and iw are the height and the width of a given image respectively, and S is written as a set $\{ s_{i,j} \mid i=1,2,\cdots,ih$ and $j=1,2,\cdots,iw \}$; T denotes a 128-bit key, and can be written as $T_1 T_2 \cdots T_{16}$, where T_i is ith character of T $(i=1,2,\cdots,16)$.

The steps of the encryption algorithm are given as follows.

Input: *a 128-bit key and the original image*

Output: *encrypted image*

Step 1: The six system parameters for the algorithm are obtained through the following computational procedures:

$$H_1 = \bigoplus_{i=1}^{ih} \bigoplus_{j=1}^{iw} s_{i,j}; \quad H_2 = (\sum_{i=1}^{ih} \sum_{j=1}^{iw} s_{i,j}) \ mod \ 256;$$

$$S = (T_1 + T_2 + \cdots + T_{16}) \ mod \ 256;$$

$$P = T_1 \oplus T_2 \oplus \cdots \oplus T_{16};$$

$$\lambda = (T_{10} + T_{11} + T_{12} + H_1 \times H_2) \ mod \ 256;$$

$$h = (T_{13} + T_{14} + T_{15} + T_{16} + II_1 \times H_2) / 256.$$

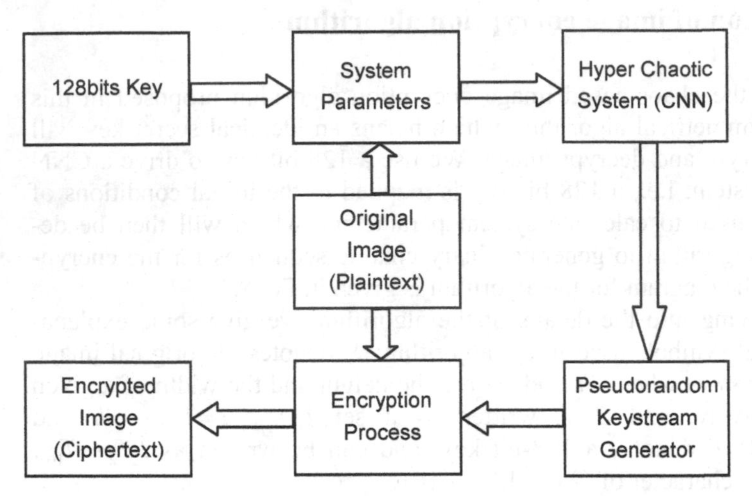

Fig. 8.4 The diagram of CNN-based image encryption algorithm.

Step 2: Determine the initial condition $(x_{1,0}, x_{2,0}, x_{3,0})$ and iteration numbers N_0 of the CNN:

$$x_{1,0} = (T_1 \times T_4 \times T_7 \times S \times P) / 256^5;$$

$$x_{2,0} = (T_2 \times T_5 \times T_8 \times S \times P) / 256^5;$$

$$x_{3,0} = (T_3 \times T_6 \times T_9 \times S \times P) / 256^5;$$

$$N_0 = (S^2 + P^2) \ mod \ 256.$$

Then iterate the CNN for $iw \times ih$ times from the initial condition $(x_{1,0}, x_{2,0}, x_{3,0})$ by using fourth-order Runge-Kutta Method (the time step size is 0.005) to solve the CNN differential equation (8.3). Here, to avoid the transient effect, the first N_0 iterations are considered.

Step 3: Let $\{(x_{1,i}, x_{2,i}, x_{3,i})\}_{i=1}^{iw \times ih}$ denotes the set of solutions of CNN. The size of the set is $iw \times ih$, and ith element has three points $(x_{1,i}, x_{2,i}, x_{3,i})$, $i = 1, 2, ..., iw \times ih$. Then we map $(x_{1,i}, x_{2,i}, x_{3,i})$ to a character that will be used to encrypt image pixel in the future, and the mapping procedure is described as follows:

Assume $v_i = h(x_{1,i}^2 + x_{2,i}^2 + x_{3,i}^2)^{1/2} + \lambda$, and u_i denotes the decimal fraction of v_i, then u_i can be represented as a binary sequence $u_i = 0.b_{i,1}b_{i,2}...b_{i,p}$, where p is a certain precision (here let $p = 32$). Let $w_i = (w_{i,0}w_{i,1}...w_{i,7})_2$, where $w_{i,j}$ denotes the jth bit of w_i, and

$w_{i,j} = b_{i,4j+1} \oplus b_{i,4j+2} \oplus b_{i,4j+3} \oplus b_{i,4j+4}$, $j = 0,1,...,7$. Hence, we obtain a character w_i from $(x_{1,i}, x_{2,i}, x_{3,i})$. (see Fig.8.5 for a schematic view). Finally, when the variable i changes from 1 to $iw \times ih$, we will obtain a pseudorandom keystream $\{w_i\}_{i=1}^{iw \times ih}$ to encrypt the original image.

Step 4: Let us suppose e_i, w_i and m_i denote the ciphertext (encrypted image), pseudorandom keystream and plaintext (original image) respectively, where $i = 1,2,...,iw \times ih$. Then the encryption process is defined as follows:

$$e_i = (w_i + m_i) \bmod 256. \tag{8.6}$$

Since our algorithm is a symmetrical one, the decryption process is similar to the encryption process except for the extend key T_E (see remark 1) used to generate w_i. The original image can be regenerated from the encrypted image through the following:

$$m_i = (e_i - w_i) \bmod 256. \tag{8.7}$$

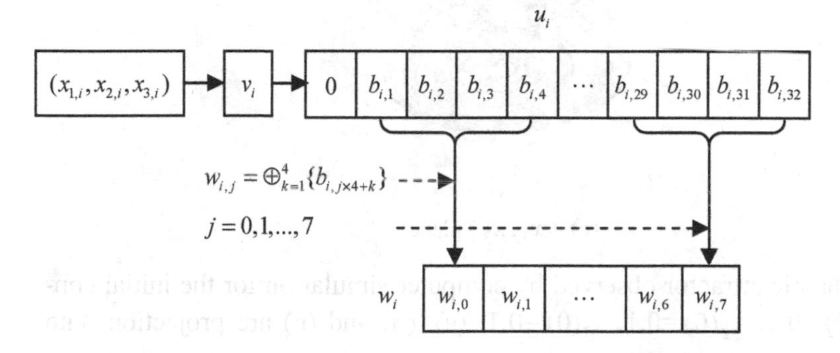

Fig. 8.5 The procedure of mapping $(x_{1,i}, x_{2,i}, x_{3,i})$ to character w_i.

Remark 1: To guarantee the sensitivity of the encryption algorithm with regard to the whole original image, we introduce two special parameters, i.e., H_1 and H_2 (see Step 1). It is clear that the generation of H_1 and H_2 depends on the plaintext S, and the keystream w_i in turn hinges on H_1 and H_2. Therefore the encryption results depend on the plaintext. This translates into the fact that the algorithm is sensitive to the plaintext, and can be verified in our experiments (see section of sensitivity analysis). If the encryption key is T, then the decryption key should be $T_E = (T, H_1, H_2)$, here we call T_E an extend key.

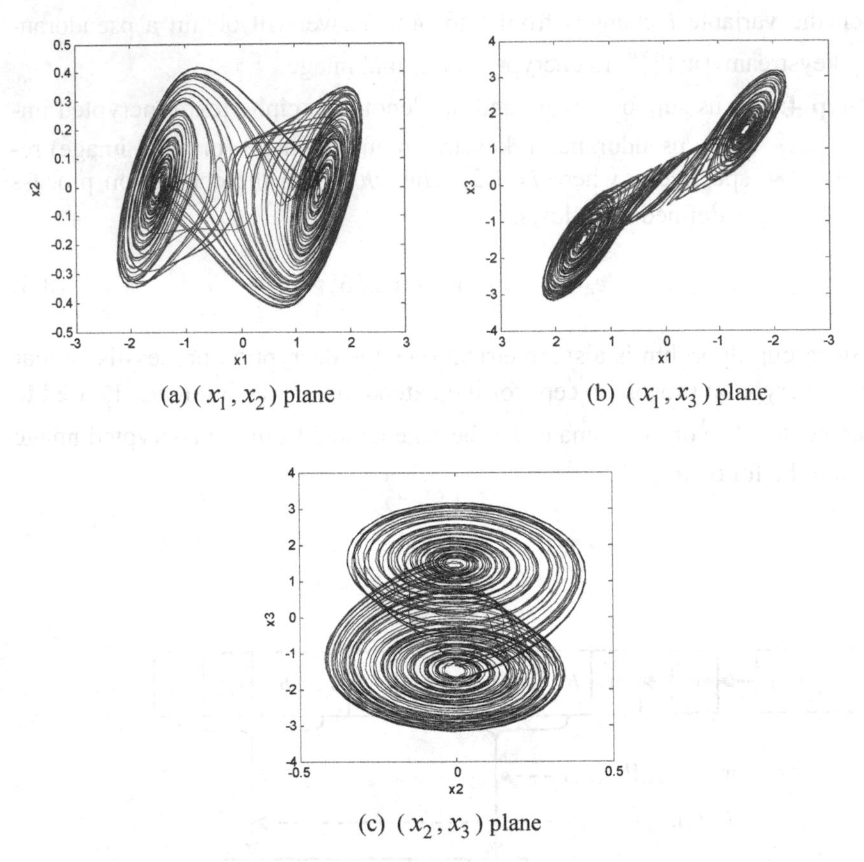

(a) (x_1, x_2) plane (b) (x_1, x_3) plane

(c) (x_2, x_3) plane

Fig. 8.6 Chaotic attractor observed by computer simulation for the initial condition $x_1(0)=0.1$, $x_2(0)=0.1$, $x_3(0)=0.1$. (a), (b) and (c) are projection onto the three planes respectively.

Remark 2: Actually, the proposed encryption algorithm mapped a 128 bits external key to parameters (S, P, h, λ), and the initial conditions of the CNN and the keystream obtained are related to the (S, P) and (S, P, h, λ), respectively. Furthermore, the control parameters of the CNN used in the algorithm are fixed as follows: $a_1 = 3.95$, $S_{11} = -1.65$, $S_{12} = 8.78$ and $S_{32} = -13.25$. The chaotic attractors obtained under these parameters are shown in Fig.8.6, and the maximum Lyapunov exponent of the sequences is 0.4684.

8.4 Security analyses

From the security standpoint, an effective encryption algorithm should withstand all kinds of known attacks. For example, it should have a large key space to thwart the brute-force attack, and it should be sensitive to the key and also to the plaintext. In this section, we present the results of several security analyses on the effectiveness of the proposed algorithm.

8.4.1 Key space analysis

It's well known that a large key space is very important for an encryption algorithm to repel the brute-force attack. Since the algorithm has a 128-bit key, the key space size is $2^{128} \approx 3.4 \times 10^{38}$. Furthermore, if we consider the four control parameters of CNN as a part of the key, the key space size will be even larger. This means if the precision is 10^{-14}, then the key space can be $2^{128} \times 10^{14 \times 4} \approx 2^{314}$. Apparently, the key space is large enough to resist all kinds of brute-force attacks. However, the extend key H_1 and H_2 should not be considered as part of the calculation of the key space. As we know, they are related to the plaintext, that means once a plaintext is given, H_1 and H_2 are constants.

8.4.2 Sensitivity analysis

We also perform experiments on the algorithm's sensitivity with respect to the key and the plaintext. The sensitivity means the change of a single bit in the secret key or the original image should result in a completely different encrypted image. The image 'Lena' and 'Pepper' with size 512×512 are used as the original images, and two keys *key*1 and *key*2 are randomly selected as follows, except for the fact that there is only one character difference between them.

*key*1 = 'hT4#6RuJha&3m%B8'
*key*2 = 'hT4#6RuJha&3m%B9'

The results are shown in Fig.8.7. We can see that, in Fig.8.7 (b) and (c), the encrypted images (ciphertext) are very different when using a slightly different key. Besides, to test the sensitivity to the plaintext, we modified the gray-scale value (GSV) of one pixel at grid (100,100) of the original image (d) by adding one, i.e.,

$$GSV(s'_{100,100}) = (GSV(s_{100,100}) + 1) \bmod 256 . \tag{8.8}$$

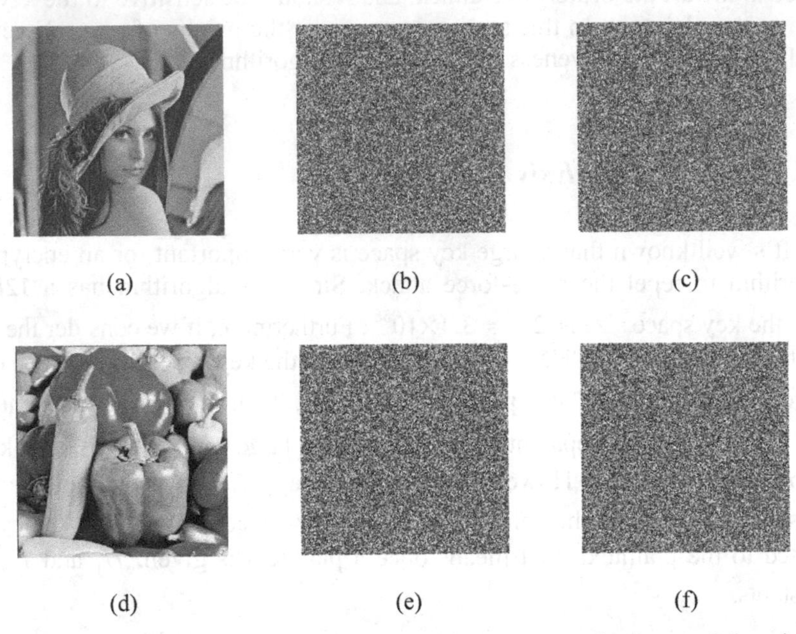

Fig. 8.7 Results of sensitivity experiment with respect to the key and the plaintext. (a) is the original 'Lena' image, (b) is the encrypted image of (a) with key1, (c) is the encrypted image of (a) with key2, (d) is the original 'Pepper' image, (e) is the encrypted image of (d) with key1, (f) is the encrypted image with key1 when only one pixel changed in (d).

The results are in Fig.8.7 (e) and (f), and obviously (e) and (f) are completely different. For the other two original images 'Boat' and 'Woman', the results of the same sensitivity experiment with respect to the key are also shown in Fig.8.8.

Furthermore, for the sake of understanding the difference between the encrypted images, two common measures (Chen et al. 2004) are used, i.e., number of pixels change rate (NPCR) and unified average changing intensity (UACI), they can be defined as:

$$\text{NPCR} = \frac{\sum_{i,j} D(i,j)}{W \times H} \times 100\% , \tag{8.9}$$

$$\mathrm{UACI} = \frac{1}{W \times H}\left[\sum_{i,j} \frac{\left|E^1_{i,j} - E^2_{i,j}\right|}{255}\right] \times 100\%. \tag{8.10}$$

where E^1 and E^2 denote two encrypted images, respectively, W and H are the width and height of E^1 or E^2, and the gray-scale values of the pixels at grid (i, j) of E^1 and E^2 are labeled by $E^1_{i,j}$ or $E^2_{i,j}$, respectively. And $D(i, j)$ is given by

$$D(i,j) = \begin{cases} 0, \text{ if } \mathrm{E}^1_{i,j} = E^2_{i,j}, \\ 1, \text{ if } \mathrm{E}^1_{i,j} \neq E^2_{i,j}. \end{cases} \tag{8.11}$$

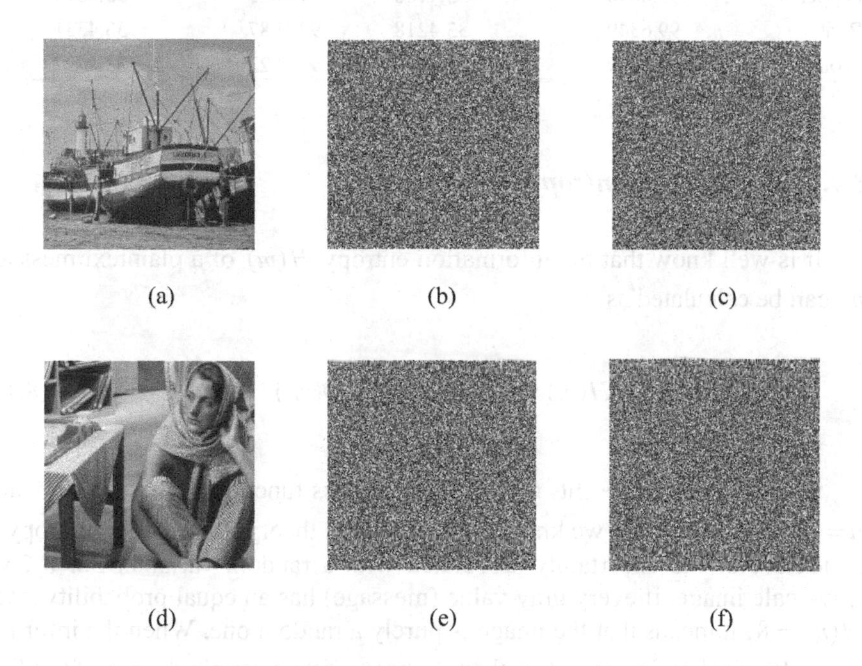

Fig. 8.8 Results of sensitivity experiment with respect to the key. (a) is the original 'Boat' image, (b) is the encrypted image of (a) with key1, (c) is the encrypted image of (a) with key2, (d) is the original 'Woman' image, (e) is the encrypted image of (d) with key1, (f) is the encrypted image of (d) with key2.

The NPCR measures the different pixel numbers between two images, and the UACI measures the average intensity of differences between two images.

We have computed the NPCR and UACI with the proposed algorithm to evaluate the sensitivity with respect to small changes in a key or in the plaintext. The calculation results for four original images are shown in Table 8.1. From the results, we have found that all the NPCR are over 99.5% and all the UACI are over 33.3%, indicating that the algorithm is very sensitive to the key and the plaintext. Thus, the algorithm can effectively resist the known-plaintext attack, through which the attacker may be able to find out a meaningful relationship between the original image and the encrypted image.

Table 8.1 Pixel difference between the encrypted images when a slight change in the key or the original image

Image (512×512)	One character change in key		One pixel change in original image	
	NPCR (%)	UACI (%)	NPCR (%)	UACI (%)
Lena	99.6105	33.4749	99.6059	33.5364
Pepper	99.6098	33.4786	99.6025	33.4314
Boat	99.6349	33.4218	99.5987	33.4331
Woman	99.6059	33.3911	99.5922	33.4422

8.4.3 Information entropy

It is well know that the information entropy $H(m)$ of a plaintext message m can be calculated as

$$H(m) = -\sum_{i=1}^{n} p(m_i)\log_2 p(m_i), \tag{8.12}$$

where $p(m_i)$ represents the probability mass function of message m_i, and $n = 256$ for image. As we know, in information theory, information entropy is a measure of the uncertainty associated with a random variable. For a 256-gray-scale image, if every gray value (message) has an equal probability, then $H(m) = 8$, it means that the image is purely a random one. When the information entropy of an image is less than 8, there exists a certain degree of predictability, which will threaten its security. Hence, to be able to stand up to the entropy attack effectively, we hope the entropy of the encrypted image is near 8.

The entropy of the four original images and their corresponding encrypted images are computed and listed in Table 8.2. From Table 8.2 we can see that all the entropies of the original images are smaller than the ideal one. This is due to the fact that practical information sources seldom generate random messages. However, the entropy of every encrypted image is very close to 8 (ideal value). This indicates that the encryption system can convert an original image to a random image, showing the system is secure against the entropy attack.

Table 8.2 Information entropy of the original and the encrypted images

Image (512×512)	Information Entropy (H)		
	Original image	Encrypted image E1 (with key1)	Encrypted image E2 (with key2)
Lena	7.4455	7.9993	7.9992
Pepper	7.5936	7.9992	7.9993
Boat	7.1238	7.9994	7.9992
Woman	7.6321	7.9993	7.9993

8.4.4 Statistical analysis

It is well know that many encryption algorithms have been successfully analyzed with the help of statistical analysis and several statistical attacks have been devised on them. Hence, an ideal image encryption scheme should be robust against any statistical attack. In order to prove the robustness of the proposed scheme, the following statistical tests are performed for four images and its corresponding encrypted one.

8.4.4.1 Histogram analysis

In statistics, an image histogram illustrates how pixels in an image are distributed by graphing the number of pixels at each gray-scale value. Fig.8.9 and Fig.8.10 show the histograms for each image in Fig.8.7 and Fig.8.8, respectively. From the figures, we can see that the histogram of the encrypted image is fairly uniform and is significantly different from the respective histogram of the original image, indicating the proposed scheme make the encrypted image look like a random one and therefore does not provide any clue to conduct any statistical attack on the suggested scheme.

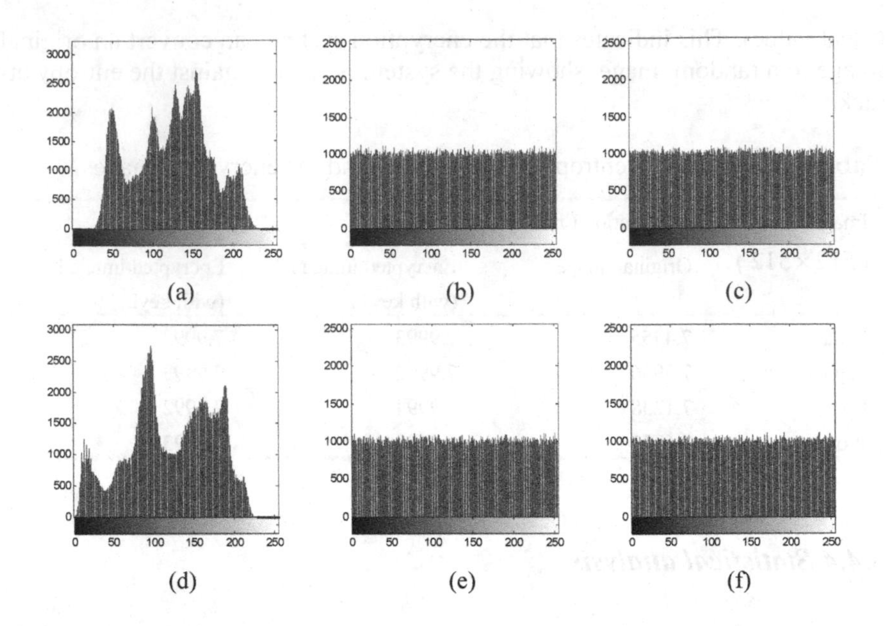

Fig. 8.9 The histograms corresponding to the images shown in Fig.8.7.

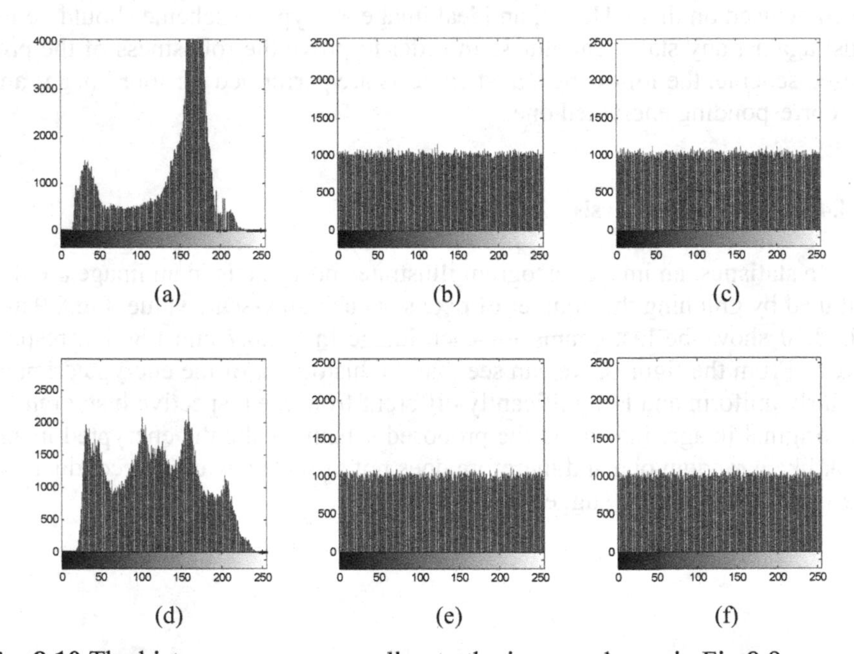

Fig. 8.10 The histograms corresponding to the images shown in Fig.8.8.

8.4.4.2 Analysis of correlation of adjacent pixels

Shannon (Shannon 1949) has suggested two basic techniques for instructing the design of practical ciphers in his classic paper: diffusion and confusion. Diffusion means to spread the influence of a single plaintext or key bits over as much of the ciphertext as possible so as to hide the statistical structure of the plaintext. Confusion means to exploit some transformations to hide any relationship between the plaintext, the ciphertext and the key, thus making cryptanalysis more difficult (Schneier 1996). These two superior properties can be demonstrated by a test on the correlations of adjacent pixels in the encrypted image (Chen et al. 2004). We examine the correlation between two vertically adjacent pixels, horizontally adjacent pixels, and two diagonally adjacent pixels, respectively. Relevant coefficient is defined as follow

$$r_{xy} = \frac{\text{cov}(x, y)}{\sqrt{D(x)}\sqrt{D(y)}}, \tag{8.13}$$

$$\text{cov}(x, y) = \frac{1}{N}\sum_{i=1}^{N}(x_i - E(x))(y_i - E(y)), \tag{8.14}$$

where $\text{cov}(x, y)$ is covariance, $D(x)$ is variance, x and y denote the gray-scale values of image. In numerical computation, the following discrete forms were used:

$$E(x) = \frac{1}{N}\sum_{i=1}^{N} x_i, \tag{8.15}$$

$$D(x) = \frac{1}{N}\sum_{i=1}^{N}(x_i - E(x))^2. \tag{8.16}$$

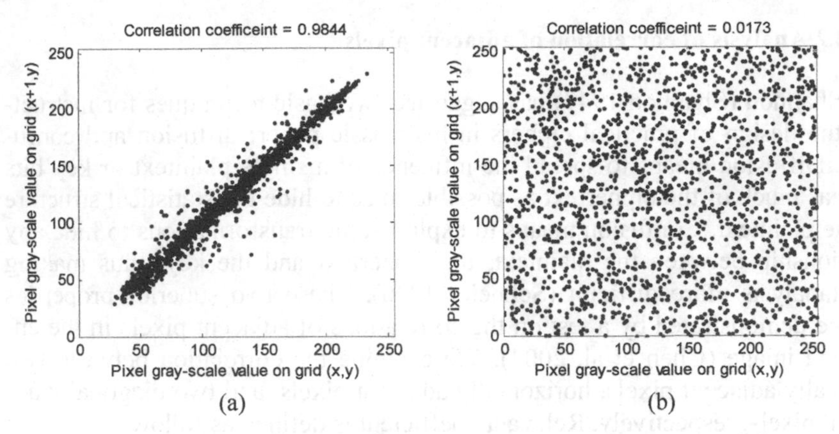

Fig. 8.11 Correlations of two horizontally adjacent pixels. (a) and (b) are correlation distribution analysis of the original 'Lena' image and the encrypted image, respectively.

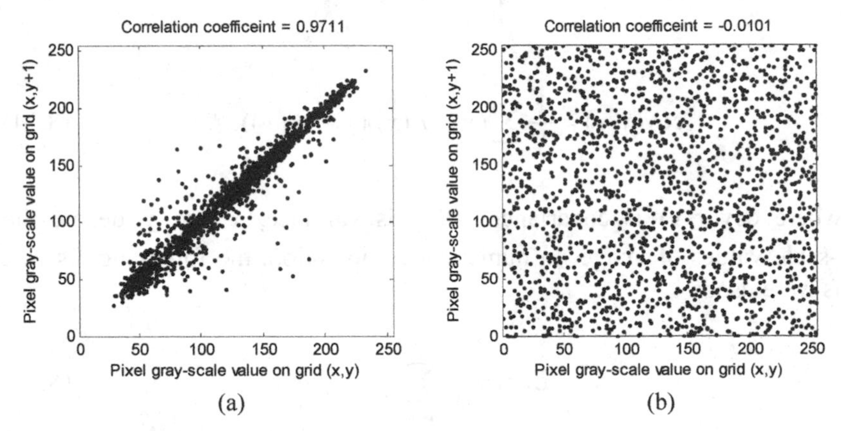

Fig. 8.12 Correlations of two vertically adjacent pixels. (a) and (b) are correlation distribution analysis of the original 'Lena' image and the encrypted image, respectively.

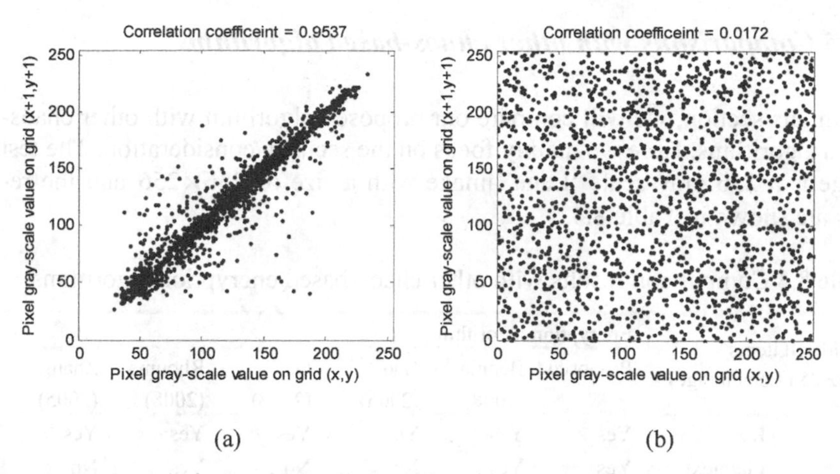

Fig. 8.13 Correlations of two diagonally adjacent pixels. (a) and (b) are correlation distribution analysis of the original 'Lena' image and the encrypted image, respectively.

We randomly select 2000 pairs of pixels from the original image and encrypted image respectively, and then compute their corresponding relevant coefficients. Fig.8.11 shows the correlation distribution of two horizontally adjacent pixels in the original image 'Lena' and its corresponding encrypted image, and the correlation coefficients are 0.9844 and 0.0173, respectively. Similar results for vertical and diagonal directions were also obtained, which are shown in Fig.8.12, Fig.8.13 and Table 8.3.

From the results of Table 8.3, we found that the correlation coefficients of the encrypted images are very small, indicating that the attacker cannot obtain any valuable information by exploiting a statistical attack.

Table 8.3 Correlation coefficients of two adjacent pixels in the original and encrypted images

Image	*Original Image*			*Encrypted Image*		
(512×512)	*Horizontal*	*Vertical*	*Diagonal*	*Horizontal*	*Vertical*	*Diagonal*
Lena	*0.9844*	*0.9711*	*0.9537*	*0.0173*	*-0.0101*	*0.0172*
Boats	*0.9740*	*0.9609*	*0.9365*	*-0.0235*	*0.0202*	*-0.0114*
Peppers	*0.9836*	*0.9787*	*0.9739*	*0.0162*	*-0.0136*	*0.0125*
Woman	*0.9598*	*0.8866*	*0.8741*	*-0.0132*	*-0.0439*	*0.0316*

8.4.5 Comparisons with other chaos-based algorithms

In this section, we will compare our proposed algorithm with other chaos-based algorithms. Here we mainly focus on the security consideration. The test image is a 256 gray-scale 'Lena' image with a size of 256×256 and the results are shown in Table 8.4.

Table 8.4 Comparison results with other chaos-based encryption algorithms

Considered items (256×256 Lena image)		Encryption algorithm					
		Proposed	Behnia (2008)	Gao (2006)	Gao (2008)	Rhouma (2008)	Zhang (2005)
Sensitivity	Key	Yes	Yes	Yes	Yes	Yes	Yes
	Plaintext	Yes	Yes	No	No	Yes	No
	NPCR (%)	99.65%	41.96%	N/A	N/A	99.58%	N/A
	UACI (%)	33.46%	33.25%	N/A	N/A	33.38%	N/A
Correlation coefficients	Horizontal	0.0016	0.0038	-0.0159	-0.0142	0.0681	0.0286
	Vertical	-0.0049	0.0023	-0.0654	-0.0074	0.0845	0.0203
	Diagonal	0.0025	0.0004	-0.0323	-0.0183	N/A	0.0197
Information entropy		7.9969	7.9968	N/A	N/A	7.9732	N/A
Key space		2^{314}	2^{260}	2^{150}	2^{233}	2^{192}	2^{129}
Chaotic system employed		CNN	Coupled chaotic system	NCA map	Chen's chaotic system	OCML	Hybrid of CNN and Logistic map

From Table 8.4 we can see that the chaotic systems used in each algorithm are different from each other. Even Zhang (Zhang et al. 2005) employed a hybrid of CNN and Logistic map, the details on the design of encryption technique are also distinct. We easily found that only our algorithm and those that were proposed in (Behnia et al. 2008, Rhouma et al. 2008) are sensitive to both the key and the plaintext, and the performance of NPCR and UACI of our algorithm are better than the algorithms in (Behnia et al. 2008), but compatible with the algorithm in (Rhouma et al. 2008). In terms of correlation coefficients, our algorithm and algorithms in (Behnia et al. 2008, Gao and Chen 2008) have equivalent performance. However, from the key space point of view, our algorithm has a larger space (2^{314}, see section of key space analysis) than the others. With a larger key space the encryption algorithm can provide a higher capability for fighting against various kinds of brute-force attacks. Generally speaking, the analyses above indicate that our algorithm has a better security performance when compared with those five algorithms.

8.5 Conclusions and discussion

In this chapter, a novel image encryption algorithm based on a hyper-chaotic CNN is proposed. One of the main motivations for using hyper-chaotic system is that we want to achieve a more sophisticated chaotic sequence to be used for encrypting the plaintext. We accomplish this through a 128-bit-long external key as the secret key of the algorithm and through a special procedure that maps the key to the system parameters. Security analyses indicate that the proposed image encryption algorithm has desirable properties from cryptographic point of view and the experiment results show that our algorithm performs better than five other similar algorithms. It should be mentioned that, although the proposed algorithm has focused on image encryption, it can also be applied in other fields such as secure information communication over the Internet.

Possible future work can be pursued in the following directions: extending the proposed algorithm to handling color image encryption and examining its security as well as speed performance. Meanwhile, other measures used in the former cryptosystems are still worthy of further research to improve the proposed algorithm's performance. For example, in order to obtain a better confusion, the following chaotic cat map

$$\begin{pmatrix} x_{n+1} \\ y_{n+1} \end{pmatrix} = \mathbf{A} \begin{pmatrix} x_n \\ y_n \end{pmatrix} \pmod N, \text{ where } \mathbf{A} = \begin{pmatrix} 1 & a \\ b & ab+1 \end{pmatrix}, \ \det(\mathbf{A}) = 1 \quad (8.17)$$

can be applied to the original image before the encryption. The corresponding reversible transformation is also like (8.17) except that $\mathbf{A}$ is replaced by $\mathbf{A}^{-1}$. As a result, the encrypted image can possess the two kinds of changes that we mentioned in the first section.

Due to the potential of the CNN architecture, as well as the application of image encryption, the CNN can be used in other information assurance and security fields. For instance, it is clear that cryptographic hash functions can provide message origin authentication between two mutually trustful parties sharing a secret key and validation of message content integrity. In 1998, Csapodi et al. had succeeded in implementing existing constructions of Cartesian authentication codes on CNN Universal Machine to provide provable (unconditional) security for message authentication (Csapodi et al.1998). In 2004, Gao and Moschytz presented a fingerprint feature-matching algorithm which is tailor-made for CNN to realize a fingerprint verification system (Gao and Moschytz 2004). Moreover, Su (Su et al. 2005) introduced a fingerprint recognition system that involves the CNN-based ridgeline thinning and feature extraction process, which also could be applied in digital signature. Recently, an anomaly detection approach for the network intrusion detection was presented (Yang and Karahoca A 2006), in which a recurrent perceptions learning algorithm (RPLA) was used to learn the templates and bias in CNN classifier. Experiments with KDD Cup 1999 network traffic connections which have been preprocessed with methods of features selection and normalization have shown that CNN model is effective for intrusion detection and has an excellent performance compared with back propagation neural network due to the higher attack detection rate with lower false positive rate.

We believe that with the development of research and practice on the information assurance and security issues using CNN, more and more cryptographic systems will be proposed to provide a secure and reliable environment or measures for a wide different of application areas via the Internet.

Acknowledgments The authors would like to acknowledge the anonymous reviewers for their valuable suggestions. The work described here was partially supported by a research funding from California State University, Sacramento, USA, the Natural Science Foundation Project of Chongqing Science and Technology Committee of China under Grant No. CSTC 2008BB2360, and the Science and Technology Research Program of the Chongqing Municipal Education Committee of China under Grant No. KJ081406.

References

[1]. Alvarez G, Li S J (2006) Some basic cryptographic requirements for chaos-based cryptosystems. Int J Bifurcation Chaos 16:2129–2151
[2]. Amigo J M, Kocarev L, Szczepanski J (2007) Theory and practice of chaotic cryptography. Phys Lett A 366: 211–216
[3]. Arthur T B, Kan Y (2001) Magic 'squares' indeed. Math Gazette 108:152–156
[4]. Baptista M S (1998) Cryptography with chaos. Phys Lett A 240:50–54
[5]. Behnia S, Akhshani A, Mahmodi H et al (2008) A novel algorithm for image encryption based on mixture of chaotic map. Chaos, Solitons, Fractals 35:408–419
[6]. Bourbakis N (1997) Image data compression encryption using G-SCAN pattern. Proc IEEE Conf SMC, 1117–1120
[7]. Bourbakis N, Alexopoulos C (1992) Picture data encryption using SCAN patterns. Pattern Recogn 25:567–581
[8]. Bourbakis N, Alexopoulos C (1999) A fractal based image processing language - formal modeling. Pattern Recogn 32:317–338
[9]. Chang H T (2004) Arbitrary affine transformation and their composition effects for two dimensional fractal sets. Image Vis Comput 22:1117–1127
[10]. Charilaos C, Athanassios S, Touradj E (2000) The JPEG2000 still image coding system. IEEE Trans Consum Electron 46:1103–1127
[11]. Chen G R, Mao Y B, Chui C K (2004) A symmetric image encryption based on 3D chaotic maps. Chaos, Solitons, Fractals 21:749–761
[12]. Chen H Z, Dai M D, Wu X Y (1994) Bifurcation and chaos in discrete-time cellular neural networks. 3rd IEEE Int Wkshp Cellular Neural Networks and Their Applicat, CNNA'94, 309–315
[13]. Cheng H, Li X (2000) Partial encryption of compressed images and videos. IEEE Trans Sig Proc 48:2439–2451

[14]. Chua L O, Hasler M, Moschytz G S et al (1995) Autonomous cellular neural networks: a unified paradigm for pattern formation and active wave propagation. IEEE Trans Circ Syst I 42:559–577
[15]. Chua L O, Yang L (1988a) Cellular neural network: theory. IEEE Trans Circ Syst 35:1257–1272
[16]. Chua L O, Yang L (1988b) Cellular neural network: applications. IEEE Trans Circ Syst 35:1273–1290
[17]. Clarke R J (1995) Digital compression of still images and video. Academic Press, New York
[18]. Cox I J, Kilian J, Leighton F T et al (1997) Secure spread spectrum watermarking for multimedia. IEEE Trans Image Process 6:1673–1687
[19]. Csapodi M, Vandewalle J, Roska T (1998) High speed calculation of cryptographic hash functions by CNN Chips. 1998 Fifth IEEE Int Wkshp Cellular Neural Networks and Their Applicat, CNNA'98, 14–17
[20]. Dang P P, Chau P M (2000) Image encryption for secure Internet multimedia applications. IEEE Trans Consum Electron 46:395–403
[21]. Fan Z, Josef A N (1991) A chaotic attractor with cellular neural networks. IEEE Trans Circ Syst 38:811–812
[22]. Fridrich J (1997) Image encryption based on chaotic maps. Proc IEEE Int Conf Syst Man Cybern, 1105–1110
[23]. Fridrich J (1998) Symmetric ciphers based on two-dimensional chaotic maps. Int J Bifurcation Chaos 8:1259–1284
[24]. Gao H J, Zhang Y S, Liang S Y et al (2006) A new chaotic algorithm for image encryption. Chaos, Solitons, Fractals 29:393–399
[25]. Gao Q, Moschytz G S (2004) Fingerprint feature matching using CNNs. Proc 2004 Int Symp Circ Syst 3:73–76
[26]. Gao T G, Chen Z Q (2008) A new image encryption algorithm based on hyper-chaos. Phys Lett A 372:394–400
[27]. Goedgebuer J P, Levy P, Larger L et al (2002) Optical communication with synchronized hyperchaos generated electro optically. IEEE J Quantum Electron 38:1178–1183
[28]. Guzman J, Astrom K J, Dormido S et al (2008) Interactive learning modules for PID control. IEEE Contr Syst Mag 28:118–134
[29]. He Z, Li K (2000) TDMA secure communication based on synchronization of Chua's circuits. J Circ Syst Comp, 10:147–158
[30]. He Z Y, Zhang Y F, Lu H T (1999) The dynamic character of cellular neural network with applications to secure communication. J Chin Inst Commun 20:59–67
[31]. Jessl J, Bertram M, Hagen H (2005) Web-based progressive geometry transmission using subdivision-surface wavelets. Proc 10th Int Conf 3D Web Technol, 29–35
[32]. Kocarcv L (2001) Chaos-based cryptography: A brief overview. IEEE Circ Syst Mag 1:6–21
[33]. Kocarev L, Jakimoski G (2001) Logistic map as a block encryption algorithm. Phys Lett A 289:199–206
[34]. Lai X J, Massey J L (1990) A proposal for a new block encryption standard. Adv in Cryptology-EUROCRYPT'90, Springer-Berlin, LNCS 473:389–404
[35]. Lee J W, Giraud-Carrier C (2007) Transfer Learning in Decision Trees. 2007 Int Joint Conf Neural Networks, 726–731
[36]. Li P, Li Z, Halang W A et al (2007) A stream cipher based on a spatiotemporal chaotic system. Chaos, Solitons, Fractals 32:1867–1876
[37]. Li S J, Zheng X (2002) Cryptanalysis of a chaotic image encryption method. Proc IEEE Int Symp Circ Syst 2:26–29
[38]. Li Z G, Li K, Wen C Y et al (2003) A new chaotic secure communication system. IEEE Trans Commun 51:1306–1312

[39]. Li X B, Knipe J, Cheng H (1997) Image compression and encryption using tree structures. Pattern Recogn Lett 18: 1253–1259
[40]. Lian S G, Wang Z Q, Li Z X (2004) Secure multimedia encoding schemes based on quad-tree structure. J Image and Graphics 9:353–359
[41]. Maniccam S S, Bourbakis N (2004) Image and video encryption using SCAN patterns. Pattern Recogn 37:725–737
[42]. Martin K, Lukac R, Plataniotis K N (2005) Efficient encryption of compressed color images. Proc IEEE Int Symp Ind Electron 3:1245–1250
[43]. Mascolo S, Grassi G (1998) Observers for hyperchaos synchronization with application to secure communications. Proc IEEE Int Conf Control App 2:1016–1020
[44]. Masuda N, Aihara K (2002) Cryptosystems with discretized chaotic maps. IEEE Trans Circ Syst I 49:28–40
[45]. Matthews R (1989) On the derivation of a chaotic encryption algorithm. Cryptologia XIII:29–42
[46]. Mozina M, Zabkar J, Bratko I (2007) Argument based machine learning. Artificial Intelligence 171:922–937
[47]. Parberry I (1997) An efficient algorithm for the Knight's tour problem. Discrete Appl Math 73:251–260
[48]. Pareek N K, Patidar V, Sud K K (2006) Image encryption using chaotic logistic map. Image Vis Comput 24:926–934
[49]. Parker A T, Short K M (2001) Reconstructing the keystream from a chaotic encryption scheme. IEEE Trans Circ Syst I 48:104–12
[50]. Pecora L M, Carroll T L (1990) Synchronization in chaotic systems. Phys Rev Lett 64:821–824
[51]. Pecora L M, Carroll T L (1991) Driving systems with chaotic signals. Phys Rev A 44:2374–2383
[52]. Perez G, Cerdeira H A (1995) Extracting messages masked by chaos. Phys Rev Lett 74:1970–1973
[53]. Pisarchik A N, Flores-Carmona N J, Carpio-Valadez M (2006) Encryption and decryption of images with chaotic map lattices. Chaos 6:033118.1–033118.6
[54]. Qiu S H, Ma Z G (2003) An image cryptosystem based on general cat map. J Commun Chin 24:51–57
[55]. Radha H, Vetterli M, Leonardi R (1996) Image compression using binary space partitioning trees. IEEE Trans Image Process 5:1610–1624
[56]. Rhouma R, Meherzi S, Belghith S (2008) OCML-based colour image encryption. Chaos, Solitons Fractals doi:10.1016/j.chaos.2007.07.083
[57]. Rivest R L, Shamir A, Adleman L (1978) A method for obtaining digital signatures and public key cryptosystems. Commun ACM 21:120–126
[58]. Scharinger J (1998) Fast encryption of image data using chaotic Kolmogorov flows. J Electron Imaging 7:318–325
[59]. Schneier B (1996) Applied Cryptography (2nd edn). John Wiley & Sons Press, New York
[60]. Shannon C E (1949) Communication theory of secrecy system. Bell Syst Tech J 28:656–715
[61]. Shiang H P, Tu W, van der Schaar M (2008) Dynamic Resource Allocation of Delay Sensitive Users Using Interactive Learning over Multi-Carrier Networks. 2008 IEEE Int Conf Commun, 2502–2506
[62]. Slavova A (1998) Dynamic properties of cellular neural networks with nonlinear output function. IEEE Trans Circ Syst I 45:587–590
[63]. Su T J, Du Y Y, Cheng Y J et al (2005) A fingerprint recognition system using cellular neural networks. 9th Int Wkshp Cellular Neural Networks App, 170–173
[64]. Quattoni A, Collins M, Darrell T (2008) Transfer learning for image classification with sparse prototype representations. 2008 IEEE Conf Comp Vis and Pattern Recogn, 1–8

[65]. Vladimir S U, Jean-Pierre G, Laurent L et al (2001) Communicating with optical hyperchaos: information encryption and decryption in delayed nonlinear feedback systems. Phys Rev Lett 86:1892–1895
[66]. Wong K W (2003) A combined chaotic cryptographic and hashing scheme. Phys Lett A 307:292–298
[67]. Wolf A, Swift J B, Swinney H L et al (1985) Detecting Lyapunov Exponents from a Time Series. Physica D 16:285–317
[68]. Yang T, Lin B Y, Chun M Y (1998) Application of neural networks to unmasking chaotic secure communication. Physica D, 124:248–257
[69]. Yang T, Wu C W, Chua L O (1997) Cryptography based on chaotic systems. IEEE Trans Circ Syst I 44:469–472
[70]. Yang Z X, Karahoca A (2006) An anomaly intrusion detection approach using cellular neural networks. Comput Inf Sci – ISCIS'06, Springer-Berlin, LNCS 4263:908–917
[71]. Yen J C, Guo J I (2000) A new chaotic key-based design for image encryption and decryption. Proc IEEE Int Conf Circ Syst 4:49–52
[72]. Zhang W, Peng J, Yang H Q et al (2005) A digital image encryption scheme based on the hybrid of cellular neural network and logistic map. Adv Neural Networks, ISNN'05, Springer-Berlin, LNCS 3497:860–867

Part III: Privacy

9 From Data Privacy to Location Privacy

Ting Wang and Ling Liu[1]

Abstract: Over the past decade, the research on data privacy has achieved considerable advancement in the following two aspects: First, a variety of privacy threat models and privacy principles have been proposed, aiming at providing sufficient protection against different types of inference attacks; Second, a plethora of algorithms and methods have been developed to implement the proposed privacy principles, while attempting to optimize the utility of the resulting data. The first part of the chapter presents an overview of data privacy research by taking a close examination at the achievements from the above two aspects, with the objective of pinpointing individual research efforts on the grand map of data privacy protection. As a special form of data privacy, location privacy possesses its unique characteristics. In the second part of the chapter, we examine the research challenges and opportunities of location privacy protection, in a perspective analogous to data privacy. Our discussion attempts to answer the following three questions: (1) Is it sufficient to apply the data privacy models and algorithms developed to date for protecting location privacy? (2) What is the current state of the research on location privacy? (3) What are the open issues and technical challenges that demand further investigation? Through answering these questions, we intend to provide a comprehensive review of the state of the art in location privacy research.

9.1 Introduction

Recent years have witnessed increasing concerns about the privacy of personal information in various data management and dissemination applications. Typically, such data is stored in a relational data model, and each record consists of three categories of information: identity attributes, quasi-identity attributes, and sensitive attributes. Individuals intend to protect their sensitive information from exposure to unauthorized parties via direct disclosure or indirect inferences [20]. Concretely, releasing of microdata with identity attributes removed may still disclose sensitive information about individuals with high probability, due to the inference attacks that link quasi-identity attributes

[1] Distributed Data Intensive System Lab, College of Computing, Georgia Tech

J.J.P. Tsai and P.S. Yu (eds.), *Machine Learning in Cyber Trust: Security, Privacy, and Reliability*, DOI: 10.1007/978-0-387-88735-7_9,

to some external knowledge. One of the most well known examples of such attacks was given in the original *k*-anonymity paper [49].

In order to alleviate such concerns, various privacy protection mechanisms have been proposed to perform transformation over the raw data before publishing. The literatures of data privacy protection techniques can be grouped into two categories: the first category of work aims at proposing privacy models and principles, which serve as criteria for measuring if the publication of a raw dataset provides sufficient privacy protection. The second category of research explores data transformation techniques that can meet the proposed privacy principles, while maximizing the utility of the resulting data.

The first part of the chapter is devoted to a close examination on the advancement of the data privacy research in the past decade, from the aforementioned two aspects. Specifically, we review the proposed data privacy principles and models, according to the types of privacy breaches identified, and analyze the implicit relationships among different types of privacy breaches. We classify the existing data anonymization algorithms and methods, based on their underlying privacy models and implementation techniques, and compare their strengths and weaknesses accordingly. In addition, we discuss various data utility optimization frameworks. Through the survey, we aim at pinpointing the existing independent research efforts on the grand map of data privacy protection.

With the ubiquitous wireless connectivity and the continued advance in mobile positioning technology (e.g., cellular phones, GPS-like devices), follows an explosive growth of location-based services (LBS). Examples include location-based store finders ("*Where is the nearest gas station to my current location?*"}), traffic condition tracking ("*What is the traffic condition on Highway 85 North?*"), and spatial alarm ("*Remind me to drop off a letter when I am near a post office.*"}). The mobile users obtain such services by issuing requests together with their location information to the service providers. While offering great convenience and business opportunities, LBS also opens the door for misuse of users' private location information. For example, the collected location information can be exploited to spam users with unwanted advertisements; Personal medical conditions, alternative lifestyles, unpopular political or religious views can be inferred by knowing users' visit to specific locations; GPS devices have even been used in physical stalking [22, 54].

Such concerns have spurred intensive research on location privacy protection recently [40]. As a special form of data privacy, location privacy features its unique characteristics and research challenges. Specifically, location privacy requirements are inherently personalized and context-sensitive. The level of privacy protection is intricately related to the quality of service delivered to the customers. Furthermore, location data is highly dynamic, and subjected to frequent updates. Such uniqueness makes it inadequate to directly apply the data privacy research results and techniques to the problem of location privacy.

In the second part of the chapter, we present a brief survey of the existing literatures of location privacy research, and examine the research challenges and opportunities, in a perspective analogous to data privacy. Within the survey, we intend to address the following three key questions: (1) Is it sufficient to apply the data privacy models and algorithms developed to date for protecting location privacy? (2) What is the current state of the research on location privacy? (3) What are the open issues and technical challenges that are worth further investigation? Specifically, we summarize the proposed location privacy principles and models, and their implicit relationships; We analyze various system architectures for implementing location anonymization; We classify the existing location anonymization tools based on their underlying architecture models and implementation techniques, and compare their strengths and weaknesses; We then discuss open issues and technical challenges for constructing complete solutions for location privacy protection. Through this survey, we intend to provide a comprehensive review of the state of the art in location privacy research.

9.2 Data Privacy

While the scope of data privacy broadly covers the topics of privacy-preserving data publication, data mining, information retrieval, etc, and involves techniques such as cryptography, perturbation, auditing, etc, this survey particularly focuses on the problem of privacy protection in data publication, using the techniques of data perturbation, as is most relevant to the location privacy protection.

The problem of controlling information disclosure in data dissemination for public-use has been studied extensively in the framework of statistical databases. Motivated by the need of publishing census data, the statistics literatures focus mainly on identifying and protecting the privacy of sensitive data entries in contingency tables, or tables of counts corresponding to cross-classification of the data.

A number of disclosure limitation techniques have been proposed [19], which can be broadly classified into *query restriction* and *data perturbation*. The query restriction family includes controlling the size of query results [21], restricting the overlap between the answers of successive queries [16], suppressing the cells of small size [13], and auditing queries to check privacy compromises [11]; The data perturbation family includes sampling data [15], swapping data entries between different cells [14], and adding noises to the data [53] or the query results [15]. Nevertheless, as shown in [1], these proposed techniques can not meet the requirement of providing high precision statistics, and meanwhile preventing exact or partial disclosure of personal information. Moreover, the perturbation-based techniques generally compromise the data integrity of the table.

Motivated by the need of publishing non-aggregated *microdata* involving personal sensitive information, e.g., medical data [49, 51], the data privacy researcher have mainly focused on ensuring that no adversary can accurately infer the sensitive information of an individual in the data, based on the published data and her background knowledge.

Typically, the microdata is stored in a relational data model, and each record in the microdata corresponds to an individual, which can be divided into three sub-categories: (1) *identifier* attribute, e.g., social security number, which can explicitly identify an individual, and therefore is usually removed from the microdata for publication; (2) *quasi-identifier* (QI) attributes, e.g., zip-code, gender and birth-date, whose values in combination can potentially identify an individual, and are usually available from other sources (e.g., voter registration list); (3) *sensitive* (SA) attribute, e.g., disease, which is the private information to be protected for the individuals.

Within this setting, a majority of the research efforts focus on addressing the *linking attacks*: the adversary possesses the exact QI-attribute values of the victim individual, and attempts to infer his/her sensitive value from the published data. Aiming at providing protection against linking attacks while preserving the information truthfulness, a set of *group-based anonymization* techniques have been proposed, which guarantees that each individual is hidden within certain group, called *QI-group* with respect to their QI-attributes values.

Two main methodologies have been proposed for achieving such group-based anonymization: *suppression* and *generalization* [51]. Specifically, for a given QI-attribute value, the suppression operation involves not releasing the value at all, while the generalization operation replaces it with a less specific, more general value that is faithful to the original. Clearly, suppression can be considered as a special form of generalization, where the QI-attribute value is replaced by a non-informative wildcard symbol '*'.

9.2.1 Models and Principles

Essentially, in linking attacks, the adversary infers the sensitive value of a victim individual, by leveraging the association between the QI attribute values of the victim, and the corresponding sensitive value (QI-SA association), as appearing in the microdata. Group-based anonymization weakens such associations by reducing the granularity of the representation of the QI-attributes values, and the protection is adequate if the weakened associations are not informative enough for the adversary to infer individuals' sensitive values with high confidence. Aiming at providing sufficient protection, various anonymization principles and models have been proposed, which can be classified, according to the type of QI-SA associations that they are designed to address, and the background knowledge that the adversary is assumed to have.

9.2.1.1 QI-SA Association

The associations between the QI attribute values and sensitive values can be categorized as *exact association* and *proximate association*. The former refers to the link between QI-attributes values and *specific* sensitive values, while the latter refers to the link between QI-attributes values and *a set of proximate* sensitive values.

Exact QI-SA Association. The exact QI-SA association is particularly meaningful for publishing categorical sensitive attribute, where different values have no sense of proximity, and it is desired to prevent the adversary from linking an individual to a specific sensitive value in the published data with high confidence. A number of principles have been proposed, with the objective of avoiding exact QI-SA association re-construction, including k-anonymity [49], l-diversity [41] and its variant (α, k)-anonymity [59].

- k-anonymity [49]. As a pioneering work on privacy-preserving data publication, Sweeney introduced the k-anonymity model to preserve information truthfulness in microdata, which paved the way for the development of group-based anonymization methodology. Intuitively, k-anonymity requires that each QI-group contains at least k tuples, therefore each individual is indistinguishable from a group of more than (k-1) others, with respect to their QI-attribute values. Clearly, k-anonymity is effective in preventing identification of an individual record.
- l-diversity [41]. It is recognized that k-anonymity ignores the characteristics of the sensitive attribute, which therefore does not prevent the disclosure of sensitive attribute values. For example, after identifying an individual belongs to a QI-group G, and 3 out of 4 tuples in G share an identical sensitive value, without further information, the adversary can infer that this sensitive value belongs to the victim individual with probability 75%. To address such "homogeneity attacks", Machanavajjhala et al. introduced the principle of l-diversity, which demands that each QI-group contains at least l "well represented" sensitive values, such that no single sensitive value is dominant in the group. A number of instantiations of l-diversity were proposed in [41], e.g., entropy l-diversity, recursive (c, l)-diversity.
- (α, k)-anonymity [59]. As a variant of l-diversity, (α, k)-anonymity essentially combines k-anonymity and l-diversity, which requires that (1) each

QI-group has size at least k, and (2) at most α percent of the tuples in each QI-group share identical sensitive values.

Proximate QI-SA Association. The principles above aim at preventing the reconstruction of exact sensitive values from the anonymized data, which is fairly reasonable for publishing categorical sensitive data. However, when publishing quantitative sensitive data is taken into consideration, the proximate QI-SA association arises as another important privacy concern, which refers to the link between QI-attributes values and a set of proximate sensitive values. Specifically, by studying the published data, the adversary may conclude with high confidence that the sensitive value of a victim individual falls in a short interval, even though with low confidence about the exact value.

To remedy this problem, several anonymization principles have been proposed recently, including (k, e)-anonymity [65], t-closeness [39], variance control [36] and (ε, m)-anonymity [38].

- (k, e)-anonymity [65]. Intuitively, (k, e)-anonymity dictates that every QI-group should be of size at least k, and in each group, the difference between the maximum and minimum sensitive values should be at least e. Essentially, (k, e)-anonymity counters the adversary from narrowing down the sensitive value of an individual to a small interval, by specifying constraints on the extreme values of each QI-group. However, it does not prevent the case that in a QI-group, a majority of tuples have nearly identical sensitive values, and the remaining few carry faraway values [38]. In such cases, the adversary can still link the victim with the set of proximate sensitive values with high confidence.
- t-closeness [39]. Li and Li separate the information gain an adversary can get from the released data into two parts: that about all the population in the data, and that about specific individuals. To limit the second kind of information gain, they proposed t-closeness, which is a further enhancement of the concept of l-diversity. Intuitively, it demands that the distribution of the sensitive values in every QI-group should not deviate from the distribution in the overall table more than a threshold t. In the paper, the EMD (earth mover distance) metric is used to measure the distance of probability distributions, which however fails to capture the probability scaling characteristics of the distribution itself.
- Variance control [37]. A natural measurement of the "concentration" of a set of numeric values is their variance. The variance control model imposes a threshold t on the variance of sensitive values in each QI-group. Nevertheless, it is proved in [38] that large variance does not ensure sufficient protection against attacks based on proximate QI-SA associations.
- (ε, m)-anonymity [38] In a recent work, Li et al. proposed the principle (ε, m)-anonymity, designed to address the proximate QI-SA-association based attacks, in publishing numeric sensitive data. Intuitively, (ε, m)-

anonymity requires that for a given QI-group, each sensitive value is similar to at most $1/m$ of all the sensitive values of the group, and the similarity is controlled by ε. One shortcoming of this principle is that since m is an integer, it only allows adjusting the level of privacy protection in a harmonic sequence manner, i.e., 1/2, 1/3, etc.

9.2.1.2 Background Knowledge

The aforementioned principles and models all assume that the adversary possesses full identification information [42], which includes (i) the identifier of the individuals in the microdata table, and (ii) their exact QI-attribute values. For scenarios with alternative background knowledge assumptions, a number of anonymization principles have been proposed, including δ-presence [46], (c, k)-safety [42], privacy skyline [8], m-invariance [62], and sequential anonymization [55].

Less External Knowledge. An implicit assumption made in the anonymization principles discussed so far is that the adversary already knows that the victim individual is definitely in the microdata. In the scenario where she has no prior knowledge regarding the presence of the individuals in the microdata, the privacy protection can be achieved from another perspective.

- δ-presence [46]. Assuming that the adversary has no prior knowledge regarding whether the victim appears in the microdata, δ-presence achieves privacy protection by preventing the adversary from inferring the presence of the individual in the data with probability no more than δ. Clearly, in this setting, if the adversary is only δ (percent) sure that the victim is in the microdata, any specific sensitive value belongs to the victim with probability no more than δ.

More External Knowledge. The problem of privacy-preserving data publishing is further complicated by the fact that in addition to the published data, the adversary may have access to other external knowledge, e.g., public records and social network relating individuals. To enforce the privacy protection against such external knowledge-armed adversary, several stricter privacy criteria have been proposed.

- (c, k)-safety [42]. Martin et al. considered the case that the adversary possesses the *implicational knowledge*, as modeled as a set of rules that if the individual o_1 has sensitive value s_1 then the individual $o_2$$ has sensitive value s_2. (c, k)-safety guarantees that even if the adversary possesses k pieces of such knowledge, the probability that she can infer the exact sensitive value of an individual is no more than c.

- Privacy skyline [8]. Besides the implicational knowledge, Chen et al. considered a set of other types of background knowledge, including that about the target individual, that about others (implicational knowledge), and that about the family of individuals sharing identical sensitive value. They proposed a multidimensional approach to quantifying an adversary's background knowledge. This model enables the publishing organization to investigate various types of privacy threats, and tune the amount of privacy protection against each type of adversarial knowledge.

Multiple Releases. The centralized-publication models discussed so far focus on the scenario of "one-time" release, while there are scenarios where the microdata is subjected to insertion and deletion operations, and need to be released multiple times. It is clear that in such cases, the adversary may potentially leverage previous releases of microdata to infer sensitive information in the current release.

- *m*-invariance [62]. To remedy this problem, Xiao and Tao proposed the principle *m*-invariance for sequential releases of microdata, which prevents the adversary from using multiple releases to infer the sensitive information of individuals. Intuitively, *m*-invariance can be considered as a stringent version of *l*-diversity, which dictates that at each release, each QI-group contains at least *m* tuples, and all of them have different sensitive values. An efficient algorithm was proposed in [62] for fulfilling *m*-invariance, via inserting counterfeit tuples.
- Sequential anonymization [55]. Wang and Fung considered a different scenario of re-publication from *m*-invariance. They assume a static microdata table that contains a large number of QI-attributes. In the first release, a subset of QI-attributes, together with the sensitive attribute are published. Later, the publisher is requested to publish an additional subset of QI-attributes. They intend to prevent the adversary from inferring the sensitive information by joining multiple publications. In [55], a technique based on lossy joins is proposed to counter such table-joining attacks. The intuition behind the approach is: if the join is lossy enough so that the adversary has low confidence in relating previous releases with the current one, then she is effectively prevented from discovering the identities of the records.

9.2.2 Techniques

While the first category of literatures on data privacy aims at proposing anonymization principles that guarantee adequate protection against the privacy attacks in question, the second category explores the possibility of fulfilling the proposed privacy principles, while preserving the utility of the resulting data to the maximum extent. They either analyze the hardness of implementing the anonymization principles, or propose efficient algorithms for computing the anonymized table under a given principle, and meanwhile minimizing the information loss. Following, we summarize the relevant literatures from these according to these two categories.

9.2.2.1 Negative Results

The techniques of fulfilling a generalization principle are usually limited by two inherent difficulties: the *hardness of optimal generalization* with minimum information loss and the *curse of dimensionality* for high-dimensional microdata.

Specifically, while there are numerous ways of generalizing the microdata table to satisfy the given privacy principle, one is usually interested in the optimal one which incurs the minimum amount of information loss, in order to preserve as much data utility as possible. However, finding such optimal generalization in general can be typically modeled as a search problem over a high-dimensional space, which is inherently difficult.

Meanwhile, when one has to deal with high-dimensional microdata, the curse of dimensionality becomes an important concern for privacy-preserving data publication. Intuitively, the combination of a large number of attribute values is so sparsely populated, that even to achieve 2-anonymity, one needs to suppress a large number of attribute values, resulting in almost useless anonymous data. Following, we present a brief survey of the existing literatures from these two perspectives respectively.

Hardness of Optimal Generalization. The first algorithm for computing generalization under the *k*-anonymity principle was proposed in [50]. It employs the domain generalization hierarchies of the QI-attributes to construct *k*-anonymous table. To minimize the information loss, the concept of *k*-minimal generalization was proposed, which requests for the minimum level of generalization, in order to maintain as much data utility as possible, for given level of anonymity. However, it is proved in [3], with the generalization height on the hierarchies as the metric of information loss, achieving the optimal *k*-anonymity with minimum information loss is NP-hard.

Meyerson et al. [43] studied the hardness of the optimal *k*-anonymity problem under the model of suppression, where the only allowed operation is to replace a QI-attribute value with a wildcard symbol '*'. As expected, it is theoretically proved that finding the optimal *k*-anonymous table with minimum number of suppressed cells is also NP-hard.

LeFevre et al. [36] proposed a multidimensional model for QI-attributes, and modeled the problem of constructing *k*-anonymous table as finding a partition of the multidimensional space. They used the *discernability measure* of attribute values [6] as the metric of information loss, and proved that finding the optimal k-anonymous partition with minimum incurred information loss is NP-hard.

Curse of Dimensionality. In [2], Aggarwal analyzed the behavior of the suppression approach for implementing k-anonymity, in the presence of increasing dimensionality. It was observed that for high-dimensional microdata, in order to meet the anonymity requirement, a large number of attribute values need to be suppressed. The situation is even worse when the adversary has access to considerable background information, which is usually the case in practice.

As a result, in anonymizing high-dimensional microdata, the boundary between QI-attributes and sensitive attribute is blurred. On one hand, it becomes extremely hard to fulfill the anonymization principle; On the other hand, a large number of attribute values are generalized to wide ranges, resulting in significant loss of data utility. So far there is no formal analysis of the curse of dimensionality regarding generalization principles other than k-anonymity. However as suggested in [41], it tends to become increasingly infeasible to implement l-diversity as the dimension of the microdata grows.

9.2.2.2 Positive Results

Despite the hardness results of finding the optimal generalized relation, it is shown that with certain constraints on the resulting generalization [35], it is usually feasible to enumerate all the possible generalizations, and find the optimal one based on a set of heuristics [6, 35]. Meanwhile, efficient greedy-manner solutions naturally are also good candidates for solving NP-hard generalization problems. To this end, extensive research has been conducted on devising generalization algorithms based on these heuristic principles [36, 56, 23].

Meanwhile, though simple to implement and efficient to operate, such heuristic methods inherently can not guarantee the quality of the resulting generalization. Recognizing this drawback of heuristic methods, another line of research has been dedicated to developing approximation algorithms [43, 3, 47], in which the quality of the solution found is guaranteed to be within a certain constant factor of that by the optimal one. Following, we summarize the existing literatures on heuristic methods and approximation methods respectively.

Heuristic Methods. Intuitively, the first set of generalization algorithms base themselves on the following principle: enumerating all possible generalizations, and using heuristic rules to effectively prune non-promising solutions along the traversal process.

In [6], Bayardo and Agrawal proposed a k-anonymization algorithm based on the technique of set enumeration tree. Specifically, they impose a total order over all the QI-attribute domains, and the values in each domain are also ordered, which are preserved by the total order. Under this model, a generalization can be unambiguously represented as a union of the generalization sets for each QI-attribute, and finding the optimal k-anonymization involves searching through the powerset of all domains of QI-attributes for the generalization with the lowest cost. While it is impossible to construct the entire set enumeration tree, they apply a systematic set-enumeration-search strategy with dynamic tree rearrangement and cost based pruning. In particular, a tree node can be pruned when it is determined that no descendent of it could be optimal.

In [35], Lefevre et al. proposed the Incognito framework for computing optimal k-anonymization, based on bottom-up aggregation along domain generalization hierarchies. Specifically, they apply a bottom-up breadth-first search strategy over the domain hierarchies: At the ith iteration, it computes the i-dimensional generalization candidates from the (i-1)-dimensional generalizations, and removes all those generalizations which violate k-anonymity, i.e., in a similar spirit of Apriori algorithm [4]. Such search continues until no further candidates can be constructed, or all the domains of QI-attributes have been exhausted.

Aiming at finding suboptimal solutions, the greedy-manner solutions usually outperform the enumeration-based methods in terms of time and space complexity. Motivated by their superior computation efficiency, a set of generalization algorithms based on the greedy heuristic have been proposed [36, 56, 23].

In [36], a multidimensional generalization model was proposed for fulfilling k-anonymity, which could also be extended to support other generalization principles. Specifically, it constructs a generalization by partitioning the multidimensional space spanned by the QI-attributes, using a kd-tree structure. At each iteration, a partitioning dimension is selected with the minimum impact with respect to the quality metric in use, and its median value is chosen to partition the space. This process continues until no further partition which obeys k-anonymity can be constructed.

In [56, 23], two complementary anonymization frameworks have been proposed, based on bottom-up generalization and top-down specialization over the domain generalization hierarchies respectively. In [56] a bottom-up heuristic is applied: starting from an initial k-anonymity state, it greedily hill-climbs on improving an information-privacy metric which intuitively measures the information loss for a given level of k-anonymity. The approach is scalable in the sense that it examines at most one generalization at each iteration for each QI-attribute. However, like all other local search methods, it may get stuck at some local optimum, if the initial configuration is not proper. Fung et al. [23] presented a complementary top-down specialization strategy, which starts from a general solution, and specializes certain QI-attributes, so as to minimize the information loss, without violating k-anonymity.

As mentioned above, the problem of finding high-quality generalization can be modeled as a search problem over a high-dimensional space, therefore a set of general heuristic tools, e.g., genetic algorithms and simulated annealing can be readily applied. In [31], Iyengar defines a utility metric in terms of classification and regression modeling, and applies genetic algorithm to optimizing the utility metric, for the given level of k-anonymity. Unfortunately, at the cost of large computational complexity, this solution offers no guarantees on the solution quality.

Approximation Methods. Along another line of research efforts, researchers have been seeking the techniques which provide guarantees on the solution quality, as measured by its deviation from that of the optimal one. A number of approximation generalization algorithms have been proposed [43, 3, 47], which guarantees the quality of the found solution to be within a constant factor or that found by the optimal one.

In [43], Meyerson and Williams present a polynomial time algorithm for optimal k-anonymity, which achieves an approximation ratio of O(k log k), independent of the size of the dataset. However, the runtime is exponential in terms of k. They further remove this constraint by proposing an O(k log m)-approximation, where m is the degree of the relation.

Aggarwal et al. [3] improved the result of [43] by showing an algorithm with an approximation ratio of O(k). They also provided improved positive results for specific values of k, including a 1.5-approximation algorithm for k = 2, and a 2-approximation algorithm for k = 3.

In a recent work [47], Park and Shim present an algorithm which achieves an approximation ratio of O(log k). Furthermore, by explicitly modeling the trade-off between solution quality and execution efficiency, they proposed an O(β log k)-approximate algorithm, which allows the users to trade the approximation ratio for the running time of the algorithm by adjusting the parameter β.

9.2.2.3 Utility Optimization

Generally, the protection for the sensitive personal information is achieved by performing certain transformation over the microdata, which necessarily leads to the loss of the data utility for the purpose of data analysis and mining. Because of this inherent conflicts between privacy protection and data utility, it is imperative to take account of information loss in the privacy preservation process, and optimize the data utility under the required privacy protection.

This is especially important when dealing with high-dimensional microdata, as shown in [2], for high-dimensional microdata, a large number of attribute values need to be generalized, even to achieve a modest level of anonymity, due to the cure of dimensionality.

To this end, extensive research has been conducted on optimizing the utility of the anonymized data, without violating the hard privacy requirement. The existing literatures can be summarized as three subcategories: The first one attempts to devise general-purpose metrics of data quality, and incorporates them in the anonymization process [6, 41, 31, 56]; The second one focuses on alleviating the impact of the anonymization operation over the data utility [33, 61, 65]; The third one targets specific applications, and tailors the anonymization to preserve maximum data utility with respect to the particular applications [56, 37].

Data Utility Metrics. Various measurements of the information loss caused by the anonymization operation have been proposed in the literatures:

- Generalization height [6]. It is the level of the generalized QI-attribute value on the domain generalization hierarchy. The problem with this notion is that not all generalization levels are of equal importance, and a generalization step on one attribute may include more tuples into an anonymous group than a generalization step on another one [41].
- Average size of anonymous groups [41] and discernability measure of attribute values [6]. Both metrics take account of the QI-group size. Specifically, discernability assigns a cost to each tuple based on how many tuples are indistinguishable from it, and depending on whether the tuple is suppressed, the cost is either the size of the microdata table, or the size of the QI-group. However, neither of the two metrics take consideration of the underlying distribution of the data, which may reveal important information.
- Classification metric [31] and information-gain-privacy-loss-ratio [56]. Both metrics take account of the distribution of the underlying microdata. The classification metric is designed specifically for the purpose of training classifier over the data, therefore may not be appropriate for measuring general-purpose information loss. The information-gain-privacy-loss-ratio is a local heuristic used to determine the next generalization step, similar in spirit to the information gain metric for deciding the splitting point in a decision tree. It is also not clear how to apply it to measure general-purpose information loss.

Utility-Based Anonymization. This category of work focuses on improving the anonymization operation itself, in order to alleviate its impact over the quality of the resulting data, without compromising the privacy protection.

Instead of anonymizing the microdata table as a whole, Kifer and Gehrke [33] advocated publishing the marginals, each of which anonymizes the projection of the microdata table on a subset of QI-attributes and sensitive attributes, in order to ameliorate the effect of the curse of dimensionality. It is shown that this approach can preserve considerable utility of the microdata, without violating hard privacy requirements.

Xiao and Tao [61] proposed a simple alternative anonymization framework, called *anatomy*, which essentially publishes QI-attributes and sensitive attributes in two separate tables. This way, the QI-attributes values need not to be generalized, since the separation already provides the same amount of protection as generalization, in the case that the adversary already knows the presence of the individuals in the microdata table. It is shown that this approach brings considerable improvement over the data quality in answering aggregation queries. A similar idea was also proposed by Zhang et al. [65].

Motivated by the fact that different attributes tend to have different utility with respect to the applications in question, a utility-based anonymization method using local-recording has been proposed in [63], which takes into consideration of the different utility weights of the QI-attributes, in performing the generalization operation.

Application-Specific Utility. Another direction of preserving data utility in the privacy protection process is to minimize the information loss in an application specific or workload-specific manner. In such cases, the utility metrics are defined depending on the underlying application or workload.

In [56], using information-gain-privacy-loss-ratio as a local heuristic, designed specifically for the classification task, a bottom-up generalization algorithm has been proposed. In [65], the anonymization algorithm is optimized for the purpose of answering aggregation queries.

An interesting workload-aware generalization algorithm has been proposed by Lefevre et al. [37]. Specifically, it differentiates the subsets of microdata according to their frequency of being requested by the users, and performs less generalization over those frequently requested subsets than others, thus achieving considerable information saving, while providing the same amount of privacy protection.

9.3 Location Privacy

Along with the great convenience of the location-based services, follow their potential threats to the users' privacy and security: the location information disclosed by the mobile users could be abused in malicious ways. Location privacy protection concerns about preventing the exposure of individuals' private location information (associated with identities) to unauthorized party.

In general, such exposures can be loosely categorized as three classes: the first one happens through direct communication, i.e., the communication channel is eavesdropped by the adversary; the second one occurs through direct observation, e.g., the LBS service provider is not trusted, therefore having direct observation over the received location and identity information; the third one takes place through indirect inference of location information combined with other properties of individual, e.g., by tracking the moving pattern of a victim individual to infer his/her identity.

As a special form of data privacy, location privacy possesses its unique characteristics and research challenges:

- First, location privacy requirements are inherently personalized, and context-specific. Different users tend to have various location privacy requirements, e.g., some users regard their positions as extremely private information, while others may care much more about the quality of the service delivered. Moreover, users tend to have different requirements with respect to the context, e.g., users may have stricter privacy requirement during night time, to reduce the risk of being stalked.
- Second, location privacy is intricately related to the service quality. The service provider processes a request based on its understanding regarding the customer' position, and more precise location information leads to higher quality of the service delivered, while ambiguous or fake location information may result in the degradation of the service quality. Therefore, there exists an implicit trade-off between location privacy and location utility.
- Third, location data is extremely dynamic, and subjected to frequent updates. Unlike that in protecting ordinary data privacy, where the microdata is typically stored in databases, in preserving location privacy, the location data is usually processed in an on-line stream manner, in order to meet the strict requirement of response time. Meanwhile, customers' location data is usually subjected to frequent updates, which opens the door for the adversary to combine multiple "snapshots" to infer individuals' current location.

In general, based the assumptions regarding the trustfulness of the LBS service providers, customers' location privacy requirements can be fulfilled using two methods: Under the model of trusted service providers where the service providers faithfully act according to its agreement with the customers, users' location privacy can be preserved using a policy-based approach: the customers specify their privacy requirements in the form of consents with the service providers. However, with the explosive growth the LBS services, it becomes extremely difficult to fulfill this model in practice.

Under the model of untrusted service providers, which reflects the current main trend, it is imperative to provide technical countermeasure against potential privacy violation by the service providers. One simple solution is that instead of using their true identities, the mobile users provide pseudonyms to request for services. However, this solution is generally insufficient for two main reasons: (i) A set of applications require to verify customers' true identifies in order to provide the corresponding services, e.g., credit-card related services. (ii) A user's identity can be potentially inferred from his/her location information. Several types of inferences are possible, as shown in [28, 7]: in precise location tracking, successive position updates can be linked together to form moving patterns, which can be used to reveal the users' identity; in observation identification, external observation is available to link a position update to an identity; in restricted space identification, a known location owned by identity relationship can link an update to an identity.

Hence, a set of location hiding techniques have been proposed to address users' location privacy concerns, including reducing the granularity of the representation of users' location information (spatial/temporal cloaking, location blurring), reporting the nearest landmark, sending false dummy locations, location obfuscation, etc. Essentially, location hiding provides protection for the location privacy of a mobile user by guaranteeing that no adversary can pinpoint this particular user to a location with high precision and confidence, through inference attacks. This system property is termed location anonymization. The rest of the section is devoted to a brief survey of the current state-of-the-art location anonymization techniques.

9.3.1 Models and Principles

Analogous to preserving data privacy, in order to provide adequate protection for mobile users' location privacy, various location anonymization principles have been proposed, including location k-anonymity [28], location l-diversity [5], and minimum spatial resolution [44]. Meanwhile, in order to guarantee the quality of the service delivered, several QoS metrics have been proposed, including maximum tolerable spatial and temporal resolutions [24, 25]. In the first part of the section, we describe in detail the location privacy profile, which captures mobile users' privacy and QoS requirements.

The existing location anonymization techniques can be roughly classified into three main categories based on their system architectures, concretely centralized trusted third party model, client-based non-cooperative model, and decentralized cooperative mobility group model. Each system architecture model is associated with its unique assumptions about the privacy threat model, and supports a different set of location anonymization mechanisms. In the second part of the section, we discuss the strengths and weaknesses of these architecture models respectively.

9.3.1.1 Location Privacy Profile Model

In location privacy profiles, mobile users specify their privacy protection and quality of service requirements. As have been mentioned earlier, location privacy protection is inherently personalized, and context-specific, which implies that the location privacy profiles need to be specified on a per-user, per message basis, and the system should support users to change their privacy profiles in an on-line manner.

Location Privacy Metrics. Borrowing from ordinary data privacy the idea of k-anonymity [51], the principle of *location k-anonymity* has been introduced in [28] to preserve the location privacy of mobile users through the use of location k-anonymization. Intuitively, location k-anonymity ensures that at a given time instance, for each LBS service request, there are at least (k-1) other messages with the same location information, each associated with a different (pseudo) identity. It guarantees that without further knowledge, the adversary can not differentiate at least k participants with respect to their location information. However, the definition of location k-anonymity in [28] is limited to a system-supplied uniform k for all users in a given LBS system. Gedik and Liu [24, 25] revised the initial definition by introducing the concept of personalized location k-anonymity, allowing variable k for different users and for different service requests of the same user. Most of the subsequent research on location privacy has adopted the concept of personalized location k-anonymity [44, 45, 5].

In [41], k-anonymity is proved insufficient for protecting data privacy in the sense that it ignores the characteristics of the sensitive attribute, and opens the door for the inference of sensitive attribute values. Interestingly, an analogous situation also happens for location privacy protection.

The first weakness of location k-anonymity is that it ignores the granularity of the reported location, i.e., under location k-anonymity, a location of fairly small area is considered as providing enough protection, as long as it contains more than k active mobile users, even though the adversary can pinpoint the users with high precision. The concept of *minimum spatial resolution* [44] has been introduced to address this problem. It allows each user to specify the minimum spatial area of his/her released location.

The second weakness of location k-anonymity is its ignorance of the number of symbolic locations or static objects associated with the reported location. Intuitively, if a location k-anonymous spatial region is associated with only one static object, e.g., church or doctor's office, then with this region as the reported location of a user, the adversary may associate the user with the specific location object with high probability. This qualifies as severe privacy threat, considering a simple example: If *Bob*'s reported location is an area containing a specific clinic as the only static object, an adversary may infer that "*Bob must be visiting the clinic with high probability*". The concept of location l-diversity [5, 40] has been introduced into the framework of location anonymization to address this weakness. Intuitively, it demands that for each LBS request, in addition to user-level k-anonymity, the released location should also be associated with at least l different symbolic objects. Mobile users can specify desired k (k-anonymity) and l (l-diversity) values in their location privacy profile [40].

Note that in order to fulfill location l-diversity, the component responsible for location anonymization is expected to support efficient access to the databases of public location objects, which makes it difficult to fulfill this principle under the client-based anonymization architecture, as will be discussed in Section 9.3.1.2.

Quality of Service Metrics. Meanwhile, in addition to the location privacy requirements, the users may also desire to specify their demands for the quality of service delivered to them. Two commonly used QoS metrics are *maximum tolerable spatial* and *temporal resolutions* [24, 25].

Specifically, maximum tolerable spatial resolution represents the threshold on the maximum area of the anonymized location. As have been mentioned above, depending largely on the location information provided by the client, the quality of a LBS service tends to degrade as the level of anonymization increases. Consider the example of querying the nearest gas station: if the client reports his/her location as a wide area, a large number of candidate results will be returned, which not only incurs heavy computation overhead of filtering false positive information, but also wastes the precious wireless bandwidth. The maximum tolerable resolution bounds the level of anonymization, therefore indirectly guaranteeing the quality of the delivered service.

Another important QoS metric is the response time of the service request. As will be discussed in Section 9.3.2, under the centralized location anonymization architecture, caching the service requests temporarily can improve the throughput of the location anonymization component, at the cost of delaying the processing of the requests. Therefore, mobile users are inclined to specify threshold for the allowed delay time as the maximum tolerable temporal resolution.

9.3.1.2 Location Anonymization: Alternative System Architectures

Three alternative system architectures have been proposed for implementing location anonymization [40]: centralized architecture with trusted third-party, non-corporative client-based architecture, and decentralized peer-to-peer corporative architecture. In this section, we briefly describe each of these system architecture models, including their privacy threat models, as well as their advantages and drawbacks. We defer the discussion of the concrete techniques developed under each architecture model to the next section.

Centralized Architecture. In a centralized location anonymization architecture, a trusted third party, called *location anonymizer component*, acts as a proxy (middleware) for all communications between mobiles users and LBS service providers.

The overall communication between the clients, the location anonymizer, and the service provider can be loosely divided into four phases: (1) The location anonymizer receives the service request, plus the position information from the mobile client, and performs spatial and temporal anonymization over the location information, based on the user-specified privacy profile. Meanwhile, according to the performed transformation, it generates the filtering condition that will be used later to prune the false positive information from the candidate results as returned by the service provider, to produce the exact result to the original request; (2) The location anonymizer then relays the request associated with anonymized location information to the corresponding service provider. Upon receiving the anonymized request, the service provider invokes the anonymous request processing module, which produces the set of candidate results; (3) After receiving the candidate results from the service provider, the location anonymizer performs the filtering operation to remove the false positive results, by applying the corresponding filtering condition; (4) Finally, the exact results to the original request are delivered to the mobile user who issues it.

Note that under this architecture, the location anonymizer is responsible for both anonymizing the LBS service requests, and filtering false positive information from the candidate results, which could potentially become the bottleneck of the system. An alternative design suggested in [5] is to perform the filtering at the client side: for each service request received at the location anonymizer, the anonymized request will be relayed to the service provider, and the filtering condition is returned to the mobile user who issues the request. The service provider then directly passes the candidate results to be filtered at the client side. This design however introduces additional communication and processing overhead for the mobile clients.

The privacy threat model commonly addressed by this centralized location anonymization architecture can be specified as follows: (i) The true identity of the mobile user is hidden from the service provider, as specified by the security policy, and the LBS application is assumed to accept pseudo-identity; (ii) The LBS service provider is considered as hostile observer, therefore setting constraints on what information can or cannot be revealed from one service provider to anther is usually insufficient; (iii) The location sensing infrastructure (e.g., GPS, WiFi, Cricket) and the location anonymizer are trusted by the mobile users.

Since the location anonymizer gathers the location information from all the active users, and is usually able to provide efficient access to the database of public location objects, the strength of this architecture is its being able to support all the location anonymization requirements as mentioned in Section 9.3.1.1 (strong privacy protection). Meanwhile, since all the communications pass through the location anonymizer component, which could potentially become the bottleneck of the whole architecture.

Non-Cooperative Architecture. In a non-cooperative architecture, the mobile users maintain their location privacy based on their knowledge only, without involving any centralized trusted authority.

In this client-based location anonymization model, the communications between the mobile client and the LBS service providers follow the model as: (1) the client obfuscates its location information, which is sent, in conjunction with the service request to the service provider. Meanwhile, the client generates the filtering condition according to the location perturbation performed; (2) Upon receiving the anonymized request, the service provider invokes the anonymous request processing module, which produces the set of candidate results, to be delivered to the mobile client; (3) After receiving the candidate results from the service provider, the client obtains the exact answer to the original request by filtering out the false positive information from the candidate results.

The privacy threat model commonly addressed in the client-based non-cooperative location anonymization model is as follows: (i) The true identities of the users are hidden from the service provider, and the LBS application is considered to accept pseudonyms; (ii) No third authority (e.g., LBS service provider, location anonymizer) are trusted, who are interested in intruding mobile users' location privacy; (iii) The location sensing infrastructure (e.g., GPS, WiFi, Cricket) and the location anonymizer resided on mobile clients are trusted.

Under this model, all the communications are inherently distributed, leading to its strength of high throughput and fault-tolerance, and no centralized trusted authority is needed. Meanwhile, since the location perturbation is performed at the mobile clients, without knowing the location information of other clients, and it is usually not affordable for the clients to directly access the databases of public location objects (one of the main LBS applications), this model does not support location k-anonymity or location l-anonymity, i.e., weak privacy protection.

Peer-to-Peer Cooperative Architecture In a peer-to-peer location anonymization framework, a group of mobile users collaborate with each other to provide location privacy protection for each single user, without the interleaving of a centralized trusted entity. Specifically, within this architecture, the communications between the mobile clients and the service providers follow the model as: (1) the request-issuing mobile client communicate with neighboring clients to collect the location information of peers, and performs informed location perturbation to meet its privacy requirement. It (or a peer) then sends the service request, plus the anonymized location information to the LBS service provider, meanwhile generating the filtering condition; (2) upon receiving the anonymized request, the service provider invokes the anonymous request processing module, and produces the set of candidate results, to be delivered to the mobile client who sends the request; (3) The results are routed to the client who issues the request (if the issuer and the sender are different), which then obtains the exact results by filtering the false positive information from the candidate results.

Commonly, the peer-to-peer cooperative location anonymization architecture is applied to address the privacy threat model as follows: (i) the true identity of the mobile user is hidden from the service provider, and the LBS service provider is considered to accept pseudonyms; (ii) all third parties (e.g., LBS service providers, location anonymizer) are considered as hostile observers, who are interested in intruding mobile users' privacy; (iii) the location sensing infrastructure, the location anonymizer resided on mobile clients, and the peers in the mobility group are trusted.

This architecture requires no centralized trusted authority, therefore can scale to a large number of mobile clients. However, it is shown in [27] that it is naturally computationally expensive to support location anonymity under this architecture, since each client has to communicate with each other to construct anonymous location. Also, to support this model, a mobile peer-to-peer communication infrastructure is needed. Clearly, this architecture offers support for location *k*-anonymity; nevertheless, it is inherently difficult to support location *l*-diversity since it is usually not feasible for mobile clients to directly access the databases of public location objects. In conclusion, this architecture provides protection stronger than the client-based non-cooperative architecture, but weaker than the centralized architecture.

9.3.2 Location Anonymization Techniques

We have given an overview of location privacy research, with focus on privacy models and system architectures. In this section, we present a brief survey of the representative techniques developed for aforementioned system architectures, and analyze their advantages and drawbacks.

9.3.2.1 Centralized Architecture

As a pioneering work on location k-anonymization, Grusteser and Grunwald [28] introduced a quadtree-based spatial cloaking method, under the centralized location anonymization model. Intuitively, for a given set of service requests, it recursively divide entire geographical space of interest into quadrants until a quadrant has less than k users, and then returns the previous quadrant, i.e., the minimum quadrant that meets k-anonymity, as the anonymous location for the mobile users within it. An example is shown in Fig.9.1, where the dashed box is the minimum quadrant that contains more than k = 4 users.

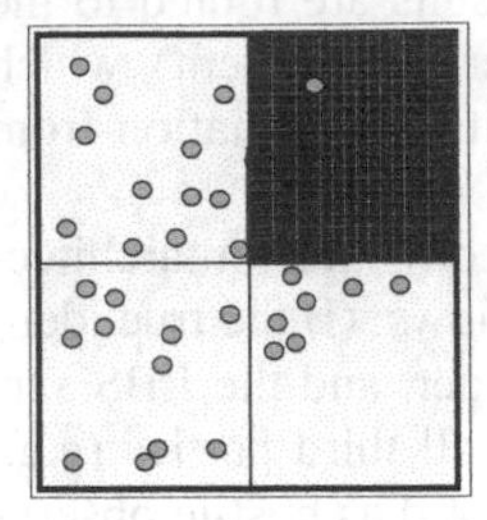

Fig. 9.1 Quadtree based spatial cloaking

Using a quadtree-based spatial partitioning technique, this method provides efficient support for *universal location k-anonymity*, where a system-supplied uniform k is used and thus the same level of k-anonymity is applied to all the users. However, as a straight-forward use of the spatial partitioning techniques, it is inherently difficult to support personalized location privacy in this framework.

As mentioned earlier, motivated by the need to support personalized location privacy requirements, Gedik and Liu [24, 25] introduced the CliqueCloak framework. It allows each message to specify a different k value based on its specific privacy requirement, and maximum spatial and temporal tolerance values based on its QoS requirements. The service requests are processed in an on-line stream manner: a constraint graph is constructed according to the maximum spatial and temporal tolerance settings of the requests, and a set of requests are perturbed if they form a clique in the constraint graph. An example is shown in Fig.9.2, where three messages m_1, m_2 and m_4 form a clique on the constraint graph, and are anonymized together. However, identifying cliques in a graph is an expensive operation, which severely limits the scalability of this framework. It has been empirically shown that this approach can only support fairly small k.

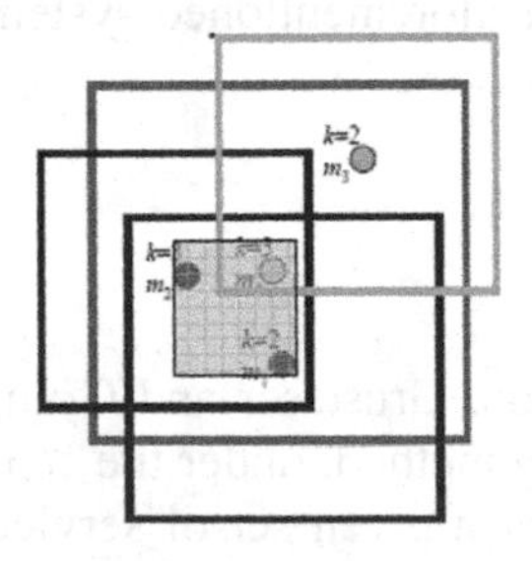

Fig. 9.2 Constraint graph based spatial cloaking.

Both frameworks above focus solely on devising efficient and effective solutions for the location anonymizer component. In [44, 45], Mokbel et al. proposed Casper, a complete location anonymization framework. It employs a hierarchical indexing structure: the universe of disclosure is represented as a complete pyramid structure, which is divided into grids at different resolution levels and each cell maintains the number of mobile users inside. It supports personalized location *k*-anonymization: on anonymizing a request, it traverses the pyramid structure bottom-up until a cell satisfying the user's privacy profile is found. They further improved this basic paradigm by introducing adaptive searching strategy, at the cost of maintaining the adaptive pyramid structure. A unique contribution of the Casper framework is its support for anonymous query processing, which extends existing spatial query processing and optimization techniques to support *k*-anonymous nearest neighbor and range queries.

It is observed in [5] that the CliqueCloak method models requests in a constraint graph and achieves anonymization by finding cliques, resulting in its low scalability. Meanwhile, the Casper method adopts a hierarchical pyramid structure, and supports anonymization by using quadrant spatial expansion. It is therefore inherently not optimal in the sense that the cloaked location is usually much larger than the minimum area satisfying *k*-anonymity. The PrivacyGrid framework developed at the DiSL group of Georgia Tech [5] addresses the drawbacks of these two frameworks, by employing a dynamic grid-based spatial index, powered with fast bottom-up, top-down and hybrid cloaking algorithms. In addition to supporting personalized location privacy and QoS requirements as defined in [24], PrivacyGrid enhances the location anonymization semantics by introducing location *l*-diversity. By adopting a flat grid index structure, PrivacyGrid achieves better efficiency than the CliqueCloak approach in terms of supporting large *k*, meanwhile higher success rate of request anonymization, and superior quality of anonymized location than the Casper approach.

9.3.2.2 Non-Cooperative Architecture

Under the non-cooperative client-based architecture, the location privacy of mobile users is usually achieved by injecting uncertainty to their location information, called location obfuscation, based solely on their own knowledge.

In [29], Hong and Landay proposed to use the nearest public landmark, instead of his/her current exact position, as a mobile user's locations in constructing a LBS request. Specifically, based on a set of public landmarks, a Voronoi diagram is constructed over the universe of disclosure, which is employed by the mobile users to identify their closest landmarks. Clearly, within this framework, the evaluation of the request is also based on the selected landmark; hence the distance between the selected landmark and the user's exact location controls the trade-off between privacy protection and service quality.

Kido et al. [32] proposed to use a set of false dummies to protect user's exact location: the reported location information consists of m locations, among which only one is the true position while other (m-1) false dummies. Only the user who issues the request knows the true location, and the service provider replies with a service for each received location. It is clear that within this model, the privacy protection is achieved at the cost of the query evaluation overhead at the server, and the communication bandwidth, since the server essentially has to evaluate m queries in order to answer one original request.

In [18], Duckham and Kulik proposed a location obfuscation approach based on a graph abstraction of the universe of disclosure. Specifically, all locations are represented as vertices in a graph with edges corresponding to the distance between two locations. A user obfuscates her location as a set of vertices, and the query is then evaluated at the server side based on the distance to each vertex in the imprecise location. Clearly, it suffers from the same problem as the false dummies approach.

In a recent work [64], Yiu et al. proposed the SpaceTwist framework to support kNN queries based on the incremental NN query processing techniques. Specifically, the mobile user reports a location different from (but near to) his/her actual position, and the nearest neighbor objects of the reported position are incrementally retrieved until the query is correctly answered by the service provider. Though simple to implement, the SpaceTwist framework incurs additional overhead on the communication bandwidth, for the protocol involves multiple exchanges of packets to ensure that all the actual results have been retrieved. Moreover, this approach offers no explicit modeling of the quality of privacy protection, which makes it difficult for users to quantitatively specify privacy requirements.

Recently, transformation-based matching techniques have been proposed to enable location privacy, which however do not offer query accuracy guarantees. In [30], a theoretical study on a client-server protocol for deriving the nearest neighbor of a query is reported. In [34], Khoshgozaran et al. defined a specific Hilbert ordering based on a key, whose value is known only by the client and a trusted entity. It is shown that without the key value, it is impossible to decode a Hilbert value into a location correctly. However, a Hilbert curve does not completely preserve spatial proximity, so the reported result can be far from the actual result. To improve the accuracy, they proposed to use two keys with orthogonal Hilbert curve.

9.3.2.3 Peer-to-Peer Cooperative Architecture

In [26, 27], Ghinita et al. proposed PRIVE, a representative framework of the peer-to-peer cooperative location anonymization model. The main idea of PRIVE is that whenever a mobile user desires to issue a LBS request, it broadcasts a group formation request to its neighbors, and a member of the group is randomly selected to act as the query sender. PRIVE provides two modes of group formation: on-demand mode and proactive mode. In on-demand mode, a mobile user invokes the group formation only when necessary; In proactive mode, mobile users periodically execute the on-demand approach to maintain their anonymous groups. Clearly, the two modes represent the two ends of the spectrum of the trade-off between maintenance cost and response time of anonymous group formation.

Specifically, in PRIVE, an anonymous location is obtained in three phases: (1) Peer searching. The user who intends for a LBS service broadcasts a multi-hop request until at least (k-1) peers are found; (2) Location adjustment. The initial anonymized location is adjusted according to the velocities of group members; (3) Spatial cloaking. The anonymized location is cloaked into a region aligned to a grid covering the (k-1) nearest peers.

9.3.3 Open Issues and Challenges

While a plethora of work has been done on preserving the location privacy of mobile users in LBSs, the techniques developed to date are far from being adequate in providing comprehensive solutions to the location privacy problem in general. In this section, we discuss open issues and possible research directions towards constructing a complete end-to-end solution for location privacy protection.

Enhancement of Location Anonymization Techniques. As a special form of data privacy, location privacy exhibits its unique characteristics and challenges. Thus it is inadequate to directly apply the extensive collection of data privacy techniques proposed to date for location privacy protection. To the best of our knowledge, the successful extensions of data privacy preserving techniques to location privacy are still limited to *k*-anonymity [49, 51], *l*-diversity [41] and distance-preserving space transformation techniques [9, 10]. One possible direction is to further extend the available data privacy techniques for enhancing existing location anonymization tools.

For example, the *m*-invariance [62] principle has been proposed for sequential releases of microdata, which can be possibly adapted to protecting location privacy for continuous location-based queries. For such continuous location updates/queries, an alternative solution can possibly be the extension of the output perturbation techniques [57], originally developed for countering inferences over the mining results of consecutive stream windows.

The aforementioned three location anonymization architectures possess their own strengths and weaknesses. Though the centralized architecture provides the strongest privacy protection, and extensive research efforts so far have been focused on developing anonymization techniques under this architecture, with the explosive growth of the scale of mobile clients and LBS services, the client-based and decentralized architectures are anticipated to become the main trends. Therefore it is imperative to put forward the research effort on remedying their weaknesses by introducing stronger privacy guarantees.

Potential Attacks and Countermeasures. A variety of inference attack models have been studied in the area of ordinary data privacy, which show that by leveraging certain external knowledge and her understanding regarding the anonymization techniques, an adversary can potentially penetrate the protection mechanism and infer the sensitive information of individuals. For example, in [62, 55], it is shown that in the scenario of multiple releases of microdata, the adversary may potentially exploit previous releases to infer sensitive information in the current one. In [60], it is shown that knowledge of the mechanism or algorithm of anonymization can also lead to extra information that assists the adversary and jeopardizes individual privacy.

Most existing location anonymization tools have not taken full account of the possible attacks to the anonymized solutions, resulting in their weak resilience to inference attacks.

For example, analogous to the sequential releases of microdata, continuous updates/queries in LBSs can be exploited to infer sensitive location or identity information of individuals. In [17], a mobility model based attack is considered: The adversary utilizes the knowledge about mobile users' motion parameters such as maximum velocity, known trajectory, frequent travel path to perform inference attacks. For instance, if the adversary knows the maximum velocity of the user, and obtains its consecutive cloaked location updates, by computing the maximum movement boundary of the same identity (pseudonym), she can potentially locate the user at the intersection of two cloaked spatial regions. In [12], an overlapping window based inference attack model is considered. Intuitively, if the adversary knows some locations of the targeted victim, even different pseudonyms are used in different updates/queries, by analyzing the overlapping spatial or temporal windows of two consecutive cloaked location updates/queries, she can infer the linkage of location with the targeted victim.

As another example, analogous to the minimality attack in publishing microdata [60], it is shown in [58] that knowing the principles applied in the location anonymization algorithms, e.g., optimality of service quality for given privacy requirement, and the underlying background of users' movement, e.g., road network, the adversary can pinpoint a victim individual with high precision.

Therefore, a promising future direction is to study such potential attack models, and to devise effective countermeasures to inject attack resilience into current location anonymization tools.

General Framework for Anonymous Query Processing The anonymous location query processing modules in literatures, e.g, [44], are designed to support only limited types of spatial queries, e.g., range queries, or *k*NN queries, and are constructed over the top of existing spatial query processing techniques. As mobile users' location privacy becomes a paramount concern in numerous LBS services, general location query processing modules designed specially for anonymous spatial queries are expected to bring significant impact over the industries. One future research direction is to extend the existing spatial query processing techniques, e.g.,[52, 48], to support more types of queries, e.g., continuous *k*NN queries.

9.4 Summary

We have presented an overview of the advances in the area of data privacy research. We discussed a variety of privacy principles and models, and analyzed their implicit relationships. We also summarized the representative algorithms and methods for implementing the proposed principles, and discussed their fundamental limitations. In a perspective analogous to data privacy, we examined the state of the art of the location privacy research, and compared the strengths and weaknesses of the proposed models, architectures, and algorithms. We concluded the chapter with a discussion of open issues and promising research directions towards providing comprehensive end-to-end location privacy solutions.

Acknowledgments This work is partially supported by grants from NSF CyberTrust program, an AFOSR grant, and an IBM SUR grant.

References

[1]. N. Adam, and J. Wortman. Security-control methods for statistical databases. *ACM Computing Surveys*, 21(4), 1989.
[2]. C. Aggarwal. On *k*-anonymity and the curse of dimensionality. In *VLDB*, 2005.
[3]. G. Aggarwal, T. Feder, K. Kenthapadi, R. Motwani, R. Panigraphy, D. Thomas, and A. Zhu. Anonymizing tables. In *ICDT*, 2005.
[4]. R. Agrawal, and R. Srikant. Fast algorithms for mining association rules. In *VLDB*, 1994.

[5]. B. Bamba, L. Liu, P. Pesti, and T. Wang. Supporting anonymous location queries in mobile environments with PrivacyGrid. In *WWW*, 2008.
[6]. R. Bayardo, and R. Agrawal. Data privacy through optimal *k*-anonymization. In *ICDE*, 2005.
[7]. A. Beresford. Location privacy in ubiquitous computing. PhD thesis, University of Cambridge, 2005.
[8]. B. Chen, R. Ramakrishnan, and K. LeFevre. Privacy skyline: privacy with multidimensional adversial knowledge". In *VLDB*, 2007.
[9]. K. Chen, and L. Liu. A random rotation perturbation approach to privacy preserving data classification. In *ICDM*, 2005.
[10]. K. Chen, and L.Liu. Towards attack-resilient geometric data perturbation. In *SDM*, 2007.
[11]. F. Chin, and G. Ozsoyoglu. Auditing and inference control in statistical databases. *IEEE Trans. Softw. Eng.*, SE-8(6), 1982.
[12]. C. Chow, and M. Mokbel. Enabling private continuous queries for revealed user locations. In *SSTD*, 2007.
[13]. L. Cox. Suppression methodology and statistical disclosure control. *J. Am. Stat. Assoc.*, 75(370), 1980.
[14]. T. Dalenius, and S. Reisss. Data swapping: a technique for disclosure control. *J. Stat. Plan. Infer.*, 6, 1982.
[15]. D. Denning. Secure statistical databases with random sample queries. *ACM TODS*, 5(3), 1980.
[16]. D. Dobkin, A. Jones, and R. Lipton. Secure databases: Protection against user influence". *ACM TODS*, 4(1), 1979.
[17]. J. Du, J. Xu, X. Tang, and H. Hu. iPDA: enabling privacy-preserving location-based services". In *MDM*, 2007.
[18]. M. Duckham, and L. Kulik. A formal model of obfuscation and negotiation for location privacy. In *Pervasive*, 2005.
[19]. G. Duncan, S. Fienberg, R. Krishnan, R. Padman, and S. Roehrig. Disclosure limitation methods and information loss for tabular data. *Confidentiality, Disclosure, and Data Access: Theory and Practical Applications for Statistical Agencies*, pp 135-166, Elsevier, 2001.
[20]. C. Farkas, and S. Jajodia. The inference problem: a survey. *SIGKDD Explor. Newsl.*, 4(2), 2002.
[21]. I. Fellegi. On the question of statistical confidentiality. *J. Am. Stat. Assoc.*, 67(337), 1972.
[22]. Foxs News. Man accused of stalking ex-grilfriend with gps. http://www.foxnews.com /story/0293313148700.html.
[23]. B. Fung, K. Wang, and P. Yu. Top-down specialization for information and privacy preservation. In *ICDE*, 2005.
[24]. B. Gedik, and L. Liu. Location privacy in mobile systems: a personalized anonymization model". In *ICDCS*, 2005.
[25]. B. Gedik, and L. Liu. Protecting location privacy with personalized *k*-anonymity architecture and algorithms. *IEEE Transactions on Mobile Computing.*
[26]. G. Ghinita, P. Kalnis, and S. Skiadopoulos. MOBIHIDE: a mobile peer-to-peer system for anonymous location-based queries. In *SSTD*, 2007.
[27]. G. Ghinita, P. Kalnis, and S. Skiadopoulos. PRIVE: anonymous location based queries in distributed mobile systems. In *WWW*, 2007.
[28]. M. Gruteser, and D. Grunwald. Anonymous usage of location-based services through spatial and temporal cloaking. In *MobiSys*, 2003.
[29]. J. Hong, and J. Landay. An architecture for privacy-sensitive ubiquitous computing. In *MobiSys*, 2004.
[30]. P. Indyk, and D. Woodruff, Polylogarithmic private approximations and efficient matching. In *TCC*, 2006.
[31]. V. Iyengar. Transforming data to satisfy privacy constraints. In *KDD*, 2002.

[32]. H. Kido, Y. Yanagisawa, and T. Satoh. An anonymous communication technique using dummies for location-based Services. In *ICPS*, 2005.
[33]. D. Kifer, and J. Gehrke. Injecting utility into anonymization databases. In *SIGMOD*, 2006.
[34]. A. Khoshgozaran, and C. Shahabi. Blind evaluation of nearest neighbor queries using space transformation to preserve location privacy. In *SSTD*, 2007.
[35]. K. LeFevre, D. Dewitt, and R. Ramakrishnan. Incognito: efficient full-domain k-anonymity. In *SIGMOD*, 2005.
[36]. K. LeFevre, D. DeWitt, and R. Ramakrishnan. Mondrian multidimensional k-anonymity. In *ICDE*, 2006.
[37]. K. LeFevre, D. DeWitt, and R. Ramakrishnan. Workload aware anonymization. In *SIGKDD*, 2006.
[38]. J. Li, Y. Tao, and X. Xiao. Preservation of proximity privacy in publishing numerical sensitive data. In *SIGMOD*, 2008.
[39]. N. Li, T. Li, and S. Venkatasubramanian. t-closeness: privacy beyond k-anonymity and l-diversity. In *ICDE*, 2007.
[40]. L. Liu. From data privacy to location privacy. In *VLDB*, 2007.
[41]. A. Machanavajjhala, J. Gehrke, D. Kifer, and M. Venkitasubramaniam. l-diversity: privacy beyond k-anonymity. In *ICDE*, 2006.
[42]. D. Martin, D. Kifer, A. Machanavajjhala, J. Gehrke, and J. Halpern. Worst-case background knowledge in privacy. In *ICDE*, 2007.
[43]. A. Meyerson, and R. Williams. On the complexity of optimal k-anonymity. In *PODS*, 2004.
[44]. M. Mokbel, C. Chow, and W. Aref. The new casper: query processing for location services without compromising privacy. In *VLDB*, 2006.
[45]. M. Mokbel. Privacy in location-based services: state of art and research directions. In *MDM,* 2007.
[46]. M. Nergiz, M. Atzori, and C. Clifton. Hiding the presence of individuals from shared databases. In *SIGMOD*, 2007.
[47]. H. Park, and K. Shim. Approximate algorithm for k-anonymity. In *SIGMOD*, 2007.
[48]. S. Saltenis, C. Jensen, S. Leutenegger, and M. Lopez. Indexing the positions of continuously moving objects. In *SIGMOD*, 2000.
[49]. P. Samarati, and L. Sweeney. Protecting privacy when disclosing information: k-anonymity and its enforcement through generalization and suppression. Technical Report SRI-CSL-98-04, SRI Computer Science Laboratory, 1998.
[50]. P. Samarati. Protecting respondents' identities in microdata release. *IEEE Trans. Knowl. Data Eng.*, 13(6), 2001.
[51]. L. Sweeney. K-anonymity: a model for protecting privacy. *Int. J. Uncertain. Fuzz.*, 10(5), 2002.
[52]. Y. Tao, D. Papadias, and Q. Shen. Continuous nearest neighbor search. In *VLDB*, 2002.
[53]. J. Traub, Y. Yemini, and H. Woznaikowski. The statistical security of a statistical database.. *ACM TODS*, 9(4), 1984.
[54]. [54] USA Today. Authorities: Gps systems used to stalk woman. http://www.usatoday.com /tech/news/2002-12-30-gps-stalker_x.htm.
[55]. K. Wang, and B. Fung. Anonymizing sequential releases. In *KDD*, 2006.
[56]. K. Wang, P. Yu, and S. Chakraborty. Bottom-up generalization: a data mining solution to privacy protection". In *ICDM*, 2004.
[57]. T. Wang, and L. Liu. Butterfly: protecting output privacy in stream mining. In *ICDE*, 2008.
[58]. T. Wang, and L. Liu. Location privacy protection for road network basedmobile computing system. CS Technical Report, Georgia Tech, 2008.
[59]. R. Wong, J. Li, A. Fu, and K. Wang. (alpha, k)-anonymity: an enhanced k-anonymity model for privacy preserving data publishing. In *SIGKDD*, 2006.
[60]. R. Wong, A. Fu, K. Wang, and J. Pei. Minimality attack in privacy preserving data publishing. In *VLDB*, 2007.

[61]. X. Xiao, and Y. Tao. Anatomy: simple and effective privacy preservation. In *VLDB*, 2006.
[62]. X. Xiao, and Y. Tao. *m*-invariance: towards privacy preserving re-publication of dynamic datasets. In *SIGMOD*, 2007.
[63]. J. Xu, W. Wang, J. Pei, X. Wang, B. Shi, and A. Fu. Utility based anonymization using local recording. In *KDD*, 2006.
[64]. M. Yiu, C. Jensen, X. Huang, and H. Lu. SpaceTwist: managing the trade-offs among location privacy, query performance, and query accuracy in mobile services. In *ICDE*, 2008.
[65]. Q. Zhang, N. Koudas, D. Srivastava, and T. Yu. Aggregate query answering on anonymized tables. In *ICDE*, 2007.

10 Privacy Preserving Nearest Neighbor Search

Mark Shaneck, Yongdae Kim, Vipin Kumar[1]

Abstract Data mining is frequently obstructed by privacy concerns. In many cases data is distributed, and bringing the data together in one place for analysis is not possible due to privacy laws (e.g. HIPAA) or policies. Privacy preserving data mining techniques have been developed to address this issue by providing mechanisms to mine the data while giving certain privacy guarantees. In this chapter we address the issue of privacy preserving nearest neighbor search, which forms the kernel of many data mining applications. To this end, we present a novel algorithm based on secure multiparty computation primitives to compute the nearest neighbors of records in horizontally distributed data. We show how this algorithm can be used in three important data mining algorithms, namely LOF outlier detection, SNN clustering, and kNN classification. We prove the security of these algorithms under the semi-honest adversarial model, and describe methods that can be used to optimize their performance. Keywords: Privacy Preserving Data Mining, Nearest Neighbor Search, Outlier Detection, Clustering, Classification, Secure Multiparty Computation

10.1 Introduction

Privacy advocates and data miners are frequently at odds with each other. In many cases data is distributed, and bringing the data together in one place for analysis is not possible due to privacy laws (e.g. HIPAA [18]) or policies. Privacy preserving data mining techniques have been developed to address this issue by providing mechanisms to mine the data while giving certain privacy guarantees. Research in this field typically falls into one of two categories: data transformation to mask the private data, and secure multiparty computation to enable the parties to compute the data mining result without disclosing their respective inputs.

[1] Mark Shaneck, Liberty University, Lynchburg, VA e-mail: mshaneck@liberty.edu
Yongdae Kim, University of Minnesota, Minneapolis, MN e-mail: kyd@cs.umn.edu
Vipin Kumar, University of Minnesota, Minneapolis, MN e-mail: kumar@cs.umn.edu

J.J.P. Tsai and P.S. Yu (eds.), *Machine Learning in Cyber Trust: Security, Privacy, and Reliability*, DOI: 10.1007/978-0-387-88735-7_10,

In this chapter we address the problem of privacy preserving *Nearest Neighbor Search* using the cryptographic approach. The problem of finding nearest neighbors is as follows: for a given data point x, and a parameter k, find the k points which are closest to the point x. In a distributed setting, with data that is horizontally partitioned (i.e. each party has a collection of data for the same set of attributes, but for different entities), the main difference is that the points in the neighborhood may no longer be entirely located in the local data set, but instead may be distributed across multiple data sets.

Many data mining algorithms use nearest neighbor search as a major computational component [47]. In this chapter, we show how to incorporate our search algorithm into three major data mining algorithms, namely *Local Outlier Factor* (LOF) outlier detection [6, 5], *Shared Nearest Neighbor* (SNN) clustering [12, 13, 26], and *k Nearest Neighbor* (kNN) classification [9]. These are an important set of data mining algorithms. For example, kNN classification is highly useful in medical research where the best diagnosis of a patient is likely the most common diagnosis of patients with the most similar symptoms [36]. SNN clustering provides good results in the presence of noise, works well for high-dimensional data, and can handle clusters of varying sizes, shapes, and densities [13, 47]. LOF outlier detection provides a quantitative measure of the degree to which a point is an outlier and also provides high quality results in the presence of regions of differing densities [5, 47].

To the best of our knowledge, this is the first work that directly deals with the issue of privacy preserving nearest neighbor search in a general way. Privacy preserving approaches for kNN classification [7, 29] also require finding nearest neighbors. However in these works, the *query point* is assumed to be publicly known, which prevents them from being applied to algorithms such as SNN clustering and LOF outlier detection. Previous work on privacy preserving outlier detection [51] required finding the number of neighbors closer than a threshold. Although this is related to the finding of nearest neighbors, it also cannot be directly adopted to compute SNN clusters or LOF outliers as it is limited in the amount of information it can compute about the points. Since our approach directly deals with the problem of finding nearest neighbors, it can be used by any data mining algorithm that requires the computation of nearest neighbors, and thus is more broadly applicable than the previous related works. For more detail on how our work differs from these previous works, we refer the reader to Section 10.6.

This chapter makes the following contributions: we design a novel cryptographic algorithm based on secure multiparty computation techniques to compute the nearest neighbors of points in horizontally distributed data sets, a problem which, to the best of our knowledge, has never been dealt with previously. This algorithm is composed of two main parts. The first computes a superset of the nearest neighborhood (which is done using techniques similar to [51]). The second part reduces this set to the exact nearest neighbor set. We show how to extend this algorithm to work in the case of more than two parties, and we prove the security of this algorithm. In addition, we show how this search algorithm can be used to compute the LOF outlier scores of all points in the data set, to find SNN clusters in the data, and to perform kNN classification with the given data sets taken as the training data, thus showing how the nearest neighbor search can be practically applied. We show how all of these can be done while giving guarantees on the privacy of the data. We also analyze the complexity of the algorithms and describe measures that can be taken to increase the performance. We stress that while this work makes use of existing primitives, these are extended and combined in novel ways to enable the computation of the k nearest neighbors, as well as three major data mining algorithms that rely on nearest neighbor search.

The rest of this chapter is organized as follows. Section 10.2 includes a formal description of the problem addressed, an overview of the secure multiparty primitives used in this work, as well as a brief discussion on provable security in the secure multiparty computation setting. The secure protocol for nearest neighbor search is described in Section 10.3. Section 10.4 explains the application of this protocol for higher level data mining algorithms. Next we discuss the complexity of the scheme in Section 10.5, outline the areas of related work in Section 10.6, and we conclude and outline some areas of future work in Section 10.7.

10.2 Overview

10.2.1 Problem Description

The objective of this work is to find the k nearest neighbors of points in horizontally partitioned data. The basic problem of k-nearest neighbor search is as follows. Given a set of data points S, and a particular point $x \in S$, find the set of points $N_k(x) \subseteq S$ of size k, such that for every point $n \in N_k(x)$ and for every point $y \in S$, $y \notin N_x(x) \Rightarrow d(x,n) \le d(x,y)$, where $d(x,y)$ represents the distance between the points x and y. In a distributed setting, the problem is essentially the same, but with the points located among a set of data sets, i.e. for m horizontally distributed data sets $S_i (1 \le i \le m)$, and a particular point $x \in S_j$ (for some j, $1 \le j \le m$), the k-nearest neighborhood of x is the set $N_k(x) \subseteq S = U_{i=1}^{m} S_i$ of size k, such that for every $n \in N_k(x)$ and $y \in S$, $y \notin N_k(x) \Rightarrow d(x,n) \leq d(x,y)$. If a distributed nearest neighbor search algorithm is privacy preserving, then it must compute the nearest neighbors without revealing any information about the other parties' inputs (aside from what can be computed with the respective input and output of the computation). In our case, we compute some extra information in addition to the actual nearest neighbor sets. While this is not the ideal solution, we argue that this work provides an important stepping stone for further work in this area. We also provide a description of what this information reveals in Section 10.3.1.1. For the sake of simplicity, everything is described in the setting of two parties. For each algorithm that we describe, there is a description of what needs to be done to extend it to the more general multiparty case.

10.2.2 Definitions

Throughout the discussion in this work, the data is assumed to be horizontally partitioned, that is, each data set is a collection of records for the same set of attributes, but for different entities. This would be the case, for example, in medical data, where the same information is gathered for different patients, or for network data, where the same set of attributes are gathered for different IP addresses in different networks. We describe the algorithms in this work from the perspective of one of the participants, and use the term "local" to indicate data contained within the data set of the party from whose perspective we are discussing, and the term "remote" to indicate data that is not "local". Also, all arithmetic is done using modular arithmetic in a sufficiently large field F (e.g. mod p for the field $\mathbb{Z}_p$). We refer to this throughout the discussion as "mod F". Note that in order to preserve distances, this element should be larger than the largest pairwise distance of the points in the set.

In addition to the normal notion of nearest neighbor sets, we introduce the notion of an *Extended Nearest Neighbor Set*, in the context of distributed nearest neighbor search. This term is used to represent a logical extension of the nearest neighbor set, and can be described as follows. Imagine a point x in the local data set. We can find the k nearest neighbors from the local data set easily. Select one such local neighbor n. Now we can define a Partial Nearest Neighbor Set to include the point n and all local and remote points that are closer to x than n. Thus the Extended Nearest Neighbor Set would be the smallest Partial Nearest Neighbor Set that has a size greater than or equal to k. This notion will be used in our algorithm for privately finding the k nearest neighbors.

10.2.3 Secure Multiparty Computation Primitives

In this section, we give a brief overview of the secure multiparty computation primitives used in our algorithm, along with their requirements, and some examples of existing implementations.

10.2.3.1 Distance

An important part of computing the nearest neighbors of a point is to find the distance between the points securely. The requirements of the protocol are that the answer should be shared between the two parties involved, such that each share is uniformly distributed over a sufficiently large field F when viewed independently of each other, and that the sum of the shares (mod F) equals the distance between the two objects. One such distance measure has been suggested [51], in which the square of the Euclidean distance is computed, by making use of a secure dot product computation [19]. Note that this formula computes the square of the distance, however this does not affect the results of the nearest neighbor algorithm, since the square of the distance preserves order and we are interested in the nearest neighbors. Also, any distance computation can be used that computes the distance securely according to the above requirements, including any similarity or non-Euclidean distance measures.

10.2.3.2 Comparison

Another cryptographic primitive needed in this algorithm is a protocol for secure comparison, i.e. if two parties have numbers a and b respectively, how can they determine which is larger without revealing the actual values to each other. This was originally formulated as Yao's Millionaire Problem [52] and can be accomplished by the general circuit evaluation protocol [20], or by other more efficient protocols [17, 23, 31]. At various points in our algorithms we make use of two variations of existing solutions for this problem. In the first variation the result is given to only one of the parties in question. In the second variation the result should be split between the two parties, such that the sum equals 1 (mod F) if the answer is true, and the sum is 0 if the answer is false (this requirement is the same as is needed in [51]). We also make use of a secure equality computation. It should be clear from the context which different versions are used.

10.2.3.3 Division

In Algorithm 4.1.1 in Section 10.4.1, we need to make use of the following primitive for secure division. In the two party case, one party has r_1 and r_3, and the other party has r_2 and r_4 and together they compute shares z_1 and z_2, such that

$$z_1 + z_2 = \frac{r_1 + r_2}{r_3 + r_4}$$

This problem has existing solutions for both the two party [11] and multiparty [4, 34] case.

10.2.4 Provable Security

Each of the algorithms presented below are privacy preserving, which is a notion that can be proven. For an algorithm to be privacy preserving, it is enough to prove that the algorithm is secure in the *Secure Multiparty Computation* sense. This notion of security is defined in [20], and we will use the notion of *semi-honest* behavior for our proofs. A semi-honest party is one that follows the protocol as specified, but may use the results and any intermediate knowledge to try to extract additional information about the other party's data. This is a realistic model, since in the target application scenarios all parties would have a mutual interest in the correct data mining output, while at the same time would desire guarantees that the other parties cannot learn extra information about their data. This also allows us to focus on more efficient computations, since protocols that are secure under malicious adversaries require the use of expensive bit commitments and zero knowledge proofs. However, it is interesting to note that all protocols that are secure in the semi-honest model can be converted into protocols that are secure in the malicious model [20].

In addition, we make use of the composition theorem [20], which states that for functions f and g, if g is privately reducible to f and there is a protocol for privately computing f, then there is a private protocol for computing g. In order to prove that an algorithm is secure, it is sufficient to show that given the final output and the respective initial input, a party is able to simulate a valid transcript for the protocol, and that the simulated transcript is computationally indistinguishable from a transcript of a normal computation. In other words, if a party can use its own input and the final output to simulate the messages that it views during the protocol, then it is proven that the party can not learn anything extra from these messages.

10.3 Nearest Neighbor Algorithm

In our scenario, two parties each have a set of data points that are gathered for the same set of attributes, that is, the data is horizontally partitioned. They want to compute the nearest neighbors of their points using the union of their data sets. For a local point x, the steps to find the k nearest neighbors are as follows.

1. First, the local nearest neighbor set for x is computed.
2. The second step is to securely compute the distance from x to each of the remote points. The individual shares from this calculation are stored for use in the following steps.
3. Next, starting with the closest local neighbor of x to the farthest, the distances to all remote points are compared with the distance to the current local neighbor. Using this information, party A can count the number of total neighbors that are closer than the local neighbor.
4. This is repeated until the number of neighbors exceeds the parameter k.
5. The set of identifiers of the remote points which are closer than this local neighbor are then discovered. This set forms the Extended Nearest Neighbor Set (see definition is Section 10.2.2).
6. Finally, the Find Furthest Points algorithm is used to remove the remote points within this set that should not be in the actual nearest neighbor set.

The following section describes the algorithm that performs these steps. The Find Furthest Points algorithm, used in the final step, is described in Section 10.3.2.

10.3.1 Nearest Neighbor Search

In this description we will assume that we are calculating the neighborhood of a point x which is owned by the party A, A has a set of points $X = \{x_1, \cdots, x_n\}$, and that the other party is B, who has a set of points $Y = \{y_1, \cdots, y_n\}$. What we intend to find is the set of points $nn\_set \subset X \cup Y$ of size k, such that all points in nn_set are closer than all points within $(X \cup Y) \setminus nn\_set$.

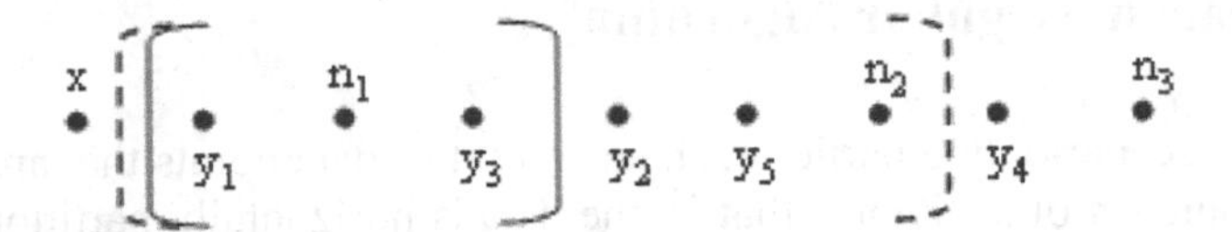

Fig. 10.1 Example nearest neighbor scenario

This is done by first computing the ordered list $local\_nn$, the k nearest neighbors in the local data set X. Next the distances to all remote points are computed and stored. Then, from the closest local neighbor to the farthest, the distance to the remote points are compared with the distance to the local neighbors, in order to count the number of remote and local points that are closer than this local neighbor. This is illustrated in Figure 10.1. In this figure the point x and all n_i are local points, and all y_i are remote points, and $k = 3$. Thus the distance from x to each point y_i would be computed and compared first to n_1. The number of points in this case would be 1. Thus, the next local neighbor n_2 would need to be checked. This would result in a count of 5. Since this is larger than k, then we have found that the Extended Nearest Neighbor Set consists of all points closer than the point n_2 (including the point n_2), which is depicted by the points between the dashed braces in Figure 10.1. Since we know the size of the Extended Nearest Neighbor Set, we know how many points need to be excluded to get a set of the correct size, that is k. Thus, we just need to find the correct number of the furthest points from x from within the Extended Nearest Neighbor Set and remove them. In our example, the number of remote points to remove is 2, and the set of furthest points would be the set $\{y_2, y_5\}$. If these are removed (along with the furthest local neighbor n_2), then the Nearest Neighbor Set will be found, which is depicted in Figure 10.1 by the points within the solid braces.

In order to make this algorithm privacy preserving, we need to make use of secure distance computations and secure comparisons. When the distance between x and a remote point y is computed, as shown in step 3 of Algorithm 1, the result will be shared between the two parties, such that the sum of the shares is equal to the distance between them (modulo F). When the distance from x to y is compared with the distance to a local neighbor in step 9 of Algorithm 1, the secure comparison algorithm must be used. The result of this comparison will be shared (as c_{iy}^A and c_{iy}^B) in the same manner as the distance shares, with the sum of the shares equal to 1 (mod F) if the point y is closer than the local neighbor, and 0 if it is not. Once all points have been compared in this way, the sum of all these shares ($c_i^A = \sum_y c_{iy}^A$ and $c_i^B = \sum_y c_{iy}^B$) become shares of the number of remote points which are closer than the local neighbor n_i. This number, plus the number of local neighbors i, can be securely compared to the value k, to see if this neighbor is the final neighbor in the Extended Nearest Neighbor Set by checking if $c_i^A + c_i^B + i \geq k$. Next the identifiers of the remote points in the Extended Nearest Neighbor Set are gathered, and the entire Extended Nearest Neighbor Set is passed to the Find Furthest Points algorithm (described in Section 10.3.2) with the appropriate number of remote points which need to be removed from the Extended Nearest Neighbor Set, which is the size of the Extended Nearest Neighbor Set minus 1 (for the final local neighbor) minus k. This is only necessary, of course, if the size is greater than $k+1$, since if it equals $k+1$ then the final local neighbor is the only point that needs to be removed. Also, if it equals k, then the Nearest Neighbor Set is already found. The rest of the details of the algorithm are shown in Algorithm 1. Note in the algorithm that when a comparison is made, it is understood that the secure comparison protocol is used in the underlying implementation.

Require: A local point x
Require: A local data set X and a remote data set Y
Require: A field F and a global parameter k
1: $local\_nn \leftarrow$ sorted k nearest neighbors locally (from X)
2: **for all** $y \in Y$ **do**
3: Compute (and store) distance(x, y) s.t. A gets d_{xy}^A and B gets d_{xy}^B where $d_{xy}^A + d_{xy}^B = \text{distance}(x, y) \bmod F$
4: **end for**
5: **for** $i = 1, \cdots, k$ **do**
6: $n_i \leftarrow local\_nn[i]$
7: $d_{xni} \leftarrow \text{distance}(x, n_i)$
8: **for all** $y \in Y$ **do**
9: Compute $d_{xy}^A + d_{xy}^B < d_{xni}$ where A gets c_{iy}^A and B gets c_{iy}^B such that $c_{iy}^A + c_{iy}^B = 1 \bmod F$ if $d_{xy}^A + d_{xy}^B < d_{xni}$, and 0 otherwise
10: **end for**

11: A computes $c_i^A = \sum_y c_{iy}^A$ and B computes $c_i^B = \sum_y c_{iy}^B$ where $c_i^A + c_i^B$ is the number of remote points closer than n_i
12: $g \leftarrow c_i^A + c_i^B + i \geq k$ where only A gets g
13: **if** g is TRUE **then**
14: $I \leftarrow i$
15: Break out of loop
16: **end if**
17: **end for**
18: $nn\_set \leftarrow \{n_1, \cdots, n_j\}$ for $j \leq I$
19: $d_{xn_I} \leftarrow \text{distance}(x, n_i)$
20: **for all** $y \in Y$ **do**
21: $g \leftarrow d_{xy}^A + d_{xy}^B < d_{xn}$ such that A gets g
22: **if** g is TRUE **then**
23: $nn\_set \leftarrow nn\_set \cup \{y\}$
24: **end if**
25: **end for**
26: **if** $|nn\_set| = k+1$ **then**
27: $nn\_set \leftarrow nn\_set \setminus \{n_1\}$
28: **else if** $|nn\_set| > k+1$
29: $\pi \leftarrow |nn\_set| - 1 - k$
30: $nn\_far \leftarrow$ FindFurthestPoints(x,π,nn_set)
31: $nn\_set = nn\_set \setminus nn\_far$
32: **end if**
33: Return nn_set

Algorithm 1: Nearest Neighbor Search

10.3.1.1 Security Analysis

In this algorithm, A learns the nearest neighbors and extended nearest neighbors of all its points, even if it includes points from the other party's set. Thus, for example, if A computes the Extended Nearest Neighborhood of its point x, and the resulting neighborhood has n_i as the furthest local neighbor, then A learns that the remote points in the set are closer than the point n_i. Also, A knows which points are not in the Nearest Neighbor Set, and thus knows that those points are further from the next closest local neighbor n_{i-1}. Thus, the points not in the Nearest Neighbor Set (but in the Extended Nearest Neighbor Set) lie in a particular hypershell constrained by the distances from x to n_{i-1} and n_i. This hypershell is most constrained when the distances from x to n_{i-1} and n_i are close together, which allows A to infer some information regarding the distribution of B's points (i.e. A could discover that B has many points at a distance between the distances to these two local neighbors). This ability to estimate the distribution of B's points decreases greatly in areas where A's points are sparse, and also greatly decreases as the number of dimensions increases. Thus the information that is leaked (from the specified output) does not give an unreasonable amount of information to the other party.

Given that the nearest neighborhood is the specified output, then this information leakage is expected. Note that the information in the Nearest Neighbor Set does not include any information about the points themselves, just an identifier (that is, an ID that can be used to reference a particular point in the set). However, since these identifiers can be used to infer some extra information about the distribution of the other party's data set (as described above), it would be better if this information could be securely shared. This would allow futher computation to be performed without releasing this intermediate information, and we intend to pursue this line of research in future work.

If the Extended Nearest Neighbor Set is included as part of the output for party A, along with the Nearest Neighbor Set, then it can be proven that this algorithm is privacy preserving.

Theorem 1. The Nearest Neighbor calculation (Algorithm 1) privately computes the Nearest Neighbor Set and the Extended Nearest Neighbor Set of a point x in the semi-honest model.

Proof. The private inputs of each party are their respective sets of points, with one point in one of the sets as the point whose neighborhood is being calculated, and the output is the Nearest Neighbor Set (with the Extended Nearest Neighbor Set). As can be seen from the algorithm, all communication takes place in steps 3, 9, 12, 21, and 30. In step 3 the distance is calculated, with each party receiving a random share of the distance, where each share is uniformly distributed over the field when viewed independently of each other. Thus these messages can be simulated with a value randomly selected from the field. In step 9, the distance from the point x to a remote point is compared to the distance between x and a local neighbor. This procedure is secure, and the results are again random shares, which can also be simulated as before. In step 12, the shared count is compared to the value k. Only A

gets the result of this comparison, and thus can be simulated, since A knows from the Extended Nearest Neighbor Set which n_i is the furthest local neighbor, and thus when to output true and stop the comparisons. In step 21, A simply gets the identifiers of the remote points which are in the Extended Nearest Neighbor Set, which is part of its output. Finally the Extended Nearest Neighbor Set is passed to the Find Furthest Points algorithm in step 30, with the results being the points which are not in the Nearest Neighbor Set, which A can compute since it knows both the Extended Nearest Neighbor Set and the Nearest Neighbor Set. Since the Find Furthest Points algorithm is secure, as will be shown in 3.2, then applying the composition theorem to this algorithm, the Find Furthest Points subroutine, the secure distance protocol, and the secure comparison protocol proves that the nearest neighbor computation is secure in the semi-honest model.

10.3.2 Find Furthest Points

In this portion of the algorithm, we start with a local point x, and a set of points Z which contains n remote points, and we need to find the set $Z_{far} \subset Z$ of size π such that all the points in Z_{far} are further than all the points which are in $Z_{close} = Z \setminus Z_{far}$. In other words we need to find the π furthest remote points from x, but without revealing any extra information that cannot be derived from the output, such as the ordering of the points. Note that initially Z would contain both remote and local points, but the local points can be easily filtered out by A to match the description above, and thus we can assume without loss of generality that all points in Z are remote.

This can be done in the following way. First we test each point, to see if it should be placed in the set Z_{close} or Z_{far}. Since we have already computed the distances from x to each point in Z, we can compare the distance to a given point in Z with the distances to all other points in the set. If it turns out that the point is closer than enough other points, then it can be added to Z_{close}, otherwise is should be added to Z_{far}. Since the size of the set Z is n, and we are trying to determine which π points are furthest, then a given point must be closer than at least π other points. All points which pass this test will be added to Z_{close}, and those which do not pass are added to Z_{far}.

In order to make this algorithm privacy preserving, the following must be done. In this case, the distances are shared between the two parties, such that the distance to z_i from x is shared as $d^A_{xz_i}$ and $d^B_{xz_i}$, where the sum of the two shares equals the distance (modulo F), and the distance between x and z_j is shared as $d^A_{xz_i}$ and $d^B_{xz_i}$. Thus to see if the distance to z_i is less than the distance to z_j, then we need to compute $d^A_{xz_i} + d^B_{xz_i} < d^A_{xz_j} + d^B_{xz_j}$. Note that, without loss of generality, this algorithm assumes that there are no two points of equal distance away from x. The case of equal distance can be handled by breaking ties by means of the point identifiers.

This distance comparison is done in such a way that the answer is randomly split between the two parties A and B as c^A_{ij} and c^B_{ij}, such that $c^A_{ij} + c^B_{ij} = 1 \bmod F$ if z_i is closer to x than z_j and 0 otherwise. Once all the points have been compared, each party can sum all their shares ($c^A_i = \sum_j c^A_{ij}$ and $c^B_i = \sum_j c^B_{ij}$), such that the sum of these two shares is equal (mod F) to the number of points which are farther from x than z_i. If this number is greater than π then it belongs in Z_{close}, which can be computed by comparing $\pi < c^A_i + c^B_i$. Otherwise the point is added to the set Z_{far}. Then this process is repeated for all $z_i \in Z$. Once this loop completes, Z_{far} contains the π points which are furthest from x, and Z_{close} contains the rest. The pseudocode for this algorithm is found in Algorithm 2.

Require: x basis (local) point
Require: π is number of farthest points to return
Require: $Z = \{z_0, \cdots, z_n\}$, (remote points) with $n \geq \pi$
Require: A and B share $d^A_{xz_i}$ and $d^B_{xz_i}$, for $1 \leq i \leq n$, such that
$d^A_{xz_i} + d^B_{xz_i} = \text{distance}(x, z_i)$
1: $Z_{close} \leftarrow \{\}$, $Z_{far} \leftarrow \{\}$
2: **for** $i \leftarrow 1$ to n **do**
3: **for** $j \leftarrow i$ to n, $i \neq j$ **do**
4: Compute $d^A_{xz_i} + d^B_{xz_i} < d^B_{xz_j} + d^B_{xz_j}$, where A gets c^A_{ij} and B gets c^B_{ij}, such that $c^A_{ij} + c^B_{ij} = 1 \bmod F$ if $d^A_{xz_i} + d^B_{xz_i} < d^B_{xz_j} + d^B_{xz_j}$ and 0 otherwise
5: **end for**
6: A computes $c^A_i = \Sigma_j c^A_{ij}$
7: B computes $c^B_i = \Sigma_j c^B_{ij}$

8: $g \leftarrow \pi < c_i^A + c_i^B$, where only A gets g
9: **if** g is TRUE **then**
10: add z_i to Z_{close}
11: **else**
12: add z_i to Z_{far}
13: **end if**
14: **end for**
15: Return Z_{far}

Algorithm 2: Find Furthest Points

10.3.2.1 Security Analysis

Theorem 2. The Find Furthest Points calculation (Algorithm 2) privately computes the π furthest points from x in the set Z in the semi-honest model.

Proof. In this algorithm, the private inputs are the local point x and the set of remote points Z, along with the distance shares from x to each point in Z, and π, the number of furthest points to find. The output is the set Z_{far} of π furthest points from the set Z. The only communication in this algorithm takes place in steps 4 and 8. In step 4, two distances are compared, using the secure comparison protocol. The answer is returned as two uniformly distributed shares in the field F, such that they sum to 1 if the answer is true, and 0 if the answer is false. Since these shares are uniformly distributed when viewed independently of each other, they can also be simulated with random values from the field F. The other communication takes place in step 8, which is another secure comparison. The results of this comparison can be simulated since Z_{far} is the output and the answer will be true if the point is in the set $Z_{close} = Z \setminus Z_{far}$ and false if it is in the final set Z_{far}. Thus applying the composition theorem to this algorithm and the two invocations of the secure comparison protocol, this algorithm is secure in the semi-honest model.

10.3.3 Extension to the Multiparty Case

So far, we have concentrated on the case where two parties are involved in the protocol. In many applications, however, this algorithm would be more useful if it was extended to the case when more than two parties are involved. This is straightforward for all computations which involve at most two parties (e.g. distance computations) and thus can be performed serially, except for the cases where comparisons involve all parties. One such comparison occurs in step 12 in Algorithm 1 where the counts are shared across all parties. The other case where multiparty comparison occurs is in the Find Furthest Points algorithm (Algorithm 2), where party A has the point x and all the other parties have sets of points. In this case, the comparisons in steps 4 and 8 can involve multiple parties, and the result of the comparison (in step 4) needs to be shared. However the general solution, which is used to compute the comparisons, is applicable for the multiparty case, and thus could be used for the multi-party comparisons.

10.4 Applications

In this section we discuss ways in which the privacy preserving nearest neighbor search can be used, by showing three data mining algorithms that utilize nearest neighbor search.

10.4.1 LOF Outlier Detection

LOF outlier detection [6, 5] is a method of outlier detection that relies on relative densities of data points. This approach works well in cases where there are clusters of points with differing densities, and provides a measure of the degree to which a point can be considered an outlier. Note that in our algorithm we compute the simplified version of LOF described in [47], which computes as the LOF score the ratio of the density of a point to the average density of its neighbors. The original LOF is a more complicated function that takes the distance to all k nearest neighbors as the distance to the k-th nearest neighbor (i.e. the reachability distance for each point in the k-distance neighborhood is simply the k-distance), and allows for variable sized nearest neighbor sets (when there are multiple points whose actual distance is equal to the k-distance). Extending our algorithm for simplified LOF to the original LOF calculation is relatively straightforward, but is omitted due to space constraints.

Before we discuss the privacy preserving algorithm for LOF outlier detection, we must first discuss what the requirements for a solution are, and how a distributed version of LOF outlier detection should work. To do this, let us first recall how the local LOF algorithm works. The LOF algorithm computes as the outlier score of a point x the ratio of the density of x times the parameter k, and the sum of the densities of the k nearest neighbors (i.e. the ratio of the density of x and the average densities of its nearest neighbors). In a distributed version of LOF, the computation of the outlier score, and thus also the computation of the densities of the points, must take into account the fact that the k nearest neighbors may not be contained within the local data set.

For an ideal secure computation (that is, one in which all inputs are sent to a trusted third party who performs the calculations locally, and returns each party's respective output), the only information that would be revealed would be the final outlier scores for each point. Also, since each party does not need to know the outlier scores of the other party's points, only the scores for the local points should be in the output for each party. Since each party can run the algorithm locally, and then compare the scores from the secure computation to the local computation, it is important to examine what information would be leaked from the secure computation. By comparing the two computed scores, a participant can infer some information about the locations of the other party's points. For example, if the other party has a dense cluster of points within the neighborhood of a local point, and the local neighborhood of that point has a relatively low density, then the outlier score could increase. If that dense cluster is very close to the local point, and it is denser than the local neighborhood, then the outlier score can decrease. However this leakage is unavoidable, since this information is leaked even in the case of the ideal computation.

10.4.1.1 Protocol

For the simplified version of LOF that were are using, the LOF score of a point x is

$$LOF_k(x) = \frac{density_k(x) \cdot k}{\sum_{n \in N_K(x)} density_K(n)} \quad (1)$$

where $density_k(x) = \sum_{n \in N_K(x)} \text{distance}(x, n)$. Since the distance between x and each of its neighbors is shared between the parties A and B, then shares of the density of the point x can be computed by simply summing the shares of the distances. In other words, A can compute its share of the density of the point x as $\delta_x^A = \sum_{n \in N_k(x)} d_{xn}^A$. Also, B can compute its share of the density of

x as $\delta_x^B = \sum_{n \in N_k(x)} d_{xn}^B$. The shares of the sum of the densities of the neighbors (the numerator in Equation 1), $\Delta_{N_x}^A$ and $\Delta_{N_x}^B$, can be computed in the same way. In order to compute the LOF score for the point x, we just need to compute

$$LOF_k(x) = \frac{density_k(x) \cdot k}{\sum_{n \in N_K(x)} density_K(n)} = \frac{\delta_x^A k + \delta_x^B k}{\Delta_{N_x}^A + \Delta_{N_x}^B}$$

This can be computed securely by means of the secure division primitive discussed in Section 10.2.3.3. The full details of this algorithm are found in Algorithm 3.

Require: Two sets of points X and Y owned by the two parties A and B respectively
1: **for all** $x \in X$ **do**
2: Compute the nearest neighbors for x, using Algorithm 1.
3: Using the distance shares from the last step, A and B compute shares of the density of x: δ_x^A and δ_x^B respectively.
4: **end for**
5: Repeat steps 1-4 for all $y \in Y$
6: **for all** $x \in X$ **do**
7: $N_x \leftarrow k$ nearest neighbors for x (for which density has already been computed)
8: A computes $\Delta_{N_x}^A = \sum_i \delta_{n_i}^A$ for all $n_i \in N_x$
9: B computes $\Delta_{N_x}^B = \sum_i \delta_{n_i}^B$ for all $n_i \in N_x$
10: $LOF_x \leftarrow divide(\delta_x^A k, \delta_x^B k, \Delta_{N_x}^A, \Delta_{N_x}^B)$
11: **end for**
12: Repeat steps 6-11 for all $y \in Y$

Algorithm 3: Privacy Preserving LOF Outlier Detection

10.4.1.2 Security Analysis

As in the case of all the applications of the basic method, the entire algorithm is only provably secure if the Nearest Neighbor Set and the Extended Nearest Neighbor Set are included in the specified output. In this case, the Nearest Neighbor Set needs to be in the output of both parties.

Theorem 3. The Privacy Preserving LOF algorithm (Algorithm 3) privately computes the LOF scores of each party's points, as well as the Nearest Neighbor Sets and Extended Nearest Neighbor Sets, in the semi-honest model.

Proof. The private inputs in this case are the respective sets of data points, and the output is the outlier score of each point in the respective sets of points. In addition, the output contains the Nearest Neighbor Set along with each point's Extended Nearest Neighbor Set. The communication that takes place in this algorithm occurs in steps 2, 5, 10, and 12. In step 2 and 5, the nearest neighbors are being computed for each point. This was shown to be a secure computation, and the output of this protocol is the Nearest Neighbor Set (and the Extended Nearest Neighbor Set) for each point, which is part of the specified output. In steps 10 and 12, the LOF score for a point is computed using the secure division algorithm. The output of these steps is the LOF score of the point, which is part of the output. Thus applying the composition theorem to this algorithm, the density calculation, and the secure division protocol, this algorithm is secure in the semi-honest model.

10.4.1.3 Extension to Multiparty Case

To extend this algorithm to the multiparty case, all that is needed is a multiparty version of the division algorithm. The nearest neighbor detection algorithm can be extended to the multiparty case, as was shown above. Once the neighbors are found using this method, the density will be shared among multiple parties, and thus in order to get the LOF score, all that is needed is a multiparty case of the division algorithm, where there are m parties, the i-th party holds s_i and t_i, and the LOF score is

$$\frac{s_1 + \cdots + s_m}{t_1 + \cdots + t_m}$$

This can be computed in the three party case [34] and in the more general m-party case [4].

10.4.2 SNN Clustering

SNN Clustering is a technique for clustering that defines similarity based on the number of nearest neighbors that two points share. The shared nearest neighbor similarity for two points x and y is defined to be the number of neighbors that appear in the nearest neighbor lists for both x and y, if x and y are in each other's nearest neighbor lists. If they are not in each other's nearest neighbor lists, then their similarity is 0. A graph that uses this notion of similarity is constructed and is called the SNN graph. This graph can be constructed based on information computed using our nearest neighbor search algorithm if the parties share their nearest neighbor lists (as was required for LOF outlier detection in Section 10.4.1).

A description of SNN Clustering techniques can be found in [47]. There are two main methods of SNN Clustering, the Jarvis-Patrick method [26] and SNN density based clustering [12, 13]. The Jarvis-Patrick clustering method is to compute the nearest neighbors of each point, find the similarity between each point, extrapolate this into a sparse graph where points are connected if their similarity exceeds a certain threshold, and designate as clusters the connected components of this graph. For SNN density based clustering, a notion of density is computed for each point that is based on the nearest neighbor similarity of the point, and then a DBSCAN [14] like algorithm is run to identify connected components of high density points. In both of these algorithms, no additional private information is required beyond what is needed to construct the similarity graph. Also, both of these can be extended to the case of more than two parties since the nearest neighbor search algorithm was shown to be extensible to the multiparty case.

10.4.3 kNN Classification

The problem of kNN classification [9] is as follows. Given a set of labeled records, and a given query instance, find the k nearest neighbors, and output as the predicted class label the majority class label of those neighbors. In a distributed setting, the nearest neighbors can be discovered among all the various sites. Previous approaches to solving this problem in a privacy preserving way [7, 29] assume that the query point is public knowledge for all the sites involved. However, it would be more desirable if the query point was only known to the party to which it was submitted. Using our basic nearest neighbor search algorithm, these goals can be achieved.

10.4.3.1 Protocol

Given the k nearest neighbors of a point x, our task in this algorithm is to find which class is represented by the most neighbors. Once the k nearest neighbors have been determined, each party knows which of its points are in the nearest neighbor set, and thus knows the class labels for these points. Thus party B, for example, knows how many of the neighbors it owns are members in each class c_i (where $1 \leq i \leq m$, and m is the number of class labels). So each party has a list of class labels with the number of neighbors in each class. In order to find the majority class, the sum of the number of neighbors in each class needs to be compared to the sum of the number of neighbors in all the other classes. The class for which the sum is greater than all other sums is the majority class.

In order to accomplish this, the only hurdle that needs to be overcome is the comparison. In the two party case this is fairly straightforward. Using the two party comparison protocol described in Section 10.2.3.2, a sum can be tested to see if it is the largest by comparing it to each other sum individually. If the result is true for each comparison, then it is the largest and thus represents the majority class. This can be repeated for each class label until the largest is found[2]. The answers to these comparisons are shared securely between the parties, and the sum is compared for equality after all comparisons are made (the sum of the shares should equal $m-1$). Also, the answers to the equality tests should be received only by the party who owns the point in question, so that the other parties do not learn the predicted class label of the point. This algorithm also assumes that no two classes have the same counts, and as mentioned in 3.2, ties can be broken using class identifiers. The pseudocode for this algorithm is found in Algorithm 4.

Require: A query point x given to party A
Require: A has a set of labeled data points X
Require: B has a set of labeled data points Y
Require: m is the number of total class labels
1: Run Algorithm 1 to find the nearest neighbors for the point x
2: For each class c_i, sum locally the number of neighbors in each class as s_i^A and s_i^B for party A and B respectively
3: **for all** class c_i **do**
4: **for all** class c_j, where $j \neq i$ **do**
5: Compute $s_i^A - s_j^A \geq s_j^B - s_i^B$, where A gets r_{ij}^A and B gets r_{ij}^B such that $r_{ij}^A + r_{ij}^B = 1 \bmod F$ if $s_i^A + s_i^B \geq s_j^A + s_j^B$, and 0 otherwise.
6: **end for**
7: A computes $r_i^A = \Sigma_j r_{ij}^A$
8: B computes $r_i^B = \Sigma_j r_{ij}^B$
9: Compute $r_i^A - m + 1 = r_1^B$ such that only A gets the answer
10: **end for**

[2] If the answer of true is obtained before the last class is checked, the party A should not terminate this loop, as it will reveal the identity of the majority class to the other party.

11: For one of the classes, A will have received an answer of true for the equality test, and this can be output as the majority class

Algorithm 4: Privacy Preserving kNN Classification

10.4.3.2 Security Analysis

As in the case of all the applications of the basic method, the entire algorithm is provably secure if the Nearest Neighbor Set and the Extended Nearest Neighbor Set are included in the specified output.

Theorem 4. The Privacy Preserving kNN Classifier (Algorithm 4) privately computes the majority class of the query point, as well as the Nearest Neighbor Set and Extended Nearest Neighbor Set, in the semi-honest model.

Proof. The private inputs in this case are the respective sets of data points, and the query point (which is known only to the local party A). The output of the algorithm is the majority class of the query point, along with the Nearest Neighbor Set and the Extended Nearest Neighbor Set of the query point. The nearest neighbors are found securely (as was shown in Section 10.3) in step 1. The output of this step can be simulated with the nearest neighbor set, which is part of the output. The next communication takes place when the classes are compared to each other in step 5. The result of these comparisons are shares that are uniformly distributed when viewed independently, and thus can be simulated with a uniform random number from the field F. Finally, the sums of the shares are compared for equality. The output of this (from A's perspective) will be true for the majority class (which is part of A's output) and false for all the others. A can simulate the answers from these comparisons with the output. B on the other hand, gets no information about the comparisons, including which class causes the loop to terminate, since the loop will not be terminated early. Thus, by applying the composition theorem to this algorithm, the nearest neighbor algorithm, and the secure comparison primitives, this algorithm is secure in the semi-honest model.

10.4.3.3 Extension to Multiparty Case

In the case of two parties, the total sum of neighbors for a given class is shared between two parties. Thus the two sums can be compared using each party's local shares as shown in step 5 of Algorithm 4. However, in the case of three or more parties, the sums are spread across all parties, and thus the two-party comparison cannot be used. In this case a multi-party comparison algorithm could be used to compute the comparison. Alternatively, this comparison can be accomplished in the following manner, using a two-party comparison protocol. Let us take the case of 3 parties, A, B, and C. We will also consider 2 classes, C_1 and C_2. The total counts for these two classes will be c_1 and c_2, respectively. Now the count c_1 is shared among the three parties as c_1^A, c_1^B, and c_1^C, so that $c_1^A + c_1^B + c_1^C = c_1$. Similarly, the count c_2 is shared among the three parties as c_2^A, c_2^B, and c_2^C, so that $c_2^A + c_2^B + c_2^C = c_2$. To perform the comparison, C simply chooses at random two values r_1 and r_2 from the field F, and sends r_1 and r_2 to A, and sends $c_1^C - r_1 \bmod F$ and $c_2^C - r_2 \bmod F$ to B. At this point A and B can compare the counts to see if $c_1 > c_2$ by computing if $c_1^A + c_1^B + (c_1^C - r_1) + r_1 \bmod F > c_2^A + c_2^B + (c_2^C - r_2) + r_2 \bmod F$. This can also be extended to an arbitrary number of parties by picking one of the parties to play the role of B (A would be the party with the query instance), and having each other party in turn play the role of C. Thus the parties playing the role of A and B would accumulate random shares of each other parties counts and at the end would be able to compute the comparison. In this case the comparison results would be shared between A and B, and thus step 9 of the algorithm would remain unchanged.

This extension maintains the security of the kNN Classification protocol. Since the parties playing the role of C select the r value uniformly at random from the field, the shares that A and B get of C's values are uniformly distributed when viewed independently. Thus the shares give no information as to C's individual counts. Also, since all parties have a share of the total (even if a party has 0 neighbors for that class), no information is leaked as to how many neighbors each party has within each class. Thus the multiparty extension is as secure as the two party version.

10.5 Complexity Analysis

The overall complexity of the algorithm is dominated by the secure multiparty computation protocols (this algorithm involves $O(n^2)$ secure computations). There are two major components to the complexity of these protocols: communication and computation. If we assume that we use homomorphic encryption [19] as the underlying primitive for the distance computations, we can concretely analyze the computational and communication complexity of this part of the protocol. For this analysis we will assume that only two parties are involved, and that each party has n data records, for a total of $2n$ records. Let us also assume that each data point is of dimension d, that each record contains 32 bit integers, and thus the field in which we perform the arithmetic is the integers modulo a $64+d$ bit integer (in order to contain the largest possible dot product, where each of the d components in the two vectors are 2^{32}). Also, the homomorphic encryption scheme that we use is Paillier's encryption scheme [43], with a 1024 bit RSA modulus, and thus the encryption involves 2048 bit exponentiation and multiplication. Finally, we assume that Paillier's encryption scheme has the appropriate values precomputed (the amount of time required for the precomputation will be discussed later in the section).

Now a total of n^2 pairwise distances need to be computed as the first step in the algorithm. This involves the computation of n^2 dot products. As per the dot product algorithm in [19], each component must be encrypted by the first party and sent to the second party. The second party computes the n^2 dot products and sends back encrypted random shares, which the first party must decrypt. Thus, the first party must compute dn encryptions, and send these dn values to the second party. For each dot product, the second party must compute d exponentiations and d multiplications, followed by an encryption and a multiplication. Finally, these n^2 encrypted values must be sent back to the first party, who must decrypt them all. Thus, the total computation required by the first party is dn encryptions and n^2 decryptions. If we ignore all multiplications and divisions in the encryption and decryption routines, we have the first party computing $dn+n^2$ 2048 bit exponentiations. The first party then must send $2048dn$ bits to the second party. The second party must then compute $(d+1)n^2$ exponentiations. These n^2 2048 bit values must be sent back to the first party, who must compute n^2 exponentiations. Thus, the total bandwidth required is $2048(n^2+dn)$ bits, and the total number of 2048 bit exponeniations is $dn+n^2+(d+1)n^2+n^2=dn+(d+3)n^2$. However, it is interesting to note that these encryptions can be highly parallelized, and thus can be greatly sped up through the use of high powered computers with multiple processors.

In addition to this, there are $O(n^2)$ comparisons that need to be computed. It is not entirely straightforward what the most efficient mechanism is to perform this computation, and thus providing a concrete implementation for this protocol is left for future work. It is important that the end solution is not only as efficient as possible, but also easy to implement.

10.5.1 Optimizations

We can see from the previous section that the algorithm is somewhat heavyweight, in that it is quadratic in terms of secure multiparty computation primitives, and may be difficult to execute in the case when the number of points is very large. Unfortunately, this is the case of many Privacy Preserving Data Mining solutions based on secure multiparty computation. Thus, in this section, we propose a few mechanisms that can be used to speed up the execution of the protocol, as a tradeoff in terms of the accuracy of the results.

The first is the use of sampling, which is a common mechanism used in data mining. The parties involved could select a subset of their data that will be used to find the neighbors. Thus, the neighbors can be found for each point, but the neighbors will only be located in the sample[3]. Thus, if the sample is of size m, with $m << n$, then the protocols would run in time $O(mn)$ for outlier detection and clustering and time $O(m)$ for classification.

The next possibility to increase performance of the algorithm is to only run it for a restricted portion of the dataset. This would not be applicable for the classification protocol, since there is only one query point for which neighbors are found. However, the outlier detection could be run only on the local points that are anomalous locally. Since points that are in dense clusters would be both locally and jointly non-anomalous, finding the joint anomaly score for these points would give no extra information. Similarly, the clustering could be run on points that are not a part of dense clusters, since those points would likely remain in those clusters even with the joint data. It is difficult to quantify how much accuracy would be lost in this way, however it would enable the parties involved to be able to run the algorithm and receive some results, hopefully resulting in better results than if the algorithm was only run locally. Related to this idea would be the use of preclustering in order to prearrange the data, so that neighbors are only searched for in the closest cluster. This would affect the accuracy of the results in a somewhat unpredictable way, but would greatly reduce the complexity of the algorithm.

A final possibility to increase performance, which is left as an area of future work, is to investigate the field of approximate nearest neighbor search. Much work has been done in this field [3, 21, 33], and applying those principles to the field of privacy preserving nearest neighbor detection could result in faster algorithms that privately find the approximate nearest neighbor sets. Some initial work has been done in this area [22] for approximate private distance computation over points in $\{0,1\}^d$, with promising results.

---

[3]This technique has been used quite effectively to identify anomalous net-flows in large collections of network data using the LOF scheme [10].

10.6 Related Work

Much work has been done in the field of Privacy Preserving Data Mining. There are two main approaches to this field: the data modification approach, and the cryptographic approach. The data modification approach consists of two major angles as well, one being the randomization method and the other being k-anonymization. The randomization method was initially proposed by Agrawal and Srikant [2], with their work on reconstructing approximations of distribution of the original dataset from randomly perturbed values. The randomization approach has also been applied to association rule mining [15, 44] and clustering [41, 40, 42], in addition to further work on decision trees [1]. This randomization approach was shown to have some limitations [30], where under certain circumstances, the original data points were able to be recovered with fairly high accuracy, thus greatly reducing the privacy guarantees of some randomization methods. Algorithms have also been developed to compute clusters by exchanging higher level information, such as generative models [37], without exchanging the actual data and without relying on randomization.

The k-anonymization approach was largely established by the works of Sweeney and Samarati [45, 46]. The idea of k-anonymization is simply to modify the records of the data such that for every record, there are $k-1$ other records that are indistinguishable from it. This is effective at preventing re-identification, which is the process of using fields that do not obviously identify an individual (e.g. zip code or data of birth), and combining them with some external information (e.g. a phone book), to link the identity to the record, thus breaking the privacy of that individual. The main technique to obtain a k-anonymous data set is generalization and suppression. Various techniques have been proposed in [45, 46, 24], and they essentially work by reducing the granularity of the data, making the attributes less specific. This is typically done on a selected subset of the attributes that are deemed potentially identifying, as achieving optimal k-anonymization was found to be NP-hard [38]. Anonymization has also been used for IP Addresses [16, 39].

The cryptographic approach primarily makes use of Secure Multiparty Computation ideas from the field of Cryptography [20, 52]. In this setting, two or more parties want to jointly compute the function from the combination of their inputs, such that they learn the output, yet receive no more information about the other's input than they can discern from the output alone. Much work has been done in this area, starting with [35], where a private computation was shown for computing information gain, allowing a secure computation of ID3 decision trees. Following this, much work was done for general tools for privacy preserving data mining (including secure set operations [8, 32] and secure sum and scalar product [8]), kNN classification for horizontally partitioned data [7, 29] and vertically partitioned data [53], association rule mining [28, 48], k-means clustering [25, 27, 50], outlier detection [51], and further work on decision trees [49].

The previous work on kNN classification for horizontally partitioned data [7, 29] deserves some extra discussion, as the key step in kNN classification is to compute the k nearest neighbors of the given point to be classified. In both of these works [7, 29], the authors assumed that the query point was public knowledge (including the values for all attributes). For both works the first step for each party is to compute the k closest points from the local data sets. From this point, Chitti et al. [7] make use of a distributed randomized algorithm *PP-TopK* to compute the k smallest distances to the query point. This is followed by a secure sum algorithm to compute the totals for each class, from which the majority class can be found. In Kantarcioglu and Clifton [29], once each party computes the k closest points in their respective data sets, the distance from all mk points to the query point are compared securely to each other (where m is the number of parties involved), and the results are sent to a semi-trusted third party (after scrambling to prevent the third party from learning which party contributed which points) along with the class values for each point (encrypted using the querier's public key). The third party can then compute the k nearest points from the mk points, and engages in a secure computation with the querier to compute the majority class label.

Since in both of these works the query point is given to each party involved, neither of these schemes can be used for LOF outlier detection and SNN clustering. The reason is that each of the data points themselves are the "query" points, and releasing them would release the entire data set. For kNN classification, both of these schemes, as well as our scheme, are potentially applicable. In our work however, the query point is assumed to be private in addition to the data sets, which is a requirement in certain applications. For example, if the query point is a patient record, that patient's information should be kept private, in addition to the records of the patients in the other data sets. Note also that in Chitti et al., once the neighbors are found the secure sum algorithm is used to sum up the total counts for each class label. This effectively releases the counts for each class label, and discloses the class distribution among the nearest neighbors. In our extension for kNN classification however, only the majority class is found, and the additional class information is not leaked. Finally, for Kantarcioglu and Clifton a semi-trusted third party is required to find the nearest neighbors and the majority class label. This assumption is not required in our construction.

To the best of our knowledge, the only other work that has addressed outlier detection was by Vaidya and Clifton [51]. In that work they describe an algorithm to compute distance-based global outliers, where a point is declared an outlier if the distance to all other points (or a certain percentage p of the other points) is greater than some given threshold. As this requires finding all neighbors closer than a threshold, their kernel and our kernel are very similar.

Both run in quadratic time and are capable of detecting the nearest neighbors within a certain distance from the point in question. In their algorithm, that distance is *dt* , a predefined static distance. In our scheme however, the distance is the distance to the k-th nearest neighbor, which varies for each point. This difference is key, as it enables the computation of much more information about the points. Vaidya and Clifton's kernel allows for the computation of a binary classification for outliers - either a point is an outlier or it is not. Our kernel, however, allows for the computation of the LOF score of each point, ranking each point in its degree of outlier-ness. It also allows for outliers and clusters to be computed based on local density. Thus our algorithm is more suitable to be used as a subroutine to provide private computations of algorithms that require the computation of nearest neighbors sets, which we show in our applications to LOF outlier detection, SNN clustering, and kNN classification.

10.7 Conclusion

In conclusion, we have shown a protocol for privately computing the k nearest neighbors of points in a horizontally partitioned data set. We described this algorithm in the two party case and proved security for each of the parts of the algorithm. In addition, we showed how to extend the nearest neighbor computation to the multiparty case. We also showed how this algorithm could be used to compute LOF outlier scores, SNN Clustering, and kNN Classification, including security proofs and extensions of each to the multiparty case. Finally, we analyzed the complexity of the scheme and suggested some mechanisms to improve the performance of the protocol at the expense of accuracy.

Since this algorithm is most likely to be used as a step in a more complicated process, it would be desirable to hide the identities of the neighbors (or share this information securely to enable further computation), so that the end result is obtained without leaking this information. This would require a solution more tightly coupled with the information that is needed from the neighborhood (i.e. it would be highly application-dependent), and is left for future work. Also, this work is focused on horizontally partitioned data, and another area of future work would be extending it to vertically partitioned data. Finally, one last area of work would be to investigate the area of approximate nearest neighbor search, in order to develop faster private computations of the neighbors.

Acknowledgments We would like to thank Nicholas Hopper for his comments on the security proofs, and Karthikeyan Mahadevan, Vishal Kher, Peng Wang, Shyam Boriah, and Varun Chandola for their helpful discussions. Portions of this work were supported by NSF Grant IIS-

0308264 and NSF ITR Grant ACI-0325949, and the Army High Performance Computing Research Center under the auspices of the Department of the Army, ARL cooperative agreement number DAAD19-01-2-0014. The content of this work does not necessarily reflect the position or policy of the government and no official endorsement should be inferred. Access to computing facilities was provided by the AHPCRC and the Minnesota Supercomputing Institute.

References

[1]. Dakshi Agrawal and Charu C. Aggarwal. On the Design and Quantification of Privacy Preserving Data Mining Algorithms. *In Proceedings of the ACM SIGMOD-SIGACT-SIGART Symposium on Principles of Database Systems*, 2001.

[2]. Rakesh Agrawal and Ramakrishnan Srikant. Privacy-Preserving Data Mining. *In Proceedings of the ACM International Conference on Management of Data*, 2000.

[3]. Sunil Arya and David M. Mount and Nathan S. Netanyahu and Ruth Silverman and Angela Y. Wu. An Optimal Algorithm for Approximate Nearest Neighbor Searching Fixed Dimensions. *Journal of the ACM*, 45(6):891--923, 1998.

[4]. Mikhail Atallah and Marina Bykova and Jiangtao Li and Keith Frikken and Mercan Topkara. Private Collaborative Forecasting and Benchmarking. *In Proceedings of the Workshop on Privacy in the Electronic Society*, 2004.

[5]. M. M. Breunig and H.-P. Kriegel and R. T. Ng and J. Sander. LOF: Identifying Density-Based Local Outliers. *In Proceedings of the ACM International Conference on Management of Data*, 2000.

[6]. M. M. Breunig and H.-P. Kriegel and R. T. Ng and J. Sander. OPTICS-OF: Identifying Local Outliers. *In Proceedings of the Third European Conference on Principles of Data Mining and Knowledge Discovery*, 1999.

[7]. Subramanyam Chitti and Li Xiong and Ling Liu. Mining Multiple Private Databases using a Privacy Preserving kNN Classifier. Technical report, Georgia Tech, 2004.

[8]. Chris Clifton and Murat Kantarcioglou and Jaideep Vaidya and Xiaodong Lin and Michael Zhu. Tools for Privacy Preserving Data Mining. *SIGKDD Explorations*, 2002.

[9]. T.M. Cover and P.E. Hart. Nearest Neighbor Pattern Classification. *IEEE Transactions on Information Theory*, 13(1):21-27, 1967.

[10]. Paul Dokas and Levent Ertöz and Vipin Kumar and Aleks Lazarevic and Jaideep Srivastava and Pang-Ning Tan. Data Mining for Network Intrusion Detection. *In Proceedings of the NSF Workshop on Next Generation Data Mining*, 2002.

[11]. Wenliang Du and Mikhail Atallah. Privacy-Preserving Cooperative Statistical Analysis. *In Proceedings of the 17th Annual Computer Security Applications Conference*, 2001.

[12]. Levent Ertöz and Michael Steinbach and Vipin Kumar. A New Shared Nearest Neighbor Clustering Algorithm and its Applications. *Workshop on Clustering High Dimensional Data and its Applications, In Proceedings of Text Mine '01, First SIAM International Conference on Data Mining*, 2001.

[13]. Levent Ertöz and Michael Steinbach and Vipin Kumar. Finding Clusters of Different Sizes, Shapes, and Densities in Noisy, High Dimensional Data. *In Proceedings of the SIAM International Conference on Data Mining*, 2003.

[14]. M. Ester and H.-P. Kriegel and J. Sander and X. Xu. A Density-Based Algorithm for Discovering Clusters in Large Spatial Databases with Noise. *In Proceedings of the International Conference on Knowledge Discovery and Data Mining*, 1996.

[15]. Alexandre Evfimievski and Ramakrishnan Srikant and Rakesh Agrawal and Johannes Gehrke. Privacy Preserving Mining of Association Rules. *In Proceedings of the ACM International Conference on Knowledge Discovery and Data Mining*, 2002.

[16]. Jinliang Fan and Jun Xu and Mostafa H. Ammar and Sue B. Moon. Prefix-Preserving IP Address Anonymization: Measurement-Based Security Evaluation and a New Cryptography-Based Scheme. *The International Journal of Computer and Telecommunications Networking*, 46(2):253-272, 2004.

[17]. Marc Fischlin. A Cost-Effective Pay-Per-Multiplication Comparison Method for Millionaires. *RSA Security 2001 Cryptographer's Track and Lecture Notes in Computer Science*, pages 457-471, 2001.

[18]. 1996 Health Insurance Portability and Accountability Act. http://www.hhs.gov/ocr/hipaa/, 1996.

[19]. Bart Goethals and Sven Laur and Helger Lipmaa and Taneli Mielikëinen. On Private Scalar Product Computation for Privacy-Preserving Data Mining. *In Proceedings of the 7th Annual International Conference in Information Security and Cryptology*, 2004.

[20]. Oded Goldreich. Secure Multiparty Computation, manuscript. http://www.wisdom.weizmann.ac.il/ oded/PSBookFrag/prot.ps, 2003.

[21]. Piotr Indyk and Rajeev Motwani. Approximate Nearest Neighbors: Towards Removing the Curse of Dimensionality. *In Proceedings of the Symposium on Theory of Computing*, 1998.

[22]. Piotr Indyk and David Woodruff. Polylogarithmic Private Approximations and Efficient Matching. *Proceedings of the Theory of Cryptography Conference*, 2006.

[23]. Ioannis Ioannidis and Ananth Grama. An Efficient Protocol for Yao's Millionaire's Problem. *In Proceedings of the Hawaii International Conference on System Sciences*, 2003.

[24]. Vijay Iyengar. Transforming Data to Satisfy Privacy Constraints. *In Proceedings of the ACM International Conference on Knowledge Discovery and Data Mining*, 2002.

[25]. Geetha Jagannathan and Rebecca N. Wright. Privacy-Preserving Distributed k-Means Clustering over Arbitrarily Partitioned Data. *In Proceedings of the ACM International Conference on Knowledge Discovery and Data Mining*, 2005.

[26]. R. A. Jarvis and E. A. Patrick. Clustering Using a Similarity Measure Based on Shared Nearest Neighbors. *IEEE Transactions on Computers*, C22(11):1025-1034, 1973.

[27]. Somesh Jha and Louis Kruger and Patrick McDaniel. Privacy Preserving Clustering. *In Proceedings of the 10th European Symposium On Research In Computer Security*, 2005.

[28]. Murat Kantarcioglu and Chris Clifton. Privacy-preserving Distributed Mining of Association Rules on Horizontally Partitioned Data. *In Proceedings of the ACM SIGMOD Workshop on Research Issues in Data Mining and Knowledge Discovery*, 2002.

[29]. Murat Kantarcioglu and Chris Clifton. Privately Computing a Distributed k-nn Classifier. *In Proceedings of the 8th European Conference on Principles and Practice of Knowledge Discovery in Databases*, 2004.

[30]. Hillol Kargupta and Souptik Datta and Qi Wang and Krishnamoorthy Sivakumar. Random Data Perturbation Techniques and Privacy Preserving Data Mining. *Knowledge and Information Systems Journal*, 7(4), 2005.

[31]. Eike Kiltz and Ivan Damgaard and Matthias Fitzi and Jesper Buus Nielsen and Tomas Toft. Unconditionally Secure Constant Round Multi-Party Computation for Equality, Comparison, Bits and Exponentiation. *In Proceedings of the third Theory of Cryptography Conference*, 2006.

[32]. Lea Kissner and Dawn Song. Privacy Preserving Set Operations. *In Proceedings of Advances in Cryptology - CRYPTO*, 2005.

[33]. Eyal Kushilevitz and Rafail Ostrovsky and Yuval Rabani. Efficient Search for Approximate Nearest Neighbor in High Dimensional Spaces. *In Proceedings of the 30th Annual ACM Symposium on the Theory of Computing*, 1998.

[34]. Sven Laur and Helger Lipmaa. On Private Similarity Search Protocols. *Proceedings of the 9th Nordic Workshop on Secure IT Systems*, 2004.

[35]. Yehuda Lindell and Benny Pinkas. Privacy Preserving Data Mining. *In Proceedings of Advances in Cryptology - CRYPTO*, 2000.

[36]. Benjamin Mayer and Huzefa Rangwala and Rohit Gupta and Jaideep Srivastava and George Karypis and Vipin Kumar and Piet de Groen. Feature Mining for Prediction of Degree of Liver Fibrosis. Poster Presentation in the Annual Symposium of American Medical Informatics Association, 2005.

[37]. Srujana Merugu and Joydeep Ghosh. Privacy-preserving Distributed Clustering using Generative Models. *In Proceedings of The Third IEEE International Conference on Data Mining*, 2003.

[38]. A. Meyerson and R. Williams. On the Complexity of Optimal KAnonymity. *In Proceedings of the Twenty-third ACM Symposium on Principles of Database Systems*, 2004.

[39]. Greg Minshall. TCPdpriv Command Manual. 1996.

[40]. Stanley Oliveira and Osmar Zaïane. Achieving Privacy Preservation When Sharing Data For Clustering. *In Proceedings of the International Workshop on Secure Data Management in a Connected World*, 2004.

[41]. Stanley Oliveira and Osmar Zaïane. Privacy Preserving Clustering By Data Transformation. *In Proceedings of the 18th Brazilian Symposium on Databases*, 2003.

[42]. Stanley Oliveira and Osmar Zaïane. Privacy-Preserving Clustering by Object Similarity-Based Representation and Dimensionality Reduction Transformation. *In Proceedings of the Workshop on Privacy and Security Aspects of Data Mining*, 2004.

[43]. Pascal Paillier. Public-Key Cryptosystems Based on Composite Degree Residuosity Classes. *In Proceedings of Eurocrypt*, 1999.

[44]. Shariq Rizvi and Jayant Haritsa. Maintaining Data Privacy in Association Rule Mining. *In Proceedings of 28th International Conference on Very Large Data Bases*, 2002.

[45]. Pierangela Samarati and Latanya Sweeney. Protecting Privacy when Disclosing Information: k-Anonymity and Its Enforcement through Generalization and Suppression. Technical report, SRI International, 1998.

[46]. Latanya Sweeney. k-Anonymity: A Model for Protecting Privacy. *International Journal on Uncertainty, Fuzziness, and Knowledge-based Systems*, 2002.

[47]. Pang-Ning Tan and Michael Steinbach and Vipin Kumar. *Introduction to Data Mining*. Pearson Education, Inc., 2006.

[48]. Jaideep Vaidya and Chris Clifton. Privacy-Preserving Association Rule Mining in Vertically Partitioned Data. *In Proceedings of the ACM International Conference on Knowledge Discovery and Data Mining*, 2002.

[49]. Jaideep Vaidya and Chris Clifton. Privacy-Preserving Decision Trees over Vertically Partitioned Data. *In Proceedings of the IFIP WG 11.3 International Conference on Data and Applications Security*, 2005.

[50]. Jaideep Vaidya and Chris Clifton. Privacy-Preserving K-Means Clustering over Vertically Partitioned Data. *In Proceedings of the ACM International Conference on Knowledge Discovery and Data Mining*, 2003.

[51]. Jaideep Vaidya and Chris Clifton. Privacy-Preserving Outlier Detection. *In Proceedings of the Fourth IEEE International Conference on Data Mining*, 2004.

[52]. A. C. Yao. How to Generate and Exchange Secrets. *In Proceedings of the 27th IEEE Symposium on Foundations of Computer Science*, 1986.

[53]. Justin Zhan and LiWu Chang and Stan Matwin. Privacy Preserving K-nearest Neighbor Classification. *International Journal of Network Security*, 1(1):46-51, 2005.

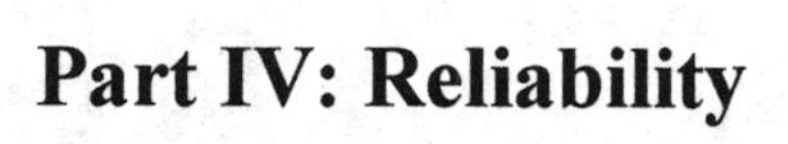

Part IV: Reliability

11 High-Confidence Compositional Reliability Assessment of SOA-Based Systems Using Machine Learning Techniques

Venkata U. B. Challagulla, Farokh B. Bastani, and I-Ling Yen[1]

Abstract Service-oriented architecture (SOA) techniques are being increasingly used for developing critical applications, especially network-centric systems. While the SOA paradigm provides flexibility and agility to better respond to changing business requirements, the task of assessing the reliability of SOA-based systems is quite challenging. Deriving high confidence reliability estimates for mission-critical systems can require huge costs and time. SOA-systems/applications are built by using either atomic or composite services as building blocks. These services are generally assumed to be realized with reuse and logical composition of components. One approach for assessing the reliability of SOA-based systems is to use AI reasoning techniques on dynamically collected failure data of each service and its components as one of the evidences together with results from random testing. Memory-Based Reasoning technique and Bayesian Belief Net-works are verified as the reasoning tools best suited to guide the prediction analysis. A framework constructed from the above approach identifies the least tested and "high usage" input subdomains of the service(s) and performs necessary remedial actions depending on the predicted results.

11.1 Introduction

Service SOA-based methods are being used increasingly for developing complex dependable systems as evident from DoD's recent projects, such as network-centric enterprise services (NCES), global information grid enterprise services (GES), and joint battle management command and control (JBMC2) (Paul 2005a). Apart from the well-defined and high levels of abstraction provided by the service interfaces, the main advantage of service oriented archi-

[1] Department of Computer Science, University of Texas at Dallas, Dallas, TX 75083-0688, USA, [uday, bastani, ilyen]@utdallas.edu

J.J.P. Tsai and P.S. Yu (eds.), *Machine Learning in Cyber Trust: Security, Privacy, and Reliability*, DOI: 10.1007/978-0-387-88735-7_11,

tecture is the loose coupling between different services that are pro-vided by the system, maximizing the system flexibility and easier reconfiguration with changes in business requirements (Gao et al. 2005). However, there exist challenges for assessing the reliability of SOA-based mission-critical software systems (Paul 2005b, Schneidewind 1998). Generally, these systems are large and complex and, hence, need more cost-effective methods of assessing their reliability.

In this chapter, we provide a framework for assessing the reliability of mission-critical SOA-based systems. The individual SOA services, that satisfy specific business requirements, are assumed to be built by logical composition of components provided by the underlying component layer. The components could be commercial off-the-shelf (COTS) / government-off-the-shelf (GOTS) software components. The business layer requests the services layer, which in-turn invokes one (its own implementation component) or more components for realizing the desired functionality. In other words, each service is exposed as an interface to which a business layer client can bind and invoke its capabilities without knowledge of its underlying implementation. A service broker/registry lists the available services. Behind services are the implementation black boxes, i.e., components. The SOA infrastructure/framework provides the mechanism for reliable, standard, and managed communication between the applications business layer and service layer. A typical SOA-based system that consists of services, applications, and the intermediate connecting infrastructure (Arsanjani 2004) is shown in Fig. 11.1.

However, there are many issues surrounding the reliability estimation of SOA-based systems. These include the following:

- How to estimate the reliability of the individual components that constitute the service?
- How to estimate the reliability of a service, both from the client and service provider view-points, from the reliability of its constituent components?
- How to estimate the overall SOA-based system reliability when the system is orchestrated using several individual services?
- How to analyze the effects of usage profile changes on the service reliability and track sensitivity factors when replacing services/components with similar ones?

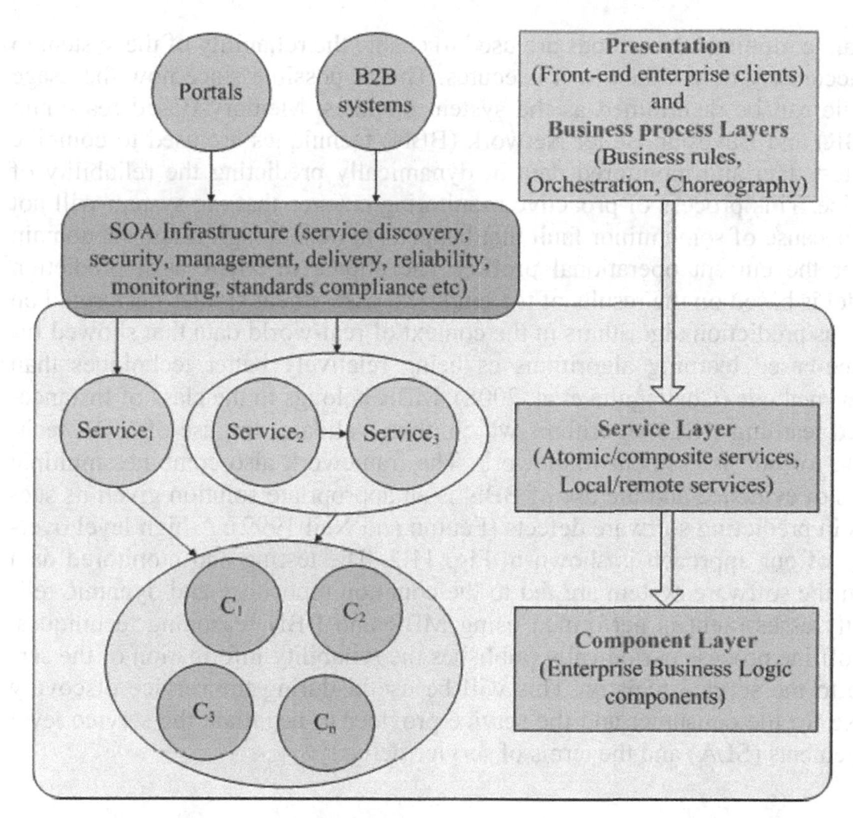

Fig. 11.1 A Typical SOA-based System

The generally used procedure for determining the reliability estimates for mission-critical SOA-based systems is to use a wide variety of testing techniques. However, achieving a sufficiently high level of confidence in the reliability estimate for these mission-critical SOA-based systems based solely on testing is impossible due to the high cost of testing resources involved. In particular, there is a possibility that some input subdomains of a service are likely to be executed much more in the actual usage than other subdomains and, hence, any latent defects in these "high usage" subdomains will impact the system reliability much more than latent defects in "low usage" subdomains. Testing methods that do not follow the operation profile cannot account for this factor, even though it is likely that good testing strategies will uncover more defects in the system than purely operational profile testing. The challenge is to combine these two methods.

One of the approaches for combining multiple evidences is to combine reliability measures from test profile based analysis and operational profile based analysis to dynamically predict the reliability of service-oriented systems. The system is first subjected to operational profile based testing and then dynamic monitoring methods are used to enable the reliability of the system to be accurately determined as it executes. This is possible since now the usage profile can be determined as the system operates. Memory-Based reasoning (MBR) and Bayesian Belief Network (BBN) techniques are used to combine the test data and monitored data in dynamically predicting the reliability of service. This process of proactive monitoring ensures that the system will not fail because of some minor fault that happens to be in a high usage subdomain under the current operational profile. The choice of MBR as a prediction model is based on the results of the empirical assessment studies performed on various prediction algorithms in the context of real-world data that showed Instance-based learning algorithms as being relatively better techniques than other methods (Challagulla et al. 2008). MBR belongs to the class of Instance-based learning (IBL) algorithms which, thus, validates the use of MBR technique for our assessment framework. The framework also combines multiple types of evidence and the use of BBN is an appropriate solution given its success in predicting software defects (Fenton and Neil 1999). A high level overview of our approach is shown in Fig. 11.2. The testing and monitored data from the software system are fed to the common repository and dynamic reliability assessment is performed using MBR and BBN reasoning techniques. An offline process periodically publishes the reliability information of the service to the service registry. This will be useful during the service discovery phase for the consumer and the service provider to negotiate the service level agreements (SLA) and the terms of service delivery.

11.2 Related Work

Though service oriented computing evolved from traditional object-oriented computing, there are fundamental differences between both of them (Paul 2005a). Dynamic discovery of services, reconfiguration and composition of services at run-time, dynamic verification, validation and monitoring are some of the most distinguishable features of service oriented systems. Because of these dynamic features combined with the abstraction provided by the services, reliability assessment of SOA-based systems is very complex. Testing a service is a combined process of testing the service by subjecting it to appropriate usage-profiles and testing the respective components with their operational profiles. However, most of the assessment techniques used for component-based systems are equally applicable for SOA-based systems with few or no modifications. To be more specific and cover the related work in a systematic way, we divide the literature survey into the following sections.

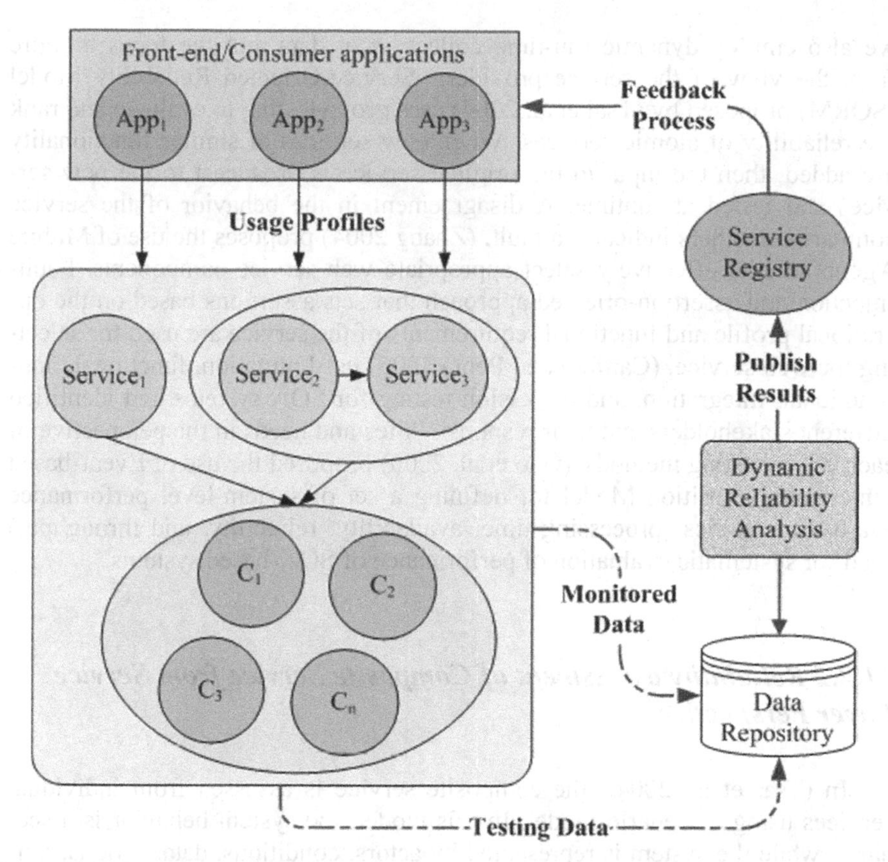

Fig. 11.2 Dynamic Reliability Assessment Framework for SOA-based Systems

11.2.1 Reliability assessment of Service from Service Layer Perspective

The factors (Tsai 2005) that need to be taken into consideration for reliability assessment of a service include a) dynamic reliability calculation based on the run-time monitoring of failure data, b) determining the validity period of failure data collected, and c) use of already existing models and data for dynamic model construction during new service composition and re-composition. The Data-driven data evaluation process (DREP) (Tsai 2005) uses a just-in-time collaborative approach where the reliability of the service is calculated from data collected from usage profile testing of different service providers. It is assumed that the source code is not available. In our analysis, we also employ dynamic run-time collection of data, but the focus is more from the view of the service provider. Service-Oriented Reliability Model (SORM) proposed by (Tsai et al. 2004) uses group testing to evaluate and rank the reliability of atomic services. When new services of similar functionality are added, then the input to the original service is broadcast to the new services and tested at runtime. A disagreement in the behavior of the service compared to others indicates a fault. (Zhang 2004) proposes the use of Mobile Agents to cost effectively select appropriate web service components. Fault-Injection and assertion-oriented approach that sets assertions based on the operational profile and functional requirements of the service are used for selecting the web service. (Canfora and Penta 2006) used mutation, functional, non-functional, integration, and regression testing for SOA systems and identified different stakeholders and their responsibilities and needs in the perspective of each of the testing methods. (Gao et al. 2006) proposed the use of Event-based Functional Transition Model for defining a set of system-level performance evaluation metrics (processing time, availability, reliability, and throughput) used for systematic evaluation of performance of SOA-based systems.

11.2.2 Reliability assessment of Composite Service from Service Layer Perspective

In (Tsai et al. 2004), the composite service is assessed from individual services using a scenario model. In this model, the system behavior is a scenario, while the system is represented by actors, conditions, data, actions, timing attributes, and events. For predicting the reliability of composite services, (Grassi and Patella 2006) identifies three policies according to the information released by the service provider regarding the reliability of the service: a) No transparency: Reveals only overall services reliability information; b) Partial transparency: Provider provides both services internal reliability and usage profile, but does not reveal how the composite service is built; and c) Total transparency: Composite-user selects the services to be assembled. Flow graphs are used to calculate the composite service reliability. (Tsai et al. 2005) proposes the use of a) Specification-based Verification and Validation (V&V): Use of Completeness and Consistency (C&C) analysis, model checking techniques and specification only based test cases for V&V, b) Collaborative V&V (CV&V): All the parties involved, clients, providers, and brokers contribute toward test case generation, and c) Group testing for effectively testing the SOA-based Web services system.

11.2.3 Reliability assessment of Services from Component Layer Perspective

For an SOA-based system built using components, the component reliability approaches described below are all relevant and usable with little or no modifications for determining the service reliabilities. However, we need to note that dynamic composition, re-configuration, and collaboration features that are specific to services are not addressed by these methods. The service is assumed to be composed of one or more components and realized by component interactions. The group of interacting components is referred to as "component system" for clarity and brevity.

Over the past three decades, several different methods have been explored for measuring the reliability of component systems. However, no single method has proved to be completely effective. (Frankl et al. 1997) compared "debug testing" and "operational testing" schemes for reliability analysis and concluded that both have advantages and the appropriate testing scheme has to be chosen depending on the specific situation. (Cukic and Bastani 1996) presented a class of techniques based on a combination of partial program proofs and program slicing to reduce the size of input domains, thereby simplifying the task of determining operational profiles. One more benefit of these techniques is the insensitivity of the component system reliability estimate to variations in the operational profile due to reduced dimensionality of the input space. In addition to the drawbacks of random testing, the complexity of software is increased with the use of component-based development, in case of mission-critical component systems. While developing software using component-based approach has its advantages in terms of reduced cost and maintenance by re-using of components, there are several drawbacks from the reliability viewpoint. (May 2002) pointed out some of the deficiencies of applying black box statistical software testing (BBSST) to component-based development. (May 2002) also pointed out that BBSST models need to explain the dependence of the reliability of software on its components, be able to estimate the overall software component system reliability from the known reliability of its re-usable software components, and be able to provide new complexity metrics based on testability that help in designing a component-based software system that is inherently reliable. (May 2002) suggested combining formal methods and statistical testing for software V&V and the use of statistical testability as a new system complexity metric. (Wohlin and Runeson 1998) uses usage based testing for reliability certification of components and lists the following steps for the certification process. First, the anticipated usage of the components should be investigated followed by the creation of usage specification of the component. Then, test cases are generated followed by the evaluation of the outcome of the test and certification of the component based on the failure data collected. One of the important issues is to take into consideration the sensitivity of the reliability estimate based on changing profiles. A similar approach for estimating the component system reliability is to average the path reliability estimates over the input test profile (Krishnamurthy and Mathur 1997). For an input test, the path reliability is computed based on the sequence of components executed. However, they also mentioned that inter- and intra-component dependence, sensitivity, associating risk parameter with operational profile, and obtaining reasonable estimates of component and interface reliabilities are some of the issues that need to be resolved for successful component system reliability estimation.

The inability to have a widely accepted technique to determine the composite component system reliability from its component reliabilities is explained by (Voas 2001). Taking an example of a sequential two component system, where the output of component A is fed to component B, some of the issues he described are a) assuming that the individual components are reliable, we cannot assume that the composite system is reliable by recognizing the fact that components can behave unreliably when connected to each other; b) there can be "non functional" component behaviors that manifest themselves only in the composite system and undermine its reliability; and c) selecting a highly reliable but incorrect component can render the component system useless.

Markov models are also used to estimate the component system reliability from constituent software components (Laprie and Kanoun 1996). (Woit and Mason 1998) suggested applying Continuation Passing Style (CPS) transformation to components to form static-independent fragments appropriate for Markov analysis. They also showed that predecessor-independent components limit the component system reliability changes with component changes/replacements. (Gokhale et al. 1998) used results from extensive testing to parameterize the discrete time Markov chain based analytical model used for architecture-based software reliability prediction. (Zhu and Gao 2003) points out that it is difficult to apply Markov models if the structure of the component based systems are very complex. (Kappes et al. 2000) modeled the component software system as a network of communicating finite state machines, with each individual component as a finite state machine with output and interactions between components captured as message queues. Using formal proofs, they show that there is no single algorithm to calculate the reliability of an arbitrary software component system using the reliability of components or the reliability of the operations in the components. (Kuball et al. 1999) used hierarchical model based on Bayesian Reasoning to estimate the probability of failure of a software system built using components. They proposed the use of a probabilistic error term to account for the changes in the component prior probability when it is shifted from an assumed operational environment to an actual deployment. Research has also been done on sensitivity analysis of reliability estimates to determine which component affects the reliability of the component system the most based on a Markov process modeling approach for reliability analysis for component based systems (Lo et al. 2003). Metrics are defined in (Lo et al. 2003) to assess the most sensitive parameter in a system and validate their analysis with different system architectural styles. Directed graphs, called component dependency graphs, have been used to represent components, component reliabilities, and link, interface, and transition reliabilities (Yacoub et al. 1999). The component transition probabilities are calculated by using component system execution scenarios. This approach is called scenario-based analysis (Yacoub et al. 1999) and the reliability of the component system is based on the usage of components in a particular scenario.

One of the approaches for obtaining better estimates of reliability is to combine all possible evidences. (Littlewood and Strigini 2000) gives a list of evidences that can be combined. Evidence from quality of the software process employed, evidence from static analysis, such as formal verification of the software product, and evidence from software components and their structure are some of the additional evidences that can be combined with results from random testing of software. (Fenton et al. 1998) provide a framework for combining diverse evidences from process and product information using Bayesian Belief Networks (BBN). (Yu and Johnson 2003) used Usability, Quality and Impact as the main nodes in a BBN framework to quantitatively calculate the component system reliability. Some researchers (Challagulla et al. 2008) have also investigated the application of machine learning techniques to software reliability prediction. (Zhang 2000, Mitchell 1997) provide a lot of references related to the use of machine learning techniques for software defect prediction.

11.3 Service Reliability Assessment Framework

The reliability assessment framework described below predicts the reliability of a component-based mission-critical service by monitoring the runtime behavior of the service. The reliability estimation for a service and its related components is initially done using the subdomain approach during the testing phase, which is then augmented by estimation using prediction techniques during the monitoring phase. The testing and monitored data collected from execution of different service components are combined to guide the reliability prediction process. The framework determines whether past data retrieved from the behavior of other "similar" service components can help to more accurately predict the reliability of the service component being monitored. Once the component reliabilities are predicted, Markov Chain Model approach is used to determine the overall service reliability from the knowledge of the reliability of its components. The monitoring and diagnosis framework thus checks whether the reliability of a particular service starts to fall below a particular threshold as predicted by the reliability model. If so, then an alarm is raised to alert the system of the problem and enable proactive recovery methods, such as switching to a different implementation of the service.

11.3.1 Component Reliability Assessment Framework

We start by describing a basic model of the subdomain approach used and the process of storing the test data during both the testing and monitoring phases. We then describe the prediction methodologies appropriate for accurately predicting the service component faults during the monitoring phase, which is termed as "dynamic component reliability"

11.3.1.1 Defining the Component Model

A service is composed of independent components that are composed together. Any input to the service invokes different components with corresponding inputs to their subdomains. It is important to note that, although generally a service is implemented by a unique corresponding component implementation, the same component could also implement different service interfaces. Hence, the input domain of a service may be considered as being different from its underlying components. The inputs that have been tested for the service and the components are represented by reliability maps that show the regions of inputs that have been successfully tested.

We briefly describe the formal definitions of the input domain model based on components. The input domain refers to the set of all possible inputs to a component. These are inputs that the component should respond to. The input could occur either during the testing or the execution phase. The input domain is further divided into a finite number of disjoint functional subdomains (Musa 1993, Hamlet et al. 2001). The operational profile of a component is the quantitative representation of component usage in terms of the inputs it is subjected to. More formally, the operational profile is defined as the likelihood that an input to the component is selected from a particular input subdomain. Hence, it is the probability distribution over the input subdomains that specifies the occurrence probabilities of each subdomain. The occurrence probabilities are also called the weights of each subdomain. It is important to note that each of the subdomains can have its own internal profile.

Let the service under investigation be represented by Sa and let a component invoked by Sa be represented as C1. Let the component input domain be defined as R and let it be divided into t subdomains, with each subdomain i containing Ri inputs, as shown in Fig. 11.3. Let the overall profile for the component be P.

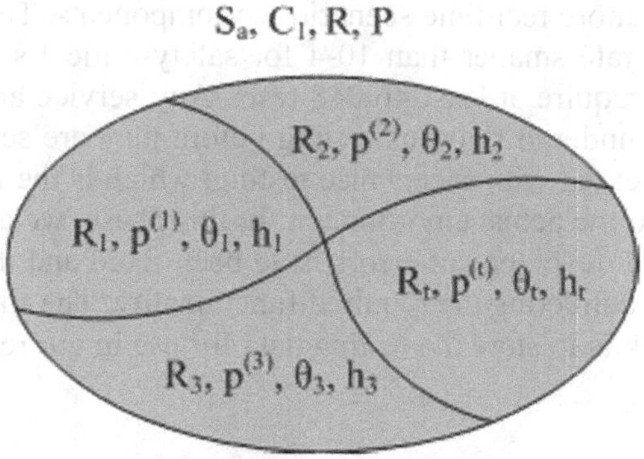

S_a, C_1, R, P

$R_2, p^{(2)}, \theta_2, h_2$

$R_1, p^{(1)}, \theta_1, h_1$

$R_t, p^{(t)}, \theta_t, h_t$

$R_3, p^{(3)}, \theta_3, h_3$

Fig. 11.3 Component Input domain.

Within each subdomain, the inputs are chosen according to some distribution, say,

$p(i), i = 1, 2, \ldots, t.$

Then, $R = R1 \cup R2 \cup \ldots \cup Rt$

and the expected probability of failure for subdomain i, Θi is,

$$\Theta i = \Sigma\, p(i)(j)\, Xij$$

where $Xij = 1(0)$ if the jth input from the ith sub-domain will result in the failure (success) of the component. If we assume that all the inputs in a subdomain have equal probabilities, then

$p(i) = 1/ |Ri|.$

Hence, $\Theta i = (\Sigma\, Xij)/ |Ri|.$

Thus, the failure rate in this case is just the proportion of inputs that lead to incorrect outputs. With the absence of actual internal profiles of the subdomain, the general process of estimating the subdomain failure rate is to use uniform random testing. The overall profile P for the component is the probability that each subdomain is invoked and let it be represented as,

$$P = \langle h1, h2, \ldots, ht \rangle$$

where hi is the probability of invocation of subdomain i.

For mission-critical services, however, we must have very low levels of failure probabilities. We calculate the number of tests required for each component sub-domain to achieve the required confidence in reliability. If Θoi denotes our hypothesized probability of failure for a given input domain, the number of test cases U to observe at least one failure with probability C (confidence level) is given by (Miller et al. 1992, Amman et al. 1994):

$$U = \ln(1 - C)/\ln(1 - \Theta oi).$$

This equation assumes testing with replacement of inputs. The distribution of the random variable that denotes the number of test cases is geometric. We choose this model as it depicts a more real-time scenario for components. This proves that to achieve a failure rate smaller than 10-4 for safety-critical services with confidence 0.99, we require at least 46,048 tests. Any service and its associated components must undergo rigorous testing before they are sent to production. We are only concerned with acceptance testing, which is the final phase before deployment into the actual environment. In this phase, we assume that most of the design and development errors have been fixed and the service component is capable of delivering the required functionality. The first step in the monitoring framework is to store the testing data for use in our reasoning to be done later.

11.3.1.2 Storing of the Testing Data

The component is first subjected to Test profile based testing and then to Operational profile based testing. The general process of estimating the operational profile for a component includes usage analysis of similar components and through surveys. The usage specification also helps to determine the probability density function associated with the sub-domains resulting in the operational profile for the component. Once the operational profile is determined for a component, random testing is conducted using inputs from the input domain. We assume that a test oracle exists for the component that determines whether or not the result of a particular test execution is correct. The results are aggregated together to yield the failure rate for the component. The failure rate and the operational profile of the component sub-domain are used to calculate the reliability for that component. The collected test data are stored in a database. The details of the stored test data are described in Sect. 11.3.1.4.

If To is the number of test cases without encountering a failure, then the question is how many more test cases are required to achieve our required reliability and confidence estimates. The prior belief of "successfully running To test cases" results in simply reducing the remaining testing effort required and, thus, the remaining number of test cases to achieve the confidence level is:

$$U = [\ln(1 - C)/\ln(1 - \Theta o_i)] - T_o.$$

For example, if 10000 test cases have been run without encountering a failure, then for achieving failure rate lower than 10^{-4} with confidence 0.99, we require at least 46048-10000 = 36048 tests However, the above equation is derived assuming failure free statistical testing only. If the component has encountered a failure during acceptance testing then the current version of the service component is not deployed. Instead, the error is fixed and the test cases are repeated again. However, the results of the entire testing process are stored in the database. Due to insufficient testing effort, it becomes imperative to monitor the real time execution of the service components and use other evidences to increase our confidence in the reliability of service. The results of testing along with the results collected from the monitoring process are used to dynamically determine the reliability of the components and, thus, the reliability of the service.

11.3.1.3 Defining the Monitoring Process

The monitoring process involves identifying the services and associated components to be monitored, and defining the monitored parameters, duration/frequency of monitoring, and format/storage of the monitored data. In the context of the current framework, the executions of all the service components provided by the service provider are monitored. The monitoring points are placed at the service component interfaces to record the input sub-domains invoked. The service components are dynamically monitored for faults. We do not cover the details of the monitoring process here. A detailed taxonomy of run-time software-fault monitoring tools can be found in (Delgado et al. 2004). The framework also ensures that the monitoring process does not alter the state of the component. The details of the stored monitored data are described in Sect. 11.3.1.4.

11.3.1.4 Combining the Testing and Monitored Data

The testing and the monitored data are combined into a central database. We developed a method of aggregating data from multiple services and their components. The following data from the testing and monitoring phase are stored: The characteristic metric information of the component, the number of times each component sub-domains are invoked along with the failure percentage (% of faults encountered) of the component. The failure date in terms of percentages is the preferred input to some of the external tools used in the framework and hence the failure data is chosen to be stored in failure percentages rather than in actual probabilities

The data that need to be stored in the database are determined based on the following key issues:

a) Use of metrics as additional evidence
Both the deterministic and probabilistic metric data from the components are collected and stored. The deterministic data involve size and complexity metrics that can be extracted from the source code. Since we are assessing the reliability from the provider's perspective, we assume that the source code for the components is available. The probabilistic data include the quality aspects of the data such as the experi-

ence of the staff, CMM level of the organization, amount of time spent on component review, number of changes since last release, and number of requirements/use-cases satisfied by this component.

b) Storing the number of inputs rather than the input itself.

There are two ways of storing the combined data for reliability estimation:

- Type I – Input specific data

  This involves storing data about each component along with its input. The execution data from the monitoring phase are just added to the existing test data. The database contains the component identifier, the input to the component, the deterministic/probabilistic metrics, and the result of the execution using that input during the testing phase and the actual usage. We use all the data for reliability estimation, except the component identifier.

- Type 2 – Input independent data

  In this case, we store the component identifier, deterministic/probabilistic metrics of the component, number of inputs to each of the sub-domains of the components, and the failure percentage of that sub-domain.

The main difference between Type 1 and Type 2 data is that we use exact input information for Type 1 data, while we just keep track of the number of input invocations for Type 2 data. We use Type 2 data for storing information about the component.

There are advantages and disadvantages for both the approaches. For Type 1 data, the new data being added into the repository of the component is equivalent to increasing the input domain for that component and, thus, the failure rate is dynamically updated for that component. This also helps in executing similar tests for a replaced component with similar interfaces. However, a disadvantage for the Type-1 approach is that memory constraints can limit us from storing all the input data. This could be very large if the service provider has a large number of components. Also, the repository is too component and input specific. The prediction analysis uses the combined data from all the components and storing input specific data hampers efficient reliability prediction. In case 2, we can easily aggregate the data from all the components, giving more data for prediction. The reason for using a combined repository is because we assume that similar components with the same set of characteristics generally also have similar defect occurrences. Sharing data from similar components is then equivalent to performing additional testing on that component. This increases our confidence in the reliability estimate. The component repository layout is shown in Fig. 11.4. The input profile to the individual component C_1 represented by, say, $\langle h_1, h_2, \ldots, h_x \rangle$ can easily be obtained from the repository data.

11.3.1.5 Combining Evidences – Use of evidences other than testing alone

When a component is invoked, the near real-time monitoring system checks if the component input domain is sufficiently tested. If it encounters a scenario where the input sub-domain is not sufficiently tested, then it tries to make a decision about its reliability using other evidences. We assume the simplest case where each service has independent components that are solely invoked by the service and are not used by any other service. While this assumption is more realistic, it also helps to be more pessimistic about the number of test cases required for assuring each component's reliability. If another service has used the same component and invoked the same sub-domain, it serves as an added bonus to the confidence in the reliability estimates. If the subdomain invocation of the service is not enough to assure reliability, then it generally translates to insufficient testing of the related components in operation. In such cases, for each component, the dynamic monitoring system tries to subjectively determine the component reliability by combining the following evidences:

a) Evidence from the corresponding input sub-domain: The frequency of executions of the sub-domain invoked by that input is used as the primary evidence.
b) Evidence from the composite repository from Sect. 11.3.1.4: The components that are given by the nearest neighbor are used to predict the reliability of components that have similar deterministic attributes. The frequency of their execution and the probability of failure are used to determine the predicted probability of the failure of the current component.
c) Combine the evidences from both steps a) and b) and determine with confidence if the reliability of the component falls within the prescribed limits. If not, necessary actions are taken

	Subdomain	Number of Inputs	Failure %
Component C_1	1		
	2		
	....		
	x		

Component	Subdomain {Num of Inputs, Failure %}	Metrics			Total Num of Inputs	Total Failure %
		M_1	...	M_n		
C_1						
C_2						
...						
C_n						

Fig. 11.4 Component Repository Layout

11.3.1.5.1 Evidence from Composite Repository

Let us first investigate the procedure of using the evidence from the composite repository. Given a component, the number of nearest neighbors to be used for intelligent analysis is determined using data from Fig. 11.4. We use Memory-based reasoning technique to determine the nearest neighbors. The memory-based reasoning technique uses Euclidean distance as the nearest neighbor distance. Standardization/normalization of metric data and assigning weights to the metrics improve the results returned by the reasoning technique. Both deterministic and probabilistic metrics are used. However, we might not always have knowledge of program quality metrics for 3rd party components. Given no prior knowledge of the quality process, we assume that the predicted reliability is solely determined by the deterministic characteristics. As the quality process of a particular component becomes available, the proportion of the predicted operational reliability affected is updated accordingly. We will not count the records pertaining to that component, as that evidence is taken care of in Step 3.1.5 (a). The optimum choice of the number of nearest neighbors chosen is continuously updated based on the effect of increasing/decreasing the numbers of nearest neighbors on the accuracy of prediction. A value equal to half the size of the repository count is generally a good initial value as an upper bound for the nearest neighbor count. Let the number of nearest neighbors chosen be represented by "N". The nearest neighbors closer to the target record are sorted according to the dissimilarity values. Once the nearest neighbor data is identified, we rank them into three categories (high, medium, low). The ranking of the component and the associated failure percentage are used to calculate the net failure percentage of a nearest neighbor component as shown by the BBN network diagram in the shaded area of Fig. 11.5. The ranking reflects the nearness of the neighbor to the actual component.

In an ideal case, instead of using ranking categorization, we could have used the distance measures as categories and assigned them a probability distribution. Given that the number of such distance categories can be very huge and with the two state (success/failure) nature of the "Failure percentage" node, we decided to go with three states for ranking each node. The sorted nearest neighbors are split into three equal groups and the top cut neighbors are placed in the "high" ranking, while the middle cut neighbors are placed in "medium" ranking and the lowest cut neighbors are assigned the "low" ranking. For example, assume that the MBR process retrieved 6 nearest neighbors for C1 and sorted them in the ascending order of distance, then the ranking assignments are as shown in Fig. 11.6.

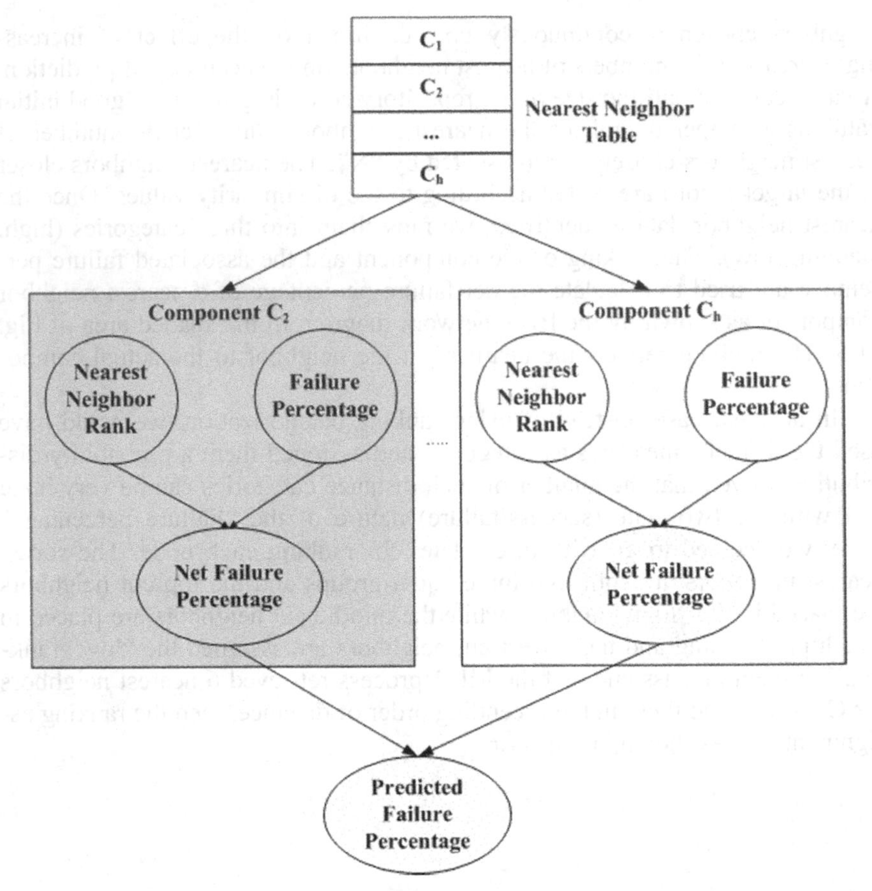

Fig. 11.5 BBN Network for calculating component operational reliability

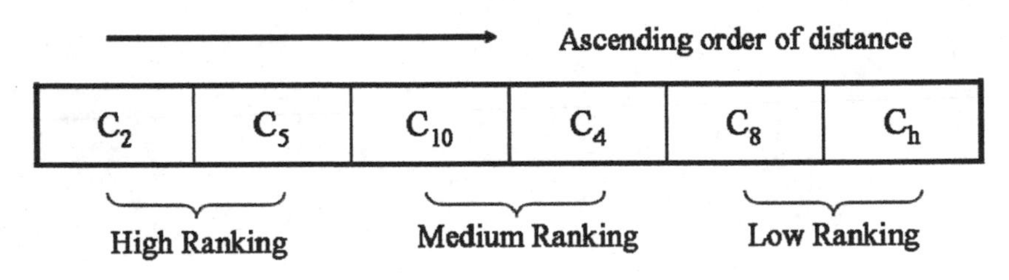

Fig. 11.6 Calculating component rankings

Calculation of the Predicted Failure Percentage of a Nearest Neighbor

Consider the BBN network diagram of a neighbor component C2 as shown in the shaded area of Fig. 11.5. The neighbor falls under the high ranking category and the node probabilities for the different nodes of the BBN network for that component are shown in Fig. 11.7.

For any BBN network, the important issues to be considered are as follows:

- Method of determining the prior probabilities of each of the states of the nodes of the Bayesian Belief network.
  For each component in the system, we have a "Nearest neighbor rank (N)" node and a "Failure Percentage (FP)" node that is dynamically updated as the component executes and participates in the nearest neighbor calculations. During the run-time calculations, update the nodes with the appropriate ranking and fault percentage. There is so much implication by adding ranking to the net operational reliability calculations. The trend derived from monitoring a component's ranking percentages tells a lot about the invocations of similar components. A very "high" percentage indicates that a set of similar components being executed is abnormally high. We can use this information to determine the sensitive components in the system.
- Method of coming up with the node probability tables.
  An example assignment of probabilities to the Node Probability Table (NPT) table for the "Net Failure Percentage" Node of component C_2 is shown in Fig. 11.7. We assume that the NPT for this node is the same for all the nearest neighbors. Initially, the NPT weights are assigned based on subjective confidence and on prior successes in assigning this type of weights. However, these are updated based on the results obtained from dynamic operation as discussed in Sect. 11.3.1.5.2.

Combining the results from all the nearest neighbor components

We need to combine the results from all the nearest neighbors to predict the operational reliability of the component. We show a unique way of combining the results, a more conservative approach that does not lead to excessive overestimation of the predicted reliability results. We assume that the component nodes are updated taking into consideration the current component rank and failure percentages. We present the combination algorithm in Fig. 11.8.

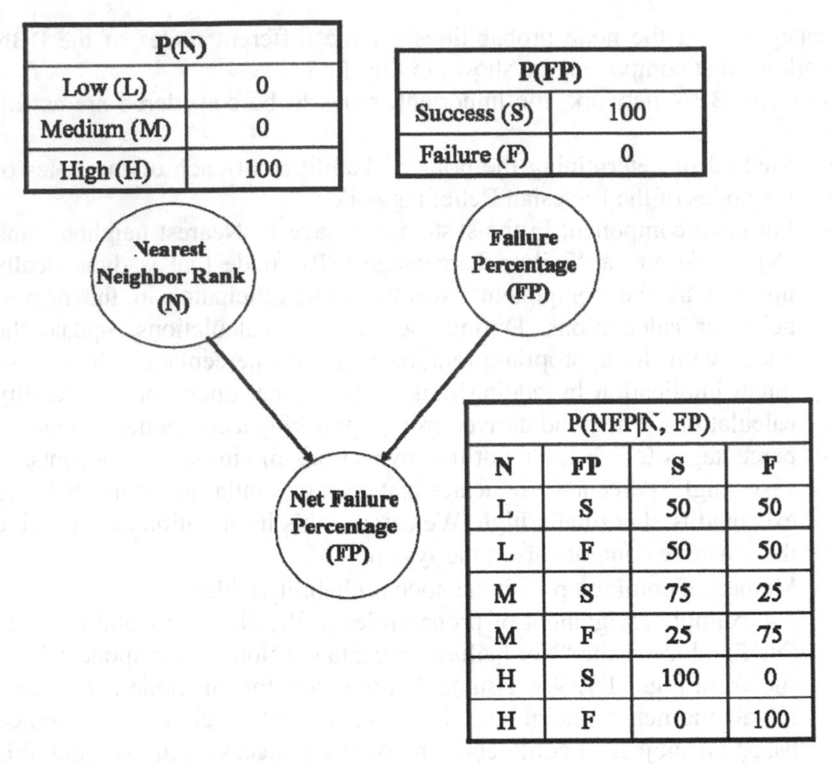

P(N)	
Low (L)	0
Medium (M)	0
High (H)	100

P(FP)	
Success (S)	100
Failure (F)	0

P(NFP|N, FP)			
N	FP	S	F
L	S	50	50
L	F	50	50
M	S	75	25
M	F	25	75
H	S	100	0
H	F	0	100

Fig. 11.7 Probabilities of BBN nodes of a component

Calculating the confidence for the combined reliability estimates

Let the failure probability calculated from Fig. 11.8 be Θip and the number of executions associated with Θip be Tp (Tp = combination of executions of all the components used in the prediction). Let the desired failure probability of the component be Θi. We have two cases:

a) Θ_{ip} is less than the desired Θ_i, which is generally not usual given the conservative nature of our prediction. Then using the formula,

$$T_p = \ln(1 - C_{ip})/\ln(1 - \Theta_{ip}),$$

we calculate the confidence C_{ip}. If C_{ip} is within our desired confidence limit, C, then we add T_p additional execution cases to the original component and calculate the reliability estimate, Θ_i, using the formula,

$$U = [\ln(1 - C)/\ln(1 - \Theta i)] - Tp.$$

$$\text{Thus, } \Theta i = 1 - e[\ln(1 - C)/(U + Tp)].$$

If C_{ip} is not within our limits, then we do not have enough confidence for assuring Θ_i. We calculate the remaining number of execution cases required to achieve the desired confidence and let it be T_q. We use this information in Sect. 11.3.1.5.2, where we combine this evidence with the actual operational failure evidence of the component in question.

b) Θ_{ip} is greater than the desired Θ_i, which is generally the case. We calculate a new failure rate Θ_{ir} that is derived by using the desired C and possible with running T_p executions using the formula specified in a). We use this information in Sect. 11.3.1.5.2.

1) For each nearest neighbor component

- Let x be the ranking state (high/medium/low) of the component
- Determine if x has the highest percentage of probability in the components "Nearest Neighbor Rank" node
- If Yes, use this component for further analysis and place it in the list of selected components
- If Not, discard this component

2) Combine the list of selected components using the inverse weighted approach.

- Let the list of selected components be $C_1, C_2, \ldots, C_n$
- Let their success percentages be $p_{s1}, p_{s2}, \ldots, p_{sn}$ as determined from the "Net failure Percentage" node.
- Let the normalized distance measures be $d_1, d_2, \ldots, d_n$
- Then The success percentage of the combined "Predicted Failure Percentage" node is calculated as,

$$\left(\sum_i p_{si} * (1/d_i)\right) \Big/ \left(\sum_j (1/d_j)\right)$$

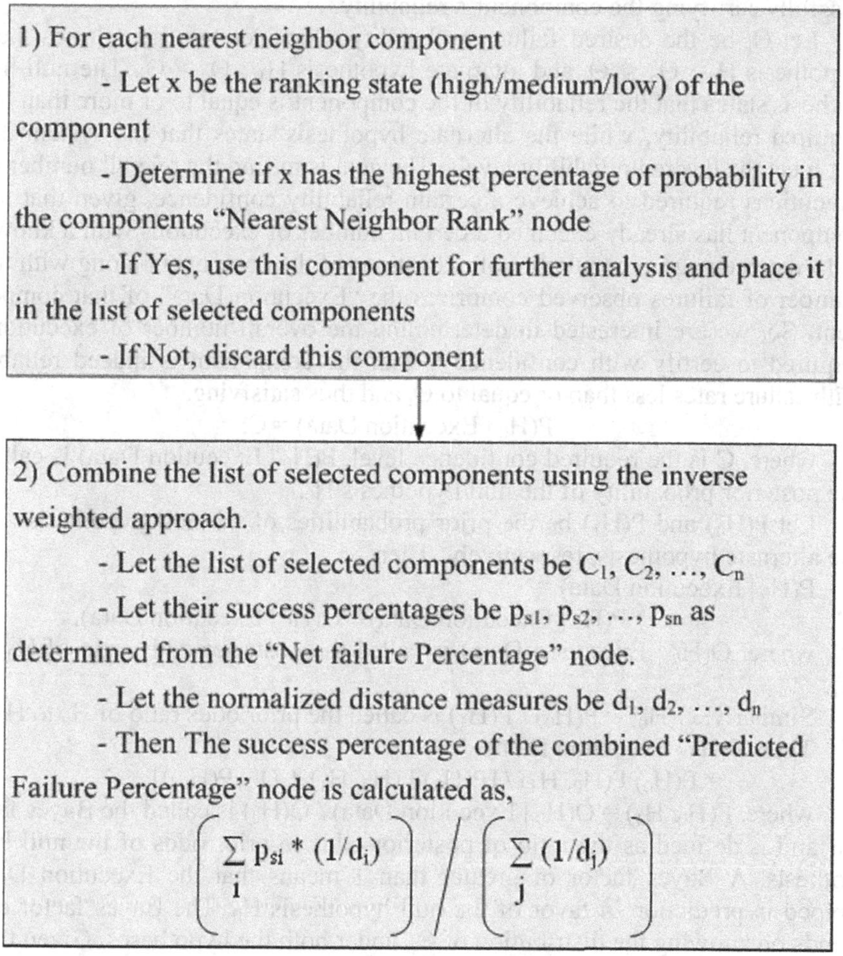

Fig. 11.8 Calculating the Combined Predicted Success Percentage

11.3.1.5.2 Dynamic Reliability Assessment from combining all the evidences

The next step is to combine the reliability measure obtained from the prediction stage with the actual reliability measures obtained from operational/testing usage of the component. We use Bayesian Hypothesis (Cukic et al. 2003) for estimating the number of remaining executions required for successfully certifying the component's reliability.

Let Θ_i be the desired failure probability. Consider two hypotheses: null hypothesis $H_0 : \Theta_o \leq \Theta_i$ and alternate hypothesis $H_1 : \Theta_o > \Theta_i$. The null hypothesis states that the reliability of the component is equal to or more than the required reliability, while the alternate hypothesis states that the system did not meet the desired reliability levels. The goal is to find the overall number of executions required to achieve a certain reliability confidence, given that the component has already executed a certain number of executions with a known failure percentage. The number of executions of the component along with the number of failures observed comprises the "Execution Data" of that component. So, we are interested in determining the overall number of executions required to certify with confidence C that the component is indeed reliable with failure rates less than or equal to Θ_i and thus satisfying,

$$P(H_0 \mid \text{Execution Data}) = C,$$

where, C is the required confidence level. $P(H_0 \mid \text{Execution Data})$ is called the posterior probability of the null hypothesis H_0.

Let $P(H_0)$ and $P(H_1)$ be the prior probabilities of the null hypothesis and the alternate hypothesis, respectively. Then,

$$P(H_0 \mid \text{Execution Data}) = P(H_1 \mid \text{Execution Data}) * O(H_0 \mid \text{Execution Data}),$$

where, $O(H_0 \mid \text{Execution Data})$ is called the posterior odds ratio of H_0 to H_1.

Similarly, $O(H_0) = P(H_0) / P(H_1)$ is called the prior odds ratio of H_0 to H1.

Then, $P(H_0 \mid \text{Execution Data})$

$$= P(H_0)\, F(H_0, H_1) / [P(H_0)\, F(H_0, H_1) + (1 - P(H_0))],$$

where, $F(H_0, H_1) = O(H_0 \mid \text{Execution Data}) / O(H_0)$ is called the Bayes factor and is defined as the ratio of posterior odds to prior odds of the null hypothesis. A Bayes factor of greater than 1 means that the Execution Data helped in prediction in favor of the null hypothesis H_0. The Bayes factor depends on knowing the distribution of Θ_o under both the hypotheses. Given that our reliability assessments are geared toward mission critical SOA systems that generally exhibit very few faults and because the desired Θ_i is very small, we assume uniform distribution of Θ_o under both the hypothesis H_0 and H_1. However, for smaller Θ_i values, assuming Uniform$(0, \Theta_i)$ for Θ_o under H_0 is a more valid assumption than assuming Uniform $(\Theta_i, 1)$ for Θ_o under H_1. The detailed conceptual foundations, assumptions, and using other distributions with the Bayesian Hypothesis framework can be found in (Cukic et al. 2003).

Calculation of $P(H_0)$

In Sect. 11.3.1.5.1, while calculating the confidence for the combined reliability estimates, we have considered two cases: Θip (calculated from Fig. 11.8) less than the desired Θi and Θip greater than the desired Θi. We revisit those scenarios to calculate P(H0) for each of the above cases

a) Θ_{ip} is less than the desired Θ_i
In this case, we calculated the remaining number of execution cases required to achieve the desired confidence to be T_q. Since, we have a very low failure rate, we assume a high probability of zero failures for T_q. Then, $P(H_0)$ translates to a simple ratio of $(U - T_q) / U$.
b) Θ_{ip} is greater than the desired Θ_i
In this case calculating $P(H_0)$ requires a more complex procedure. We calculate a new failure rate Θ_{ir} that can be achieved with confidence C using T_p executions. Consequently, Θ_{ir} will be less than Θ_{ip} and we calculate the value of $P(H_0)$ as the ratio of $\Theta_{ir} / \Theta_{ip}$.

Determining the Require Number of Component Executions

Let (r, n) represent the execution data, where "r" represents the number of failures and "n" the number of previous executions. This is easily obtained from the component repository. Then,

$$P(H_0 \mid \text{Execution Data}) = P(H_0 \mid n, r) =$$
$$P(H_0)\,F(H_0, H_1) / [P(H_0)\,F(H_0, H_1) + (1 - P(H_0))] = C,$$

So,

$$F(H_0, H_1) = [C/(1\text{-}C)]\,P(H_1) / P(H_0)$$

We consider the simplest case, when the previous executions of the component did not encounter any errors (r = 0). Assuming, uniform distribution of Θo under H0 and H1, then Bayes Factor is

$$F(H_0, H_1) = (1 - \Theta_i)\,[1 - (1 - \Theta_i)^{n+1}] / [\Theta_i\,(1 - \Theta_i)^{n+1}]$$

Hence, $[C/(1\text{-}C)]\,P(H_1) / P(H_0) = (1 - \Theta_i)\,[1 - (1 - \Theta_i)^{n+1}] / [\Theta_i\,(1 - \Theta_i)^{n+1}]$

The total number of executions required to achieve the required confidence levels is then derived as,

$$n = - \left(\ln[((C\,\Theta_i\,P(H_1)) / ((1\text{-}C)\,(1 - \Theta_i)\,P(H_0))) + 1] / \ln(1 - \Theta_i) \right) - 1$$

Thus, if we have "n" executions of the component with no failures, then we are C% confident that $\Theta_o < \Theta_i$. The confidence level should guide us to revisit the beliefs in the NPT table shown in Fig. 11.7. We increase the beliefs to be more positively oriented to predict success. However, if we encounter any failure within the n executions, we have a lower confidence in the reliability estimate. If the number of executions is less than "n", then also, we have a lower confidence in the reliability estimate. Assuming failure free operation, we calculate C at the moment of execution using Θ_i, $P(H_0)$, and n using the formula above. Depending on the C, we can make a decision of whether to continue with the component or not.

Alternatively, using the above formula, we can calculate the new predicted failure rate Θ_i of the current component that can be achieved with confidence C using its current "n" executions. As we add other evidences, the predicted failure rate will be less than the actual failure rate.

As services are executed, identifying the most sensitive service that affects the system reliability the most is critical. It helps to replace such services with new ones with similar interfaces, but having higher reliabilities. The definition of sensitivity of a component is context dependent. In our case, we define the sensitivity of a component to be dependent on the frequency of the invocations of associated services. As the component gets used by many services and is invoked many times more than other components, it is more sensitive and critical. The more sensitive the component, the more sensitive is the service that it is a part of.

11.3.2 System Reliability

The most important aspect of developing any reliability model is to define the service architecture on which the model is developed. An atomic service in an SOA-based system is a single service, which performs the task by itself without the help of other services. It is realized by composing one or more underlying components. The reliability of an atomic service depends on the reliability of those interacting components. A composite service is a higher-level service which performs its task by composing multiple atomic services. It represents a new single service that is more complete in terms of business value-addition and is often called a value-added service. The calculation of composite service reliability depends on the type of SOA-based model chosen. Some models ignore the internal services information by treating the composite service as a whole while other models calculate the composite service reliability from individual service reliabilities and the interactions between them. We concentrate on the second type of models. SOA-based systems/applications are built by using either these atomic or composite services as building blocks. Composite applications, which support specific business requirements and business processes, are created by choreographing/combining many services into a single application. While atomic services and composite services belong to the Service Layer in the SOA architecture, composite applications belong to the Business Layer.

11.3.2.1 Reliability Assessment for Atomic Services

We begin by describing the basic model of an atomic service. An atomic service is composed of independent components (for example, individual binary files) that are composed together as shown in Fig. 11.9.

Any input to the service invokes different components with corresponding inputs to their sub-domains. It is important to note that although a service is generally implemented by a unique corresponding component implementation, the same component could also implement different service interfaces. Hence, the input domain of a service may be considered different from those of its underlying components. The inputs that have been tested for the service and the components are represented by reliability maps that show the regions of inputs that have been successfully tested. For dynamic monitoring, however, the input profile constantly changes based on the mode in which the inputs are invoked in the sub-domain. The incorrect output of a service is the consequence of an incorrect output generated by one of the components used by the service. The component input sub-domains invoked by the service ultimately determine the reliability of the service.

Let the service under consideration be S_a and assume that it invokes two components, X and Y, with input sub-domains [X_1, X_2, X_3] and [Y_1, Y_2, Y_3]. For simplicity, assume that component X is the main implementation for the service and caters fully for service S_a and that X invokes Y. This simplicity enables one to assume that the input domains of S_a and X are the same. If not, one needs to probabilistically determine the inputs of X invoked by S_a for reliability calculations. The most critical part of the reliability estimation of any SOA-based system is the determination of the overall service reliability from the knowledge of the reliability of its components. We use the Markov chain model (Zhu and Gao, 2003, Hamlet et al. 2001) to model the service's reliability. A Markov chain model is constructed from service to components that show the probability of the inputs of the service to its components and then the probability of outputs from a predecessor component falling into a succeeding component. One can use this probability distribution to calculate the input profile for succeeding components and, thus, their effective reliability. We begin by formally describing the Markov model and then the overall service reliability calculation.

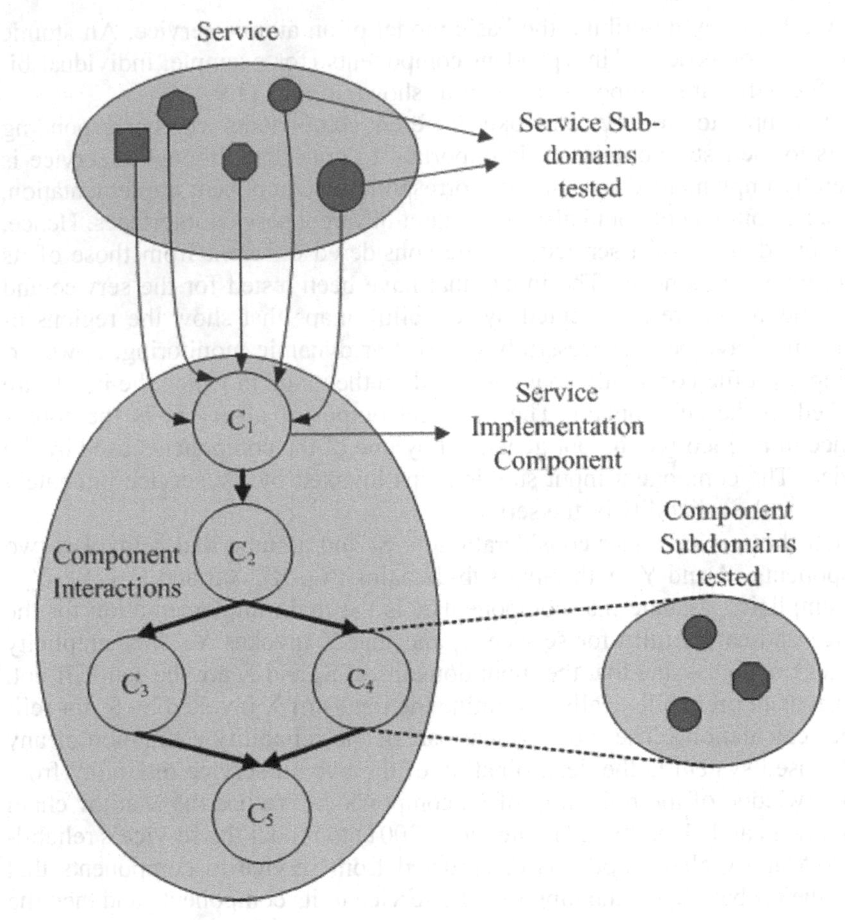

Fig. 11.9 Service Model

11.3.2.1.1 Markov Chain Description

Markov chain model is a process that is characterized by a finite set of states and transition probabilities that specify the probability of moving from one state to the next state. In general, a Markov chain has a unique initial state, where the process starts, and a unique final state, where the process terminates. The other states are called intermediate states and each transition between these states is called a step. A Markov chain is normally represented by a directed graph, consisting of a set of nodes and directed edges that link two consecutive nodes.

The directed graph representing an atomic service can be described by the following terms and their definitions:

- *Directed Graph G*: A directed graph, G = (V, E, s, t) where:
  V = {n}, is the finite set of nodes in the graph,
  E = {e}, is the set of edges representing transitions among two consecutive nodes,
  s = initial node or the invocation node, and
  t = final node.
- *Node "n"*: A node represents a state of an atomic service. In our study, a state corresponds to a sub-domain of a component. As discussed in Sect. 11.3.1.1, each sub-domain is also characterized by a corresponding failure rate. If X_i is a sub-domain of component X with failure percentage $\Theta_{x,xi}$, then node "X_i" is defined by tuple $\langle X, X_i, \Theta_{x,xi}\rangle$.
- *Edge "e"*: A directed edge models the execution flow from one component to another, specifically, from one sub-domain of a component to the sub-domain of another component. The edges could still exist between the states of a single component that represent intra-component interactions. However, from the service's reliability viewpoint, the understanding of inter-component interactions is more important than understanding intra-component interactions. The reason is that intra-component interactions are generally implemented in an ad-hoc way as compared with intra-component interactions that interact with pre-defined specific interfaces. In our study, we only consider inter-component interactions.

An edge from node "X_i" to node "Y_j" is represented by a tuple $\langle X, X_i, Y, Y_j, p_{x,xi,y,yj}\rangle$ where $p_{x,xi,y,yj}$ is the transition probability on a link from state X_i of component X to state Y_j of component Y. There can exist a maximum number of N*N directed edges in the directed graph.

- *Transition probability "$p_{x,xi,y,yj}$"*: The transition probability, $p_{x,xi,y,yj}$, is the probability that the system moves from the current state X_i to state Y_j at the next transition of the system. If there is no interaction between the two states, say, there is no transition from state X_i of component X to state Y_j of component Y, then $p_{x,xi,y,yj} = 0$ indicating the non-existence of an edge between X_i and Y_j in the directed graph. The definition of transition probabilities implies that,

$$0 \le p_{x,xi,y,yj} \le 1 \text{ and } \sum_{j=1}^{card(y)} p_{x,xi,y,yj} = 1.$$

We call this directed graph a "service control flow graph (SCFG)", since it represents the control flow among components in an atomic service. Thus, a service control flow graph is defined as,

SCFG = (V, E, s, t), V = {n}, E = {e}, s = initial node, t = final node,
$n = \langle X, X_i, \Theta_{x,xi}\rangle$, $e = \langle X, X_i, Y, Y_j, p_{x,xi,y,yj}\rangle$

$$0 \leq p_{x,xi,y,yj} \leq 1 \text{ and } \sum_{j=1}^{card(y)} p_{x,xi,y,yj} = 1.$$

Given the formal description of the Markov chain model, we proceed to determine the reliability of the simple two component model described in Sect. 11.3.2.1.

11.3.2.1.2 SCFG Construction and Atomic Service Reliability Analysis

The component architecture and SCFG of service S_a consisting of two components X and Y are shown in Fig. 11.10 and Fig. 11.11, respectively

The steps involved in construction of the SCFG graph are:

1. Estimation of the operational profiles for each of the main service components;
2. Estimation of the transition probabilities between two consecutive components;
3. Estimation of the operational profiles of the current component from the knowledge of the operational profiles of the immediate preceding component and the transition probabilities;
4. Estimation of the reliability of the service from the composition of the components and their reliabilities.

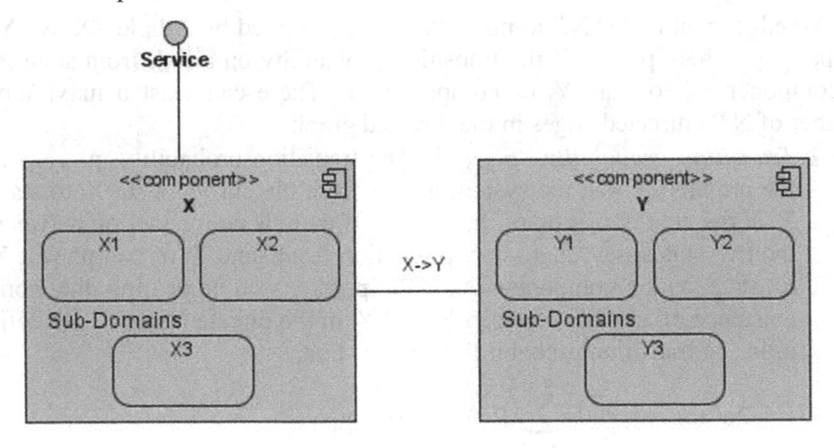

Fig. 11.10 Two Component Sequential Structure

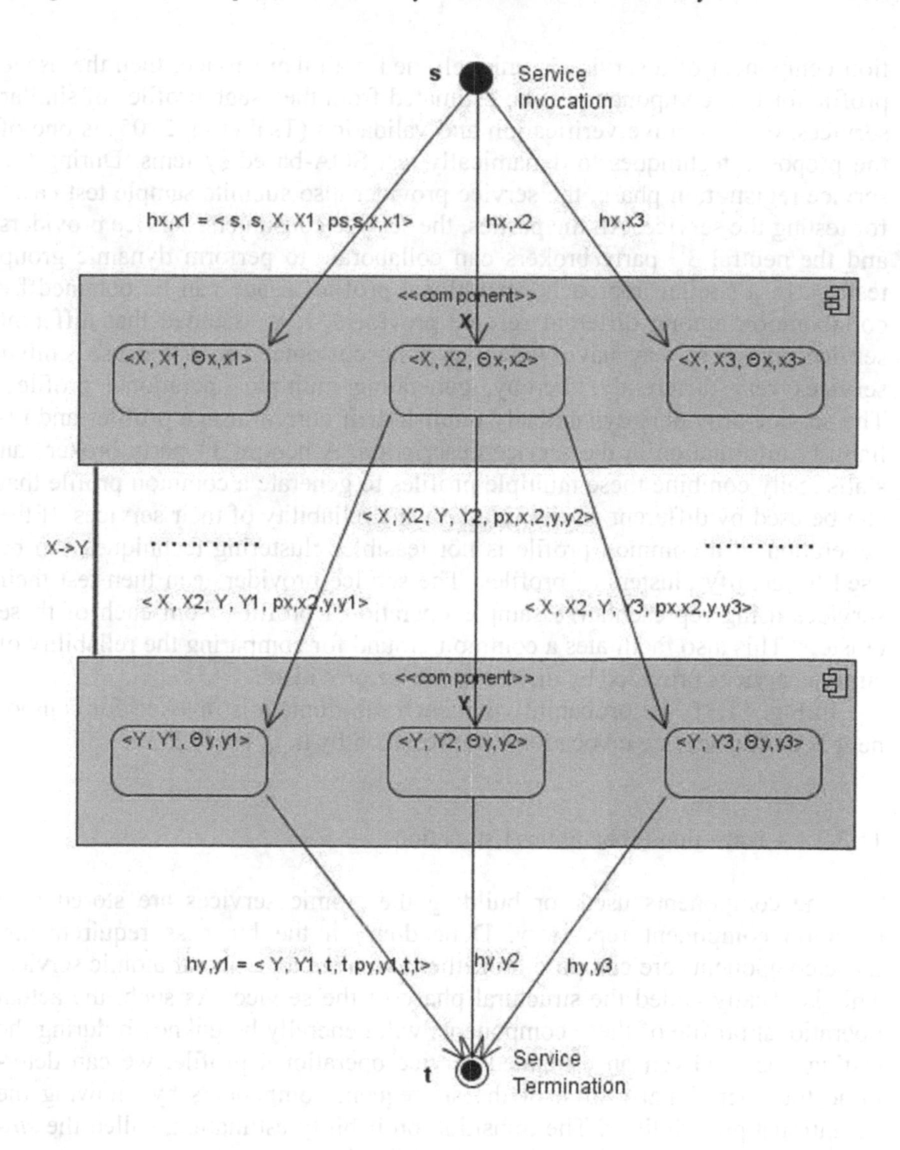

Fig. 11.11 Two Component SCFG graph

11.3.2.1.3 Operational Profiles Estimation for a Service Component

Sect. 11.3.1.2 describes some of the approaches for estimating the operational profile of a component. The most general estimation process includes usage analysis of similar components and through surveys. If the main invocation component of a service is uniquely tied with that service, then the usage profile for that component can be estimated from the usage profiles of similar services. Collaborative verification and validation (Tsai et al. 2005) is one of the proposed techniques to dynamically test SOA-based systems. During the service registration phase, the service provider also submits sample test cases for testing the service. All the parties, the service consumers, service providers and the neutral 3rd party brokers can collaborate to perform dynamic group testing. In a similar approach, operational profiles usage can be obtained by collaboration among different service providers. It is assumed that different service providers may have widely varying customer bases that use similar services very differently, thereby, generating multiple operational profiles. The service providers dynamically publish their current usage profiles and reliability information in the service description. A neutral 3rd party broker can statistically combine these multiple profiles to generate a common profile that can be used by different parties to assess the reliability of their services. If the generation of a common profile is not feasible, clustering techniques can be used to identify clusters of profiles. The service providers can then test their services using representative sample operational profiles from each of these clusters. This also facilitates a common ground for comparing the reliability of similar services provided by different service providers.

In Fig. 11.11, the probability that each sub-domain is invoked for component X during service invocation is represented by $h_{x,xj}$, j = 1, 2, 3.

11.3.2.1.4 Transition Probabilities Estimation

The components used for building the atomic services are stored in a common component repository. Depending on the business requirements, these components are composed together to realize a particular atomic service. This is usually called the structural phase of the service. As such, the actual operational profile of these components will generally be unknown during the testing phase. Given an estimated service operational profile, we can determine the operational profile of the subsequent components by knowing the transitional probabilities. The transition probability estimation, called the *statistical phase* of the service, follows the following steps:

- A random sample of inputs from each of the component's sub-domain is selected.
- The fraction of inputs falling into the succeeding component's sub-domain from each of the current component's sub-domain is the estimated transition probability between the two interacting sub-domains.

11.3.2.1.5 Operational Profile Estimation for Intermediate Components

For components X and Y, the sub-domains, input profiles, and their failure rates are shown in Table 11.1. The input profiles of component Y are unknown. Using the input profiles of component X and the transition probabilities $p_{x,xi,y,yj}$, we calculate the profile of Y as

$$h_{y,yj} = \sum_{xi \in X} h_{x,xi} * p_{x,xi,y,yj}.$$

The input profile of this component can then be used to calculate the profiles of the succeeding components.

Table 11.1 The Sub-domains, Input Profile and Failure rates of Components

	Component X	Component Y
Sub-domains	$\langle X_1, X_2, X_3 \rangle$	$\langle Y_1, Y_2, Y_3 \rangle$
Input Profile, h	$\langle h_{x,x1}, h_{x,x2}, h_{x,x3} \rangle$	$\langle h_{y,y1}, h_{y,y2}, h_{y,y3} \rangle$
Failure Rates, Θ	$\langle \Theta_{x,x1}, \Theta_{x,x2}, \Theta_{x,x3} \rangle$	$\langle \Theta_{y,y1}, \Theta_{y,y2}, \Theta_{y,y3} \rangle$

11.3.2.1.6 Service Reliability Calculation

Using the SCFG graph, we analyze the reliability of the atomic service from the reliability of the components. The calculation of service reliability involves the following steps:

- Calculation of individual component reliability: As discussed in Sect. 11.3.1.1, each component's sub-domain is subjected to uniform random testing and the failure rates are estimated. These failures rates are combined with the operational profiles calculated in Sect. 11.3.2.1.5 to calculate the component reliability under that profile. Then, the reliability of component X is calculated as,

$$Rx = \sum_{xi \in X} h_{x,xi} * (1 - \theta_{x,xi}).$$

- Calculation of atomic service reliability: The service reliability is computed from the structure of the components in the atomic service as shown by the SCFG graph. For an atomic service composed using a sequential structure of components, the operational profile for the first component is successively transformed to an operational profile of the second component. Then the atomic service reliability is just the product of the individual component reliabilities. If C_1, C_2, C_3 are the components in the sequential structure as shown in Fig. 11.12, then the reliability of the atomic service is,

$$R = R_{c1} * R_{c2} * R_{c3}.$$

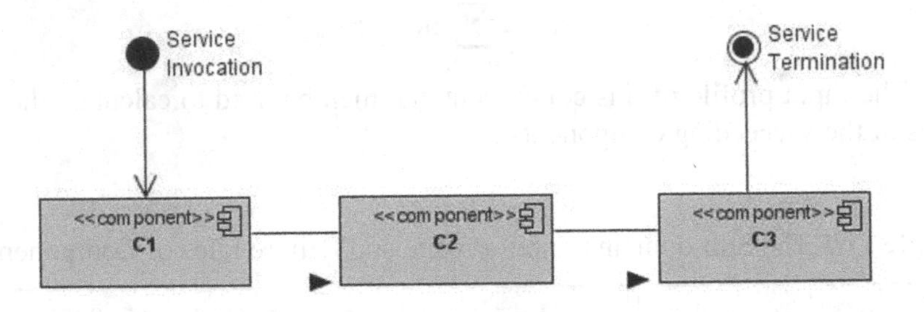

Fig. 11.12 Sequential Navigation

11.3.2.2 Reliability Assessment for Composite Services

A composite service is represented by a chart diagram, generally, called the service chart. The service chart describes, as a flow graph, the different services involved and their execution workflow process. We are only concerned with the service layer invocations from the provider's view point and do not cover the business process of the client that does these invocations. When we consider from the provider's view point, the atomic services composition to realize a composite service is not much different from the component composition that forms an atomic service. Both service and component have defined interfaces and are amenable to be reused in different contexts. However, there are very specific architectural considerations for SOA services. Generally the interaction between services is of request-response mode (similar to call-and-return style of components) and not nested like component architectures. The primary communication between services is through message exchanges/routing. This lack of continuity in the architecture of the composite service makes its overall reliability calculation very simple. The interaction can be modeled as a sequential structure of atomic services.

Thus, the composite service reliability calculation is similar to the multi component interaction described in Sect. 11.3.2.1. The reliability of composite service depends on the atomic services involved. For an input from a subdomain of the starting service in a composite service, we determine the other services involved and their sequences of invocations. Depending on the specific sequence and the input profiles of the different services involved, the reliability of the composite service is determined.

11.4 Experimental Studies

The SOA-system used in our real-world is an Enterprise Content Management (ECM) application. An ECM application generally provides basic services on the content, such as, content storage and retrieval, versioning, security, classification, search, audit trails and other administrative functionalities such as user management, reporting and export/import of bulk content. Some of the providers group a set of low level services to provide a higher level composite service. Each of these services is realized by coordination of a number of components.

11.4.1 Features of the ECM Application

Some of the features of the ECM application used in our case study are:

- Programming Language: The SOA-based services are written in Java using object-oriented methodologies. In general, components are grouped together in single package structures that represent a particular service implementation
- Services: The application predominantly consist of the following services, Content Service (that provides functionalities such as create, read, update, or download content), Authentication Service (that verifies a particular user's credentials), Authorization Service (that deals with specific user access permissions), Search Service (that provides capability to search on the stored content using different criteria), Authoring Service (that provides check-in/check-out and versioning of content), Audit Trail Service (that logs the usage of the application by different users), and Administration Service (that performs user management and bulk import/export functionalities). We have used many third party bundled libraries in our application and we do not perform monitoring of those components. Each of the services varied in the number of related components used to compose that service. The Content Service, which provided a greater functionality of the application, had nearly 3 times more components than the Authentication Service. Overall, the application has more than 200 components.
- Metrics Monitored: Table 11.2 shows the different metrics used in our analysis. These complexity and size metrics include well known met-

rics, such as Halstead, McCabe, Line count, Operator/operand count, and Branch count metrics.

Table 11.2 Metric groups used in this experiment

Metric Type	Metric	Definition
McCabe	ELOC	Executable lines of Code
Halstead	N	Length
	V	Volume
	D	Difficulty
	I	Intelligent Content
	E	Effort
	T	Programming Time
	L	Level
Line Count	BLOC	Lines of Blank
	CLOC	Lines of Comment
	LOC	Total Lines of Code including, Executable, Blank and Comments
Operator / Operand	UNOD	Unique Operands
	UNOT	Unique Operators
	NOD	Total Operands
	NOT	Total Operators

11.4.2 Dynamic Reliability Monitoring

The application is initially subjected to testing and the defect data from the services and components are stored in a common repository. The system is then subjected to dynamic reliability monitoring. The ECM application that we used lacked the logging format that is required to store the defect data in the repository. Hence, we injected our logging functionality into each of the service component classes. The application has to be run in the debug mode for it to log the required defect data into the repository. This is provided as an option for the application user to disable dynamic monitoring at any time

For each invocation of the service in the dynamically monitored environment, we do the following:

a) Determine the sequence of components invoked by the service.
b) Determine the input sub-domain for each service component. For this experiment, we assume that each component in the application has only one sub-domain.
c) For each component, determine if the input sub-domain is sufficiently tested.
d) If not, determine the net reliability using the assessment framework. In the process, update the rank of the nearest neighbors.
e) Determine the service reliabilities from component reliabilities.
f) Store the component's execution data into the repository.
g) Assess the performance of the framework.

11.4.3 Storing the Testing and Dynamically Monitored Data

After the testing data is stored in the repository, the service and components are subjected to dynamic monitoring. The number of times that each component is invoked along with its nearest neighbor rank is updated continuously. The error logs are monitored to determine if a particular component is defective or not. During the user acceptance testing phase, the percentage of faulty components noticed is 5.58%. For demonstrating our framework, we consider a specific instance of dynamic execution when, for example, the "Search Service" is invoked.

11.4.4 Dynamic Monitoring for a Specific Service Invocation

In this section, we consider the dynamic reliability of "Search Service" that performs simple text search on the stored content in the ECM application. The ECM application stores documents on which simple text search can be performed to retrieve those specific documents that either contains the search string in their text or in their file name.

During this specific invocation, we observed the following:

- The number of executions of all the components is 278454.
- The executions of the top 10 components are as follows: 119177, 24250, 14619, 11501, 10336, 9398, 8784, 8014, 7294, and 6149.
- One of the important tasks in our framework is to update the components with their ranking state (high/medium/low). We found that all the components at one point or another are involved in nearest neighbor

calculations. Of all the components, 36.8% components are ranked "high", 27.88% are ranked "medium", and 35.32% components are ranked "low".

Now, consider the invocation of the "Search Service" for performing a simple search. There are primarily four components that are being invoked:

a) Component 1 (SC_1): The Service Registry component searches the registry for the appropriate search service and returns the description of that service. This enables the clients to communicate with the service provider and acquire references to that search service.
b) Component 2 (SC_2): This component is the main implementation of the search service interface. This class queries the ECM application for getting the search results.
c) Component 3 (SC_3): This component provides the search context that enables the user to specify different search criteria such as the minimum length of the search string, the number of search results, the ranking of the results, and so on.
d) Component 4 (SC_4): A search service helper class that helps in constructing the search query according to the criteria specified by the user.

11.4.4.1 Reliability Assessment of Component SC_2

In this section, we consider the reliability of the main implementation component SC_2. The component is invoked 25 times. This is clearly insufficient for one to have a high confidence in the reliability of the component. Hence, the reliability assessment procedure presented in Sect. 11.3.1.5 is used. By using MBR, the nearest components that match the characteristics of this component are computed. The percentage of nearest neighbors that were ranked high for component SC_2 is 43.28%. The BBN's and the Euclidean distances for the first four nearest neighbors are shown in Fig. 11.13 and Table 11.3, respectively

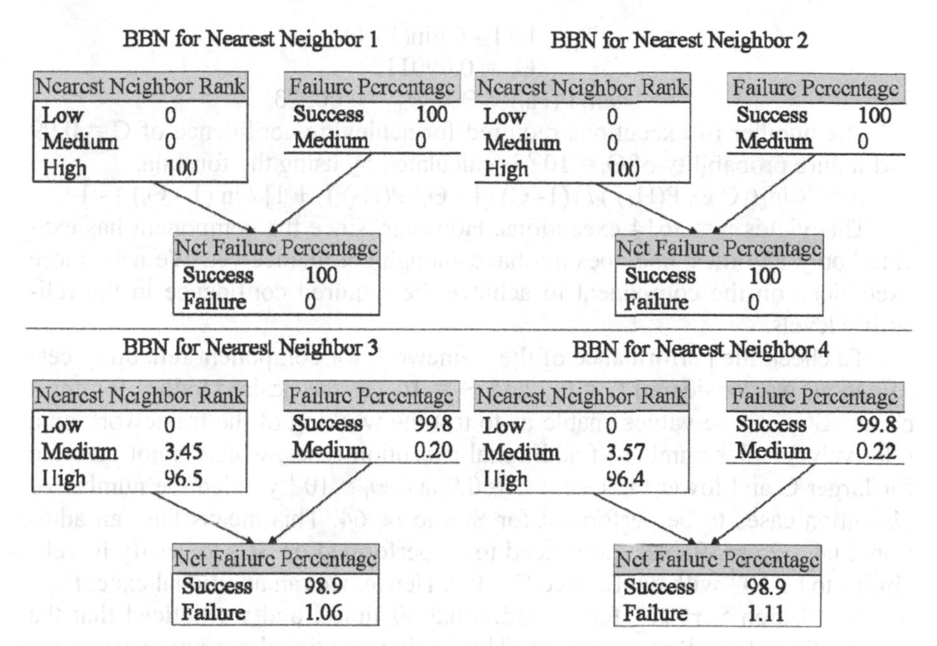

Fig. 11.13 BBN's of the four nearest neighbors of Component SC2

Table 11.3 Euclidean distance and Number of Executions for the four nearest neighbors of Component SC_2

Nearest Neighbor	Euclidean Distance	Number of Executions
1	18429.93	433
2	18922.95	218
3	25634.065	1019
4	26130.51	1824

The total number of executions performed by all the nearest neighbor components is T_p = 42076. Using the algorithm described in Fig. 11.8, the combined predicted success percentage using all the nearest neighbors is determined to be 95.97%. Hence, the combined failure probability = Θ_{ip} = 1 – 0.9597 = 0.0403. All probabilities values are rounded off to 5 decimal places and used in further calculations.

Let the desired failure probability of the component be $\Theta_i = 10^{-3}$. We notice that $\Theta_{ip} > \Theta_i$. Hence, using the formula from Sect. 11.3.1.5.2, we calculate a new failure rate Θ_{ir} that is derived by using the desired C = 0.99 and possible with running Tp executions as,

$$Tp = \ln(1 - C)/\ln(1 - \Theta_{ir}),$$
$$\Theta_{ir} = 0.00011.$$
$$\text{Then, } P(H_0) = \Theta_{ir} / \Theta_{ip} = 0.00273.$$

The number of executions required for achieving confidence of $C = 0.99$ and failure probability of $\Theta_i = 10^{-3}$ is calculated by using the formula,

$$n = - (\ln[((C\ \Theta_i\ P(H_1)) / ((1- C)\ (1 - \Theta_i)\ P(H_0))) + 1] / \ln (1 - \Theta_i)) - 1$$

This yields n = 3614 executions. However, since the component has executed only 25 times, one does not have enough confidence yet. We need more executions on the component to achieve the required confidence in the reliability levels.

To check the performance of the framework for component reliability certification, we considered $C = 0.9$ and $\Theta_i = 10^{-1}$ as the desired values for component SC_2. These values enable us to test the working of the framework on a relatively smaller number of additional execution cases, which is not possible for larger C and lower Θ_i. Using $C = 0.9$ and $\Theta_i = 10^{-1}$ yielded the number of execution cases to be performed for SC_2 to be 64. This means that, an additional 64 – 25 = 39 executions need to be performed on SC_2 to certify its reliability to be 10^{-1} with confidence $C = 0.9$. Hence, we ran additional executions on the "Search Service" for an additional 39 times, and we noticed that the system did not produce any errors. Thus, using additional execution cases, we could certify that the component obtained the necessary reliability. If the component experienced a failure during these additional executions, then more executions are required to be run to certify the component for the same reliability at the same confidence levels.

11.4.4.2 Reliability Assessment of the Search Service

Similar to SC_2, the reliability assessment calculations are performed for components, SC_1, SC_3, and SC_4. The experimental data, that shows the number of additional executions that each component requires as a function of different failure rates and confidence levels for of these four components, is shown in Table 11.4. For the search service to have a failure probability of $\Theta_i \leq 10^{-3}$ with confidence C=0.99, we need to have at least 3850 more executions of the service using random inputs.

Table 11.4 Number of additional executions required for all the four "search service" components to achieve the required reliability and confidence levels

Component	Number of Additional Executions; C=0.99; $\Theta_i = 10^{-3}$	Number of Additional Executions; C=0.9; $\Theta_i = 10^{-1}$
SC_1	792	Reliability already reached (73 required; executed 3764 times)
SC_2	3589	39
SC_3	3850	Reliability already reached (68 required; executed 147 times)
SC_4	3808	53

The most important part of the entire service certification process is the estimation of $P(H_0)$ using the MBR and BBN techniques. As is the obvious case, a system containing components with very few failures will lead to a higher estimated value of $P(H_0)$ from the nearest neighbors. This will in turn reduce the number of additional executions required to certify the service reliability. On the other hand, a system with more component failures will lead to a lower estimated value of $P(H_0)$ and, thus, will require additional executions for reliability certification. We used a very conservative approach in the estimation of $P(H_0)$ by using only highly ranked neighbors in its calculations. The number of executions calculated by the framework and the input-domain reliability formula are shown in Table 11.5 and their comparison using logarithmic-scale is shown in Fig. 11.14.

Table 11.5 Executions required for Component SC_2 as a function of Θ_i for C = 0.99

SC_2, C = 0.99; Θ_i	Number of required Component Executions calculated using framework	Number of required Component Executions calculated using, $T_p = \ln(1 - C)/\ln(1 - \Theta_i)$
10^{-1}	78	44
10^{-2}	588	459
10^{-3}	3614	4603
10^{-4}	15297	46050

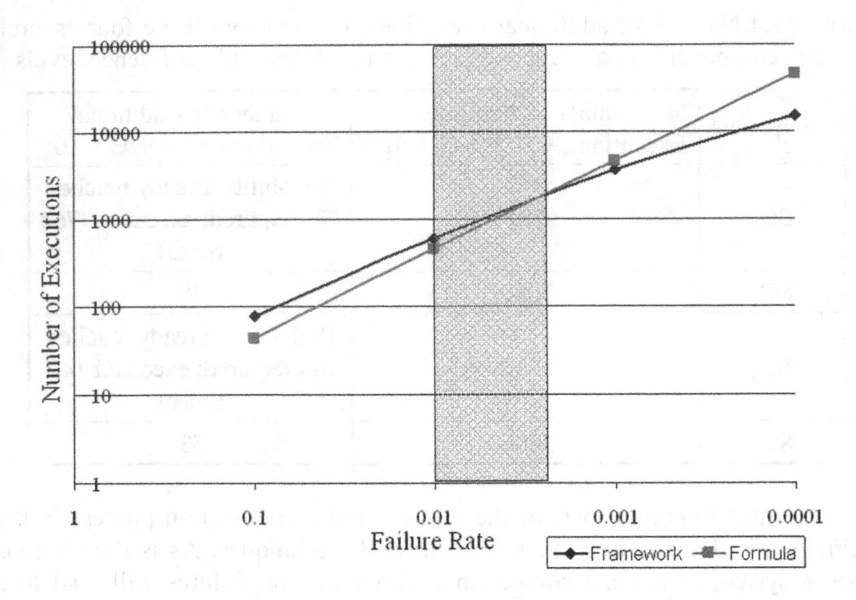

Fig. 11.14 Number of framework-based and theoretical (formula-based) executions for Component SC2 as a function of Θi for C = 0.99

For mission-critical systems, in most cases, a minimum failure rate $\Theta i < 10^{-2}$ is required. At $\Theta_i = 10^{-2}$, we notice from Fig. 11.14 that the framework estimates more execution cases than the formula. This is again due to the conservative nature of using highly ranked neighbors and node probabilities assigned in BBN calculations. The shaded rectangle in the graph shows the region between the minimum desired lower limit of failure rate and the intersection of the two execution curves. From the graph, we note that the following key points should be considered while dynamically assigning the node probabilities:

a) Overestimating $P(H_0)$: The node conditional probabilities in Fig. 11.7 could be adjusted such that the derived conditional net failure percentage given the failure percentage of the node and its nearest neighbor rank, now yields a higher percentage of success than the original value. For example, currently, we use the following conditional probability to calculate 'Net Failure Percentage of Failure', for case where nearest neighbor rank is "High" and the failure percentage is "F", as P(NFP = 'F' | N = 'H' and FP = 'F') = 1 and conversely P(NFP = 'S' | N = 'H' and FP = 'F') = 0. We could modify these probabilities to P(NFP = 'F' | N = 'H' and FP = 'F) = 0.75 and conversely P(NFP = 'S' | N = 'H' and FP = 'F) = 0.25 giving higher weightage to predicting success. This will reduce the number of execution cases required, thus moving the graph's region towards the lower limit. While this can be done, we need to be careful that the intersection point of the curves should never cross the lower limit. This will lead to dangerously under-estimating the node probabilities and, thus, overestimating the prior beliefs $P(H_0)$.

b) Underestimating $P(H_0)$: We could conversely underestimate $P(H_0)$ by adjusting node probabilities such that the derived conditional net failure percentage, given the failure percentage of the node and its nearest neighbor rank, now yields a higher percentage of failure than the original value. While this is not so detrimental compared to the overestimation case, this could lead to an increased number of execution cases for failure probability requirements bordering $\Theta_i = 10^{-2}$. At this point, however, we can use the minimum of executions cases predicted by the framework and the formula to be the ideal number of remaining execution cases required. Since our framework is more geared towards reliability assessment for mission critical systems, the assessment cost reduced by decreasing the number of execution cases always outweighs the cost of slightly underestimating $P(H_0)$.

c) Boundary of the Framework and Formula calculations: The framework calculates the number of executions required for achieving a confidence of C, and failure probability of Θ_i by using the formula, $n = -(\ln[((C\ \Theta_i\ P(H_1)) / ((1-C)(1-\Theta_i) P(H_0)))+1] / \ln(1-\Theta_i)) - 1$ However, if $P(H_0) = \Theta_i$, then the above formula reduces to, $n = (\ln(1-C) / \ln(1-\Theta_i)) - 1$, which is the input-domain reliability formula.

Due to the enormous number of execution cases required for testing the performance of framework for very lower values of Θ_i and higher C, we perform the framework validation for C=0.9; $\Theta_i = 10^{-1}$. Table 11.4 lists the number of additional executions required for various components to achieve the failure rate of $\Theta_i = 10^{-1}$, with confidence C=0.9. We subjected the components for the additional executions and noticed that no errors have been recorded. If an error has been recorded, then that means we underestimated the number of execution cases required and, thus, more executions need to be performed. With no failures from the service after additional executions, we certify with 90% confidence that the service is indeed reliable with failure rates less than $\Theta_i = 10^{-1}$.

11.5 Conclusions

We presented a methodology to combine additional evidences in the form of operational data of the services/components with testing data for better reliability estimates of a system. The number of times the input subdomains of each service/component are invoked is stored and not the actual inputs. This greatly reduces the memory needed to store the data. Software metrics are used as additional sets of information to help with the reliability diagnosis process. A technique was presented to combine data from "similar" components. The reliabilities of the components are monitored by performing dynamic reliability assessment using MBR and BBN techniques. Any decrease in the reliability estimate below a certain threshold raises an alarm. This type of proactive monitoring is very helpful for enhancing the overall dependability of mission-critical systems.

There are several interesting future research directions. Some areas of research include defining the entire monitoring process including the parameters to monitor, the frequency of monitoring the components, and developing efficient and fast retrieval techniques to determine whether an execution of a component is successful or not. Other research aspects include developing mathematical theories that can better explain the assignment of BBN node probabilities and procedures to include results of other evidences, such as verification and validation. Effective and efficient storage mechanisms for the collected dynamic data and providing mechanisms for real-time alerts in case of system failure during dynamic monitoring are some other research issues that need further investigation.

References

[1]. Amman PE, Brilliant SS, Knight JC (1994) The effect of imperfect error detection on reliability assessment via life testing. IEEE Transactions on Software Engineering 20(2):142-148

[2]. Arsanjani A (2004) Service-Oriented modeling and architecture. White paper. IBM developerWorks. http://www.ibm.com/developerworks/webservices/library/ws-soa-design1/. Accessed 24 October 2008

[3]. Canfora G, Penta MD (2006) Testing services and service-centric systems: challenges and opportunities. IT Professional 8(2):10-17

[4]. Challagulla VUB, Bastani FB, Yen IL et al (2008) Empirical assessment of machine learning based software defect prediction techniques. International Journal on Artificial Intelligence Tools 17(2):389-400

[5]. Cukic B, Bastani FB (1996) On reducing the sensitivity of software reliability to variations in the operational profile. In: Proceedings of the 7th International Symposium on Software Reliability Engineering, White Plains, October 1996, pp 45-54

[6]. Cukic B, Gunel E, Singh H et al (2003) The theory of software reliability corroboration E86-D(10):2121-2129

[7]. Delgado N, Gates AQ, Roach S (2004) A taxonomy and catalog of runtime software-fault monitoring tools. IEEE Transactions on Software Engineering 30(12):859-972

[8]. Fenton N, Littlewood B, Neil M et al (1998) Assessing dependability of safety critical systems using diverse evidence. IEE Proceedings Software Engineering 145(1):35-39

[9]. Fenton N, Neil M (1999) A critique of software defect prediction models. IEEE Transactions on Software Engineering 25(5):675-689

[10]. Frankl P, Hamlet D, Littlewood B et al (1997) Choosing a testing method to deliver reliability. In: Proceedings of the 19th International Conference on Software Engineering, Boston, May 1997, pp 68-78

[11]. Gao T, Ma H, Yen IL et al (2005) Toward QoS analysis of adaptive service-oriented architecture. In: Proceedings of the IEEE International Workshop on Service-Oriented System Engineering, Beijing, October 2005, pp 219-226
[12]. Gao J, Wu Y, Chang L et al (2006) Measuring component-based systems using a systematic approach and environment. In: Proceedings of the 2nd IEEE International Workshop on Service-Oriented System Engineering, Shanghai, May 2006, pp 121-129
[13]. Gokhale S, Wong WE, Trivedi KS et al (1998) An analytical approach to architecture based software reliability prediction. In: Proceedings of 3rd International Computer Performance and Dependability Symposium, Los Alamitos, 1998, pp 13-22
[14]. Grassi V, Patella S (2006) Reliability prediction for service-oriented computing environments. IEEE Internet Computing 10(3):43-49
[15]. Hamlet D, Mason D, Woit D (2001) Theory of software reliability based on components. In: Proceedings of the 23rd International Conference on Software Engineering, Toronto, May 2001, pp 361-370
[16]. Kappes M, Kintala CMR, Klemm RP (2000) Formal limits on determining reliabilities of component-based software systems. In: Proceedings of the 11th International Symposium on Software Reliability Engineering, San Jose, October 2000, pp 356-364
[17]. Kuball S, May J, Hughes G (1999) Building a system failure rate estimator by identifying component failure rates. In: Proceedings of the 10th International Symposium on Software Reliability Engineering, Boca Raton, November 1999, pp 32-41
[18]. Krishnamurthy S, Mathur AP (1997) On the estimation of reliability of a software system using reliabilities of its components. In: Proceedings of 8th International Symposium on Software Reliability Engineering, Albuquerque, November 1997, pp 146-155
[19]. Laprie JC, Kanoun K (1996) Software Reliability and System Reliability. McGraw-Hill, New York
[20]. Littlewood B, Strigini L (2000) Software reliability and dependability: A roadmap. In: Proceedings of the 22nd International Conference on Software Engineering, Limerick, June 2000, pp 177-188
[21]. Lo JH, Huang CY, Kuo SY et al. (2003) Sensitivity Analysis of Software Reliability for Component-Based Software Applications. In: Proceedings of the 27th Annual International Computer Software and Applications Conference, Dallas, November 2003, pp 500-505
[22]. May J (2002) Testing the reliability of component-based safety critical software. In: Proceedings of the 20th International System Safety Conference, Denver, August 2002, pp 214-224
[23]. Miller KW, Morell LJ, Noonan RE et al (1992) Estimating the probability of failure when testing reveals no failures. IEEE Transactions on Software Engineering 18(1):33-43
[24]. Mitchell T (1997) Machine Learning. McGraw-Hill Book Co., Singapore
[25]. Musa JD (1993) Operational Profiles in Software-Reliability Engineering. IEEE Software 10(2):14-32
[26]. Paul RA (2005a) DoD towards software services. In: Proceedings of the 10th IEEE International Workshop on Object-Oriented Real-Time Dependable Systems, Sedona, February 2005, pp 3-6
[27]. Paul RA (2005b) Tomorrow's needs - yesterday's technology: DoD's architectural dilemma & plan for resolution. In: The 9th IEEE International Symposium on High-Assurance Systems Engineering, Heidelberg, October 2005, pp 9-12
[28]. Schneidewind NF (1998) Methods for assessing COTS reliability, maintainability, and availability. In: Proceedings of the International Conference on Software Maintenance, Bethesda, November 1998, pp 224-225
[29]. Tsai WT (2005) Service-oriented system engineering: a new paradigm. In: Proceedings of the IEEE International Workshop on Service-Oriented System Engineering, Beijing, October 2005, pp 3-6

[30]. Tsai WT, Zhang D, Chen Y et al (2004) A software reliability model for web services. In: Proceedings of the 8th IASTED International Conference on Software Engineering and Applications, Cambridge, November 2004, pp 144-149
[31]. Tsai WT, Chen Y, Paul RA (2005) Specification-based verification and validation of Web services and service-oriented operating systems. In: Proceedings of the 10th IEEE International Workshop on Object-Oriented Real-Time Dependable Systems, Sedona, February 2005, pp 139-147
[32]. Voas J (2001) Why is it so hard to predict software system trustworthiness from software component trustworthiness?. In: Proceedings of the 20th IEEE Symposium on Reliable Distributed Systems, New Orleans, October 2001, p 179
[33]. Wohlin C, Runeson P (1998) Reliability certification of software components. In: Proceedings of the 5th IEEE International Conference on Software Reuse, Victoria, June 1998, pp 56-65
[34]. Woit D, Mason D (1998) Software component independence. In: Proceedings of the 3rd IEEE High-Assurance Systems Engineering Symposium, Washington, November 1998, pp 74-81
[35]. Yacoub SM, Cukic B, Ammar HH (1999) A scenario-based reliability analysis of component-based software. In: Proceedings of 10th International Symposium on Software Reliability Engineering, Boca Raton, November 1999, pp 22-31
[36]. Yu Y, Johnson BW (2003) A BBN approach to certifying the reliability of COTS software system reliability. In: Proceedings of the Annual Reliability and Maintainability Symposium, Tampa, January 2003, pp 19-24
[37]. Zhang D (2000) Applying machine learning algorithms in software development. In: Proceedings of the Monterey Workshop on Modelling Software System Structures in a Fastly Moving Scenario, Santa Margherita Ligure, June 2000, pp 275-291
[38]. Zhang J (2004) An approach to facilitate reliability testing of web services components. In: Proceedings of the 15th International Symposium on Software Reliability Engineering, Saint-Malo, November 2004, pp 210-218
[39]. Zhu Y, Gao J (2003) A Method to calculate the reliability of component-based software. In: Proceedings of the 12th Asian Test Symposium, Xian, November 2003, pp. 488-491

12 Model, Properties, and Applications of Context-Aware Web Services

Stephen J.H. Yang[1], Jia Zhang[2], Angus F.M. Huang[1]

Abstract Context-aware Web services is a seamless interaction between service requestors and services considering both contextual requirements of services and contextual conditions of the service requestors. We envision that providing context-aware Web services is the first step toward the ultimate goal of e-services by delivering right services from right partners at right time. This chapter aims for the models, properties, and applications of context-aware Web services by developing an ontology-based context model, and identifying context-aware applications as well as their properties. We have developed an ontology-based context model to enable formal description and acquisition of contextual information pertaining to service requestors and services. The context model is supported by context query and phrased acquisition techniques. We will also report three context-aware applications built on top of our context model as a proof-of-concept to demonstrate how our context model can be used to enable and facilitate in finding right services, right partners (collaborators), and right information (content presentation).

12.1. Introduction

One of the essential goals of the Web services technology is to help service requestors dynamically discover and locate desired services. However, the lack of taking contextual information into account usually leads to low-recall, low-precision and irrelevant results using current services discovery techniques. For example, if a service requestor is using a wireless phone as a receiver, delivering a desktop-based application may not be appropriate; if a service requestor is in a meeting, delivering an audio-based application may also be undesirable.

---

[1] Department of Computer Science and Information Engineering, National Central University, No. 300, Jhongda Rd., Jhongli City, Taoyuan County 320, Taiwan

[2] Department of Computer Science, Northern Illinois University, 1425 W. Lincoln Hwy., DeKalb, IL 60115-2825, USA

J.J.P. Tsai and P.S. Yu (eds.), *Machine Learning in Cyber Trust: Security, Privacy, and Reliability*, DOI: 10.1007/978-0-387-88735-7_12,

The objective of context-aware Web services provision is to enhance Web services provision a step further from providing services at anytime anywhere toward at the right time and right place with right services. As people are constantly on the move in nowadays heterogeneous working environment, resources (e.g., computational devices and communication network coverage) are more frequently prone to change due to physical location changes. Therefore, in the process of a service consumption, a service requestor may need to smoothly switch to different levels of services or even different services to adapt to the ever-changing environment (i.e., context), especially when a gross mismatch between resource requirements and supplies occurs. Meanwhile, people tend to continue to work on the move. Thus, more and more handheld devices (e.g., Personal Digital Assistants (PDAs) and mobile phones) have been used to access Web services via mobile communication. Nevertheless, most of the existing Web services are not originally designed for handheld devices. As a result, tools and mechanism are needed to provide users opportunities to experience transparent and seamless services provision.

Context-aware services provision is a mechanism enhancing Web service accesses based on users' varying characteristics and situated environments. In a mobile environment, users usually work with handheld devices, which are featured with good mobility but limited computational capabilities and display sizes. In addition, users' access conditions change more frequently in a mobile environment than in a traditional Web-based environment [1]. For example, users may need to access Web services while they are driving or in a meeting. Therefore, to achieve the goal of context-aware Web services, it is essential to provide personalized and adaptive services according to users' situated environments. In this chapter, the two terms "users' situated environment" and "context" are used interchangeably, both referring to surrounding information, from either service requestors or services, which may impact service execution including computational devices, communication network, lighting, noise level, location, activity, and time [2][3]. We summarize the characteristics of context-aware Web services and their requirements in the following eight aspects: mobility, location awareness, interoperability, seamlessness, situation awareness, social awareness, adaptability, and pervasiveness.

(1) Mobility: The continuousness of computing capability while moving from one position to another. Requirements include mobile computing on portable devices with embedded software.

(2) Location awareness: The capability of detecting and identifying the locations of persons and devices. Requirements include outdoor positioning and indoor positioning.

(3) Interoperability: The capability of interoperable operation between various standards of resource exchange and services composition and integration. Requirements include standards of content, services, and communication protocols.

(4) *Seamlessness: The capability of providing an everlasting service session under any connection with any device. Requirements include state transition of network roaming and service migration.*
(5) *Situation awareness: The capability of detecting and identifying person-situated scenarios. Requirements include knowing what a person is doing with whom at what time and where.*
(6) *Social awareness: The capability of knowing who are socially related? What do they know? And what are they doing at one moment? Requirements include knowing social partners' knowledge competence and social familiarity.*
(7) Adaptability*: The capability of dynamically* adjusting *services/contents depending on users' needs. Requirements include knowing people's accessibility and preferences.*
(8) *Pervasiveness: The capability of providing intuitive and transparent way of service/content access. Requirements include predicting what users want before their explicit expressions.*

The characteristics of context-aware Web services pose significant challenges. Among others, such characteristics and constraints should be formalized with requirements specification, so that they can be precisely defined and captured in order to better satisfy the demands of a mobile environment.

For better explaining why contextual information is necessary for proper services provision and better explaining our solution, let us consider an example scenario that will be used throughout this chapter. Steve is a manager, who just finished a product presentation in a trade fair and finds a "Web business meeting" service to remotely discuss some timely issues (find socially related partners). The "*availability*" of the service must be 99% or above (meet location-aware QoS requirement). During the Web meeting, Steve communicates with his colleagues using various devices for exchanging multimedia-based information (provide device-independent content adaptation). After thirty minutes of "Web business meeting," Steve needs to drive back to his office for a pre-scheduled face-to-face meeting with another customer. He wants to continue the "Web business meeting" seamlessly with his colleagues while he is driving (provide situation-aware seamless interoperability). Hence, Steve needs to automatically switch to "PDA-based Web business meeting."

Based on Steve's requirements, a published Web service entitled "Web business meeting" may be a qualified candidate, which can support both PCs and PDAs via wireless LAN and *General Packet Radio Service* (*GPRS*), respectively, with a guarantee of 99% even when users are physically moving. However, it should be noted that many other contextual information could also affect how well Steve is served. For example, who else attend the conference? Who are Steve's colleagues and what are they doing during the meeting? What if Steve's colleagues are not available at the time? Could Steve find some other information or people with proper expertise? What kind of network channels and devices will Steve use to connect to the Web? How will various situations (e.g., driving) affect Steve's device usage, network access and services delivery? As everyone knows, it is dangerous to pay attention to the video screen in video meeting while driving. Therefore, the context-aware Web service will adapt Steve's video meeting service to audio meeting service according to his contextual information such as movement or driving. The service adaptation will make the original activity continuously and will concern about the feasibility of the active user. All of these kinds of contextual information need to be considered in order to provide better services provision according to Steve's requirements and conditions at the moment.

The main theme of this chapter is the development of a context model and three context-aware applications built as proof-of-concepts. We have developed a context model to formally describe and acquire contextual information pertaining to service requestors and services. Based on the model, we have built three context-aware applications for demonstrating how to find right services, right partners (collaborators), and right information (content presentation). In this chapter, we refer the so-called "right" as "context-aware." The terms "partners" and "collaborators" are used interchangeably, and the same are the terms "information" and "content."

12.2 Related research

12.2.1 Web services and semantic Web

The emerged Web services technology provides a flexible and efficient approach to reuse existing Web-based applications and services. Web services technology concentrates not only on interoperability but also on how to describe, publish, locate, and invoke Web services. A number of standards and specifications have contributed to the development of Web services, such as WSDL (http://www.w3.org/TR/WSDL), SOAP (http://www.w3.org/TR/soap), and UDDI (http://www.uddi.org/). A service provider can thus describe a service in a standard language (e.g., WSDL) to specify what the service can do and how to invoke it. A service registry (e.g., UDDI) creates a standard interoperable platform that enables businesses and applications to quickly, easily, and dynamically locate Web services over the Internet. SOAP serves as a communication standard language. However, the above Web service infrastructure can only behave well in syntactical level and lacks semantic support for inferential capability. This motivates Web services a step further to semantic Web [4].

Semantic Web description languages, such as DAML+OIL (http://www.daml.org/) and Web Ontology Language (OWL, http://www.w3.org/TR/owl-features/), provide predicate-like and description logic-based annotation of Web services to encourage inferential procedures on annotated Web services. The combination of semantic Web's expressive power and Web service infrastructure results in the concept of semantic Web services. As the first step toward semantic Web services, OWL-S (http://www.daml.org/services/owl-s/1.1/overview/) is a specific application for Web services by applying OWL on a service's capability representation. It supports Web service providers with a core set of markup language constructs for describing properties and capabilities of their Web services in an unambiguous, machine-interpretable way. We choose OWL-S as the implementation language for our context-aware Web services.

Service oriented architecture (SOA) is a conceptual model that interrelates services through well-defined interfaces and contracts between them. The interfaces are defined in a neutral manner independent of hardware platforms, operating systems, and programming languages in which the services are implemented. This allows services built on a variety of such systems to interact with each other in a uniform and universal manner. Three fundamental roles exist in SOA - service provider, service requestor, and service matchmaker. A service provider publishes services to matchmakers. A service requestor sends requests to the matchmakers when services are needed. A matchmaker performs service matching to find available services on behalf of service requestors. While Web services are moving toward semantic Web, matchmakers in SOA also favor semantic-based service matchmaking.

LARKS is a semantic matchmaker that performs both syntactic and semantic matching [5]. It can identify different degrees of partial matching based on five filters: context matching, profile matching, similarity matching, signature matching, and constraint matching. Requestors can select any desired combination of these filters based upon different concerns, such as efficiency and computation cost. DAML-S/UDDI is another semantic matchmaker that expands UDDI to provide semantic matching [6]. It can perform inferences based on assumption hierarchy leading to the recognition of semantic matches regardless of their syntactical differences. DAML-S/UDDI adopts a flexible matching strategy based on inputs and outputs to identify similarities between requests and services' advertisements. However, although DAML-S/UDDI matchmaker expands the functionality of UDDI registry to enable capability matching, it does not take requestors' contextual information into account. Requestors may receive some invalid services, and they need to manually verify all retrieved services. In contrast, our context-aware SOA can provide automatic and better contextually matched services to meet requestors' contextual requirements.

12.2.2 Context and context-aware applications

Context can be interpreted differently from various perspectives. In this research, we will address context from the aspect of mobile and pervasive computing. Schilit, Adams, and Want [3] point out that one of the major challenges of mobile computing is how to exploit the changing environment with a new class of applications that are aware of context. They believe context encompasses more than just people's locations; instead, context should also include environment's lighting, noise level, network communication, and even people's social situation. They also believe that context-aware applications should be able to adapt to changing environment according to the location of use, a collection of nearby people, hosts, and accessible devices, as well as changes to such things over time.

Dey & Abowd [2] define *context* as "*any information that can be used to characterize the situation of an entity. An entity is a person, place, or object that is considered relevant to the interaction between a user and an application, including the user and application themselves.*" They point out that certain types of context that are, in practice, more important than others, such as identity, location, activity, and time. Given a person's identity, one can acquire many pieces of related information, such as the person's contact information and preferences. Given a person's location information, one can determine what other resources or people are nearby and what activity is occurring close to the person. Given a timestamp or a time frame, one can find out what events are taking place or what activities have occurred at the same time. Based on the four primary context types, Dey and Abowd [2] further define a context-aware system one that "*uses context to provide relevant information and/or services to the user, where relevance depends on the user's task.*" They also refer context-aware applications as applications that automatically provide information and take actions according to the user's present context as detected by sensors. In addition, context-aware applications look at *who*'s, *where*'s, *when*'s and *what*'s (that is, what a person is doing), and use this information to determine *why* the situation is occurring.

Besides the aforementioned definitions, many comprehensive survey articles and special issues on journals cover context and context-aware applications. Readers can refer to articles such as Dey and Abowd's [2] "Toward a Better Understanding of Context and Context-awareness," Korkea-aho's [7] "Context-Aware Applications Survey," and Chen and Kotz's [8] "A Survey of Context-Aware Mobile Computing Research".

There are also many research efforts on the development of toolkits supporting context-aware services provision, including HP's Cooltown project [9], Dey's Context Toolkit [10], CB-SeC framework [11], and Gaia middleware [12]. These toolkits either provide functionalities to help service requestors find services based on their contexts or enable content adaptations according to requestor's contextual information.

Mostefaoui and Hirsbrunner propose a formal definition of service context to model a service's contextual information [13]. Lemlouma and Layaida propose a framework to assert metadata information of Web contents [14]. They both use Composite Capabilities/Preferences Profiles (CC/PP) as interoperable context representation to enable communication and negotiation of device capabilities and user preferences in terms of a service invocation. Zhang et al. further propose extensions to CC/PP to enable transformation descriptions between various receiving devices [15]. Besides, several OWL-based context models are presented [16][17][18] to provide high-quality results of services discovery beyond the expressive limitations of *CC/PP*. These researchers all utilize ontology to describe contextual information including location, time, device, preference, and network.

However, existing context definitions focus on either service requestors or services alone. We argue that effective and efficient context-aware services provision needs to consider from both sides. In contrast to aforementioned related work, our approach stands out from three aspects: (1) we formalize an ontology-based context model serving both services and service requestors; (2) we provide comprehensive real-time context acquisition methods; (3) we employ a semantic matchmaking algorithm to enhance the recall and precision of context-aware applications from services discovery.

12.3 Context model

We develop a context model to support formal definition and acquisition of context descriptions. Conceiving context-awareness as an interactive model between service requestors and services, we define the term "context" from the perspectives of both parties. From a service requestor's perspective, context is defined as the surrounding environment affecting her services discovery and access, including a requestor's profiles and preferences, or network channels and devices used to connect to the Web. From a service's perspective, context is defined as the desired surrounding environment affecting services delivery and execution, such as service profiles, networks and protocols for service binding, devices and platforms necessary for service execution, and so on. In order for a service to function appropriately, its target service requestor's context must meet its required service context. Our *context model* comprises context description and context acquisition, as will be addressed as follows.

12.3.1 Context description

Based on the four primary context types (*identity*, *location*, *activity*, and *time*) defined by Dey & Abowd [2], we have developed an *ontology*-based context model, which is illustrated in Figure 12.1. Two types of context ontologies, requestor ontology and service ontology, are developed for describing requestors and services, respectively. The major differences between the two ontologies are their definition profiles. The requestor ontology contains profiles describing four primary context types: personal profile, identify profile, calendar profile, and social profile. A personal profile is used to describe a requestor's *identity* and preferences. A location profile is used to describe *location*. A calendar profile is used to describe what, when, and where an *activity* occurs as well as whom are with the requestor. A social profile is one of the important supporting mechanisms in social awareness by knowing requestors' social relations such as who are their partners? What do they know? And what are they doing at a certain time point? The service ontology is OWL-S-based and contains service profiles describing the input, output, precondition, and effect of service execution. Please be noted that the attributes listed in Figure 12.1 is selected for ease of illustration; the actual contextual information could be much more complicated in the real world.

Other than the profiles, both requestor ontology and service ontology define *Quality of Service* (QoS) and environment profiles. A QoS profile contains functional and non-functional constraints. Functional QoS constraints can be described by response time and throughput; non-functional QoS constraints can be described by reliability, availability, and cost. An environment profile is used to describe computing and communication environment that contains network constraints, situated conditions, and device constraints. Network constraints can be used to describe the types of communication protocols, such as GPRS, 3G, VoIP, and WiFi. Situated conditions can be used to describe requestors' situations, for example, whether they are in a meeting or in a driving condition. Device constraints are used to describe a device's hardware and software configurations, which can be defined by Composite Capabilities/Preferences Profile (CC/PP) and User Agent Profile (UAProf) as enabling techniques.

On the other hand, context information is often kaleidoscopic in different environment, situation even specific goal. In fact, one context model is not easy to cover overall contextual information for specific situation due to the essence of context. Therefore, it is the reason that we use ontology-based representation to define the context model. The ontology-based OWL-S proposes a powerful description capability with scalability and flexibility. We can extend concepts into the original ontology with given relationship. From the view of scalability, this context model allows users adding/deleting concepts in the Service Ontology and Requester Ontology for expanding/narrowing the description scope of this model. From the view of flexibility, this context model allows users flexibly modifying original concepts within the Service Ontology and Requester Ontology for changing the intention or meaning to achieve different expressing. After these modifications, the new OWL-S description needs to be verified by some reasoner for keeping the ontology reasonable and no conflict. We embedded the *Protégé* Ontology Editor and *RACER* Reasoner in the context-aware Web services system to let users reedit the Service Ontology and Requester Ontology for their particular requirements and to reason the new ontology with some logic checking such as consistency, subsumption, classification and unintended relationships. Correspondingly, the inference rules operating with the modified OWL-S model are needed to be changed for logical consistency. We also combine the JESS inference engine within the context-aware Web services system to verify the new rules modified.

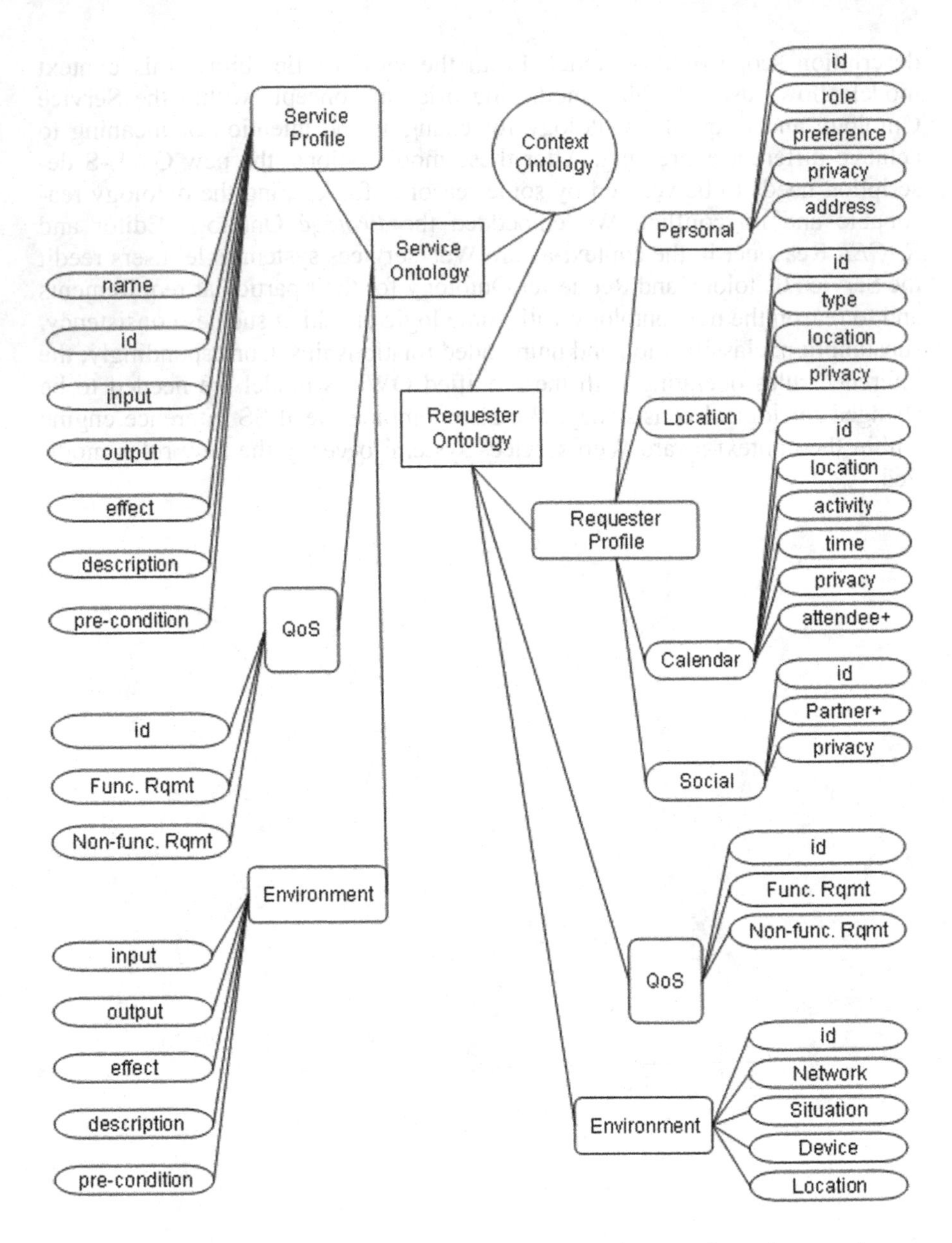

Fig.12.1 Context ontology including requestor ontology and service ontology

12.3.2 Context acquisition

With our ontology model, both service requestors and services can define their contextual information individually. We then define *context acquisition* as a process of obtaining the values of the defined properties from both sides using the requestor ontology and service ontology. By decoupling the context acquisition functions from the context-aware services, we enable the reuse of existing context acquisition functions for various services. As shown in Figure 12.2, we design a three-phase context acquisition mechanism:

(1) form-filling,
(2) context detection, and
(3) context extraction.

In the *form-filling* phase, contextual information is acquired directly from requestors' inputs. In the *context detection* phase, we utilize various sensing, recording and positioning systems (e.g., GPS, RFID, and sensor networks) for location detection. In the *context extraction* phase, contextual information is automatically derived from both requestor ontology and service ontology. In order to facilitate and automate the context acquisition process, we also developed a Context Acquisition Engine as an execution environment, as shown in Figure 12.2. It is worth noting that we designed and implemented every component of the engine as a self-contained Web service, while the entire engine is also implemented as a Web service.

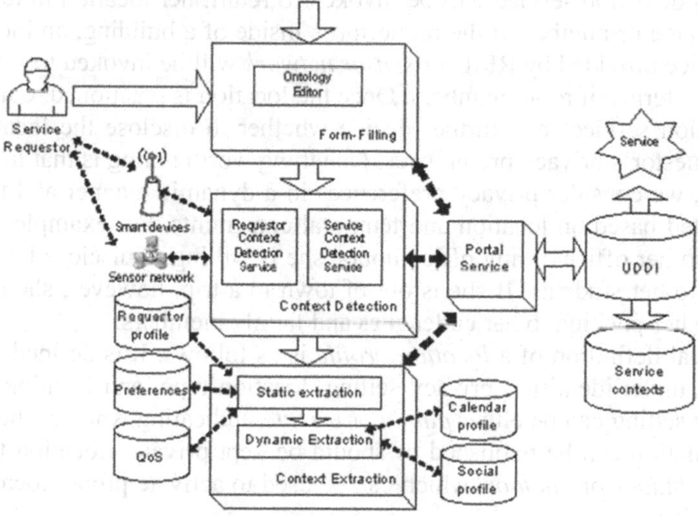

Fig.12.2 Context acquisition mechanism

The first phase form-filling is used to construct service requestor ontologies based upon user-submitted inputs, e.g., personal profile, preferences, calendar profile, and social profile. We implemented a GUI editor to prompt, receive, and organize input information. As shown in Figure 12.2, a portal Web service is designed to serve all three phases. It also keeps historical contextual information that can be used to predict requestors' requesting behaviours and patterns. In other words, the portal helps to construct requestors' preferences automatically. In the form-filling phase, obtained requestor-inputted contextual information is stored in the portal.

The goal of the context detection phase is to complement and refine contextual information during the run time. We have designed and implemented two Web services to facilitate the process: a requestor detection service and a service detection service. Their architectural design and implementation details can be found in our previous report [19]. On the service requestor side, smart devices and sensor networks are utilized to detect and react to the requestors' surrounding environments. For example, when a requestor logs in, our engine detects and analyzes the kind of devices and network channels the requestor is using so as to build device profile and environment profile for the requestor, respectively. The information retrieved is then stored in the portal. On the service side, our portal creates queries and connects to UDDI registries to retrieve contextual information of candidate services.

It should be noted that our environment is equipped with a set of location detection Web services, which can be automatically invoked based on requestors' actual contexts. For example, if a requestor is outside of a building, a GPS location detection service will be invoked to return her location in terms of building name or number; if the requestor is inside of a building, an indoor tracking service provided by RFID or *sensor network* will be invoked to return her location in terms of room numbers. Once the location is positioned, our location detection services can further decide whether to disclose the location based on requestor's privacy preferences. One thing worth noting is that in our environment, we consider privacy preferences in a dynamic manner and they can be adjusted based on location and temporal constraints. For example, if a requestor is in her office during office hours, she is willing to disclose her office number to her students. If she is out of town in a trip, however, she may only disclose her position to her colleagues and family members.

The formal definition of a *location profile* is as follows. It is defined as a four-element tuple: identifier, privacy setting, location type, and location details. Privacy setting can be either *public* or *private*, indicating whether the location information can be published or should be kept private. Location type can be either *indoor* or *outdoor*, which can be used to activate proper location detection devices.

Location_profile = {id, privacy, type, location}

```
privacy = {public | private}
type = {indoor | outdoor}
```

Under some circumstances, requestors' contexts may not be detected explicitly. Using our scenario earlier, assume at some point of time (e.g., 10:15am) when Steve is driving to his client, the connection is lost and the devices cannot directly detect his activity. By checking his *calendar profile* and social profile, it can be deducted that he is in a meeting with his colleagues. The formal definition of a calendar profile is as follows. It is defined as a six-element tuple: identifier, privacy setting, activity, time, attendee list, and location. Privacy setting can be either *public* or *private*, indicating whether the activity can be published or should be kept private. Activity is a (title, description) pair describing the brief and detailed information of the event. Time specifies the starting and ending timestamps of the activity.

```
Calendar_profile = {id, privacy, activity, time, attendee+, location}
  privacy = {public | private}
  activity = {title, description}
  time = {begin(yyyy:mm:dd;hh:mm), end(yyyy:mm:dd;hh:mm)}
```

In the third phase, *context extraction* is defined as a process to derive contextual information from requestors' preferences and profiles. Comprehensive contextual information can be extracted combining static and dynamic approaches. As shown in Figure 12.2, the static context extraction elicits a requestor's default context from her predefined preferences and personal profiles; the dynamic context extraction deduces a requestor's actual context from her calendar profiles and social profiles.

A *social profile* can be used to find the most related partners when a requestor's personal profiles do not reveal her contextual information. The strategy is to query every partner's calendar profiles to deduce the events associated with the target requestor. Note that a requestor can have various types of partners such as friends, team members, organization members, or simply community members. Partners are in turn classified by their expertise of knowledge domains with different degrees of proficiency and trust. The formal definition of a social profile is as follows. It is defined as a three-element tuple: identifier, privacy setting, and a list of partners. Privacy setting can be either *public* or *private*, indicating whether the social relationship information can be published or should be kept private. Each partner is in turn defined as a three-element tuple: partner's identifier, knowledge expertise, and social relationship with the requestor.

```
Social_profile = {id, privacy, partners*}
  privacy = {public | private}
  partners = {id, knowledge_association, social_association}
```

It is wroth noting that we consider privacy issue not only for location profiles, but also for calendar and social profiles. The profile owner can decide to make her profile public or private. In addition, the profile owner can explicitly assign who has the right to review her social interaction and trace her location and calendar. We have implemented a content acquisition system using *JESS* [20], a rule engine for the Java platform.

13.3.3 System implementation

Corresponding to the motivation example illustrated in the Introduction Section, by checking Steve's calendar profiles and social profiles, it can be inferred that at a given time, he is in a Web meeting with his colleagues remotely. Location detection services can locate Steve's position and read his calendar and social profiles and may find out where he is at a time point, what he is doing, and the person with whom he is currently working. Therefore, context-aware Web services can provide appropriate services to better match Steve's contextual changes.

Besides deriving contextual information, additional *queries* can be raised to derive more comprehensive contextual information. Let a given time be "2006:09:06:10:00", which denotes 10:00 o'clock in the morning of Sept. the 6th, 2006. We can construct JESS rules to query contextual information, as shown in the examples below.

```
;; Facts for describing calendar profile
 (deftemplate Calendar "define calendar profile"
        (slot Person-ID)
        (slot Location)
        (slot Activity)
        (slot Time)
        (slot Privacy)
        (slot Attendee))
;; Facts for describing social profile
(deftemplate Social " define social profile "
        (slot Person-ID)
        (slot Privacy)
        (slot Partner))
;; Facts for describing environment profile
(deftemplate Environment " define environment profile "
        (slot Person-ID)
        (slot Network)
        (slot Situation)
```

```
        (slot Device))
  (deftemplate QoS " define quality of service profile "
        (slot Person-ID)
        (slot FuncReq)
        (slot NonFuncReq))

  ;; context-aware inference rules
  (defrule  Q1 ((Calendar ((Person-ID ?personID&Steve) (Time ?time&
"2006:09:06:10:00")))
        =>
        (assert (Calendar (Person-ID ?personID) (Activity ?activ-
ity))))
  (defrule  Q2 ((Calendar ((Person-ID ?personID&Steve) (Time ?time&
"2006:09:06:10:00")))
        =>
        (assert (Calendar (Person-ID ?personID) (Location ?loca-
tion))))
  (defrule  Q3 ((Calendar ((Person-ID ?personID&Steve) (Time ?time&
"2006:09:06:10:00")))
        =>
        (assert (Calendar (Person-ID ?personID) (Attendee ?atten-
dee))))
  (defrule  Q4 ((Calendar ((Person-ID ?personID&Steve) (Time ?time&
"2006:09:06:10:00"))
              (Social      ((Person-ID      ?personID&Steve)
(Partner ?partnerID)))
        =>
        (assert (Calendar (Person-ID ?partnerID) (Activity ?activ-
ity))))
  (defrule  Q5 ((Calendar ((Person-ID ?personID&Steve) (Time ?time&
"2006:09:06:10:00")
             (Location ?location))
             (Environment      ((Person-ID      ?personID)
(Location ?location)))
        =>
        (assert  (Environment  ((Person-ID  ?personID)  (Net-
work ?network) (Device ?device))))
  (defrule  Q6 ((Calendar ((Person-ID ?personID&Steve) (Time ?time&
"2006:09:06:10:00")
              (Location ?location))
              (Environment      ((Person-ID      ?personID)
(Location ?location)))
        =>
        (assert (Environment ((Person-ID ?personID) (Situation ?si-
```

```
tuation))))

  (defrule  Q7  ((Calendar ((Person-ID ?personID&Steve) (Time ?time&
"2006:09:06:10:00")
                    (Location ?location))
                    (QoS ((Person-ID ?personID)))
            =>
          (assert (QoS ((Person-ID ?personID) (FuncReq ?funcReq)
(NonFuncReq ?non-funcReq))))
```

The constructed JESS *rules* can be used to answer the following questions, given the *personID* as "Steve" and the *time* as "2006:09:06:10:00". Who else is attending the conference? Who are Steve's colleagues and what are they doing during the meeting? What if Steve's colleagues are not available at the time? Can Steve find some other information or people with proper expertise? What kind of network channels and devices is Steve using to connect to the Web? What will the different situations, (e.g., driving,) affect Steve's device usage, network access and services delivery? The purposes of the defined JESS rules are summarized in the following Table 12.1, with the query numbers matched to those specified in the rules above.

Table 12.1 JESS rules for service inference

Query#	Query
Q1	What is the conference Steve attends?
Q2	Where is the conference held?
Q3	Who else attend this conference?
Q4	Who are Steve's colleagues and what are they doing during the conference?
Q5	What kind of network and device is Steve using to connect to the Web?
Q6	What is Steve's current situation?
Q7	What are Steve's default and required quality requirements?

12.4 Context-aware applications

Based on the definitions of context model and the contextual information entailed from the context acquisition, we built three context-aware *applications* for demonstrating how to find right services, right partners (collaborators), and right information (content presentation) using our proposed approach.

12.4.1 Context-aware services discovery for finding right services

Based on the requestor ontology and the service ontology defined in the context model, we have developed context-aware service oriented architecture (CA-SOA) for providing context-aware services discovery. As shown in Figure 12.3, CA-SOA consists of three components for context-aware services discovery and integration:

(1) an agent platform (comprising service agents, broker agents, and request agents),
(2) a service repository (including service profiles and service ontology), and
(3) a semantic matchmaker (including a context reasoner and a service planner).

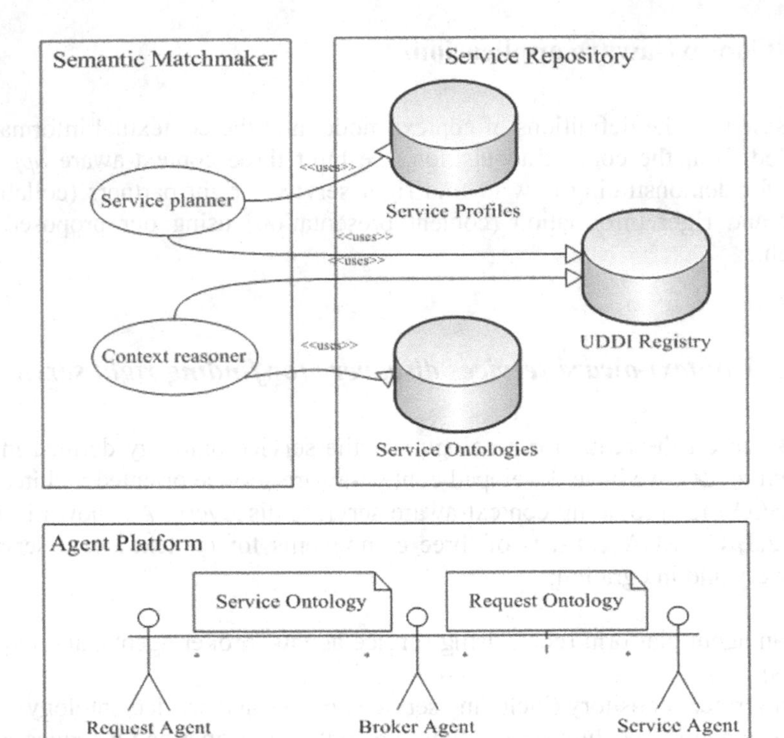

Fig. 12.3 Context-aware SOA

Three types of *agents* are identified in this CA-SOA: service agents, broker agents, and request agents. By extending the three roles from the current SOA model with context-enabled features, these agents enhance context-aware service description, publication, registration, discovery, and access. As shown in Figure 12.3, a service agent helps service providers formally describe their services and wrap the services with contextual descriptions, and then sends the services with contextual descriptions to a broker agent. A request agent helps service requestors formally define their service requests and wrap the requests with requestors' surrounding contexts, and then sends the requests with requestor's contextual descriptions to the broker agent. A *broker* agent takes the service publishing requests from the service agent and saves the service descriptions and service contextual descriptions into the service profiles and the service ontology, respectively. Then the broker agent can take the requests from the request agent and initiates a context-aware semantic matchmaking process.

Context-aware semantic matchmaking consists of two consecutive phases, capability matching and context matching, handled by a service planner and a context reasoner, respectively. As shown in Figure 12.3, our service repository is designed to encompass a general UDDI Registry associated with service profiles and service ontologies. If the required services cannot be found by the capability matching in the UDDI Registry, the semantic matchmaker will proceed context matching. The context reasoner decomposes a service request into a set of sub-requests, based on requestors' ontology sent along with the service request by the request agent. The service planner performs a capability matching process to schedule an integrated composite service based on the decomposed request. We have implemented broker agents to conduct context-aware services discovery based on the two context ontologies.

In *capability matching* phase, we utilized the DAML-S/UDDI methodology [6] that uses IN and OUT to find a set of Web services with capabilities matching the requests. In more detail, our capability matching strategy is based on inputs and outputs to identify the similarities between requests and service's advertisements. Three matching degrees are defined in capability matching - exact match, plug-in match, and subsumed match. (IN_{Req} and OUT_{Req} denote the inputs and outputs of a request, while IN_{Pub} and OUT_{Pub} denote the inputs and outputs of a published service.)

1. Exact match: A published service exactly fulfills a request's requirements. Both the inputs and outputs of the published service match the request's requirements, i.e., $IN_{Pub} = IN_{Req}$ and $OUT_{Pub} = OUT_{Req}$.
2. Plug-in match: A published service sufficiently fulfills a request's requirements. The outputs of the published service provides more detailed information than the request's requirements, or the inputs of the published service require less specific information than the request's requirements, i.e., $OUT_{Req} \subset OUT_{Pub}$ or $IN_{Pub} \subset IN_{Req}$.
3. Subsumed match: A published service only partially fulfills a request's requirements. The outputs of the published service provide less information than the request's requirements, or the inputs of the published service require more detailed information than the request's requirements, i.e., $OUT_{Pub} \subset OUT_{Req}$ or $IN_{Req} \subset IN_{Pub}$.

In *context matching* phase, we refine our queries. In more detail, we have utilized broker agents, as show in Figure 12.3, to handle the context-aware services discovery. Broker agents help service repositories to maintain references of service profiles published by service agents. Note that they only keep references while the service agents keep the original service ontology. Service agents thus need to inform broker agents whenever service profiles are updated. Broker agents also provide caching mechanisms to improve search performance when the same service is requested by multiple requestors. As shown in Figure 12.3, the algorithm of broker agents-enabled context-aware *semantic matchmaking* can be summarized as follows.

(1) A request agent sends a request wrapped with requestor's contextual information to a broker agent.

(2) The broker agent forwards the request to the semantic matchmaker to perform capability matching.
(3) If no matched service can be found, the semantic matchmaker will decompose the request into sub-requests based on requestor's contextual information, and repeat capability matchmaking in Step 2 for each sub-request.
(4) The semantic matchmaker returns matched services to the broker agent.
(5) The broker agent replies the matched services to the request agent.

Using our motivation example again, Steve's request for a context-aware "Web meeting" is formally defined as follows in Resource Description Framework (RDF). His context could be changed due to social interaction, location, calendar events, switching devices and networks. The request agent will perform context wrapping to transform Steve's request to be context-aware. Due to the page limitation, we only show part of the transformed OWL-S code with contextual description of the request. The meeting caller is *Steve* who is using either *NB* or *PDA* to access Web meeting service via *10Mbps bandwidth*. Steve's service request requires the "*availability*" of the meeting service is *99%* or above. Such a meeting requires both audio and video devices. In addition, the service is required to support both the *meeting* mode (with *Audio Off*) and the *driving* mode (with *Video off*).

```
<?xml version="1.0"?>
<rdf:RDF>
<owl:DatatypeProperty rdf:ID="Social_Owner_Name">
  <rdfs:domain rdf:resource="#Social_Owner"/>
  Steve
  <rdfs:range
rdf:resource="http://www.w3.org/2001/XMLSchema#string"/>
</owl:DatatypeProperty>
<owl:DatatypeProperty rdf:ID="Default_Value_of_Device">
  <rdfs:range
rdf:resource="http://www.w3.org/2001/XMLSchema#string"/>
  NB, PDA
  <rdfs:domain rdf:resource="#Preferences"/>
  </owl:DatatypeProperty>
  <owl:DatatypeProperty rdf:ID="Bandwidth">
  <rdfs:range
rdf:resource="http://www.w3.org/2001/XMLSchema#string"/>
  10Mbps
  <rdfs:domain rdf:resource="#Functional_Constraints"/>
</owl:DatatypeProperty>
```

```
<owl:DatatypeProperty rdf:ID="Availability">
  <rdfs:domain rdf:resource="#Non-functional_Constraints"/>
  99%
  <rdfs:range
rdf:resource="http://www.w3.org/2001/XMLSchema#string"/>
</owl:DatatypeProperty>
<Situation rdf:ID="Meeting"/> Audio Off
<Situation rdf:ID="Driving"/> Video Off
</rdf:RDF>
```

The following segment of *OWL-S code* represents one candidate service that satisfies the above requests. It is entitled "*Web business meeting,*" which can be used by either *NB* or *PDA* devices via wireless *WLAN* or *GPRS* with *99%* availability for any requestors who are out of office.

```
<owl:DatatypeProperty rdf:ID="Service_Name">
  <rdfs:range
rdf:resource="http://www.w3.org/2001/XMLSchema#string"/>
  Web business meeting
</owl:DatatypeProperty>
<owl:DatatypeProperty rdf:ID="Availability">
  <rdfs:domain rdf:resource="#Non-functional_Constraints"/>
  99%
  <rdfs:range
rdf:resource="http://www.w3.org/2001/XMLSchema#string"/>
</owl:DatatypeProperty>
<owl:DatatypeProperty rdf:ID="Network">
  <rdfs:range
rdf:resource="http://www.w3.org/2001/XMLSchema#string"/>
  Wireless_LAN, GPRS
</owl:DatatypeProperty>
<owl:DatatypeProperty rdf:ID="Platform">
  <rdfs:domain rdf:resource="#Software"/>
  NB, PDA
  <rdfs:range
rdf:resource="http://www.w3.org/2001/XMLSchema#string"/>
</owl:DatatypeProperty>
<owl:DatatypeProperty rdf:ID="OS">
  <rdfs:range
rdf:resource="http://www.w3.org/2001/XMLSchema#string"/>
  Windows NT, Windows XP
  <rdfs:domain rdf:resource="#Software"/>
```

```
  </owl:DatatypeProperty>
  <owl:DatatypeProperty rdf:ID="Browsers">
    <rdfs:range
rdf:resource="http://www.w3.org/2001/XMLSchema#string"/>
    Microsoft IE, Mozilla Firefox, Netscape
    <rdfs:domain rdf:resource="#Software"/>
  </owl:DatatypeProperty>
```

12.4.2 Context-aware social collaborators discovery for finding right partners

As indicated by Cross, Rice, and Parker [21], social networks significantly effect how people seeking information to do their work. They pointed out that people often receive some combination of five benefits when seeking information from other people (*collaborators*), solutions, meta-knowledge, problem reformulation, validation of plans or solutions, and legitimation from contact with a respected person [21].

Based on the requestor's social profiles defined in the context model, we can derive a *personalized* social network regarding the requestors' social relations such as who are the requestors' partners? What do they know? And what are their relationships? As the social profiles defined in the context model, we are interested to find a requestor's partners (or collaborators) based on their knowledge association and social association. A person's knowledge association can be described by the person's knowledge domain along with her proficiency pertaining to this domain. A person's social association indicates the person's degree of social familiarity to her partners appeared in the social profile.

We use ACM Computing Classification System to classify domain of knowledge, and use *Bloom taxonomy matrix* [22] to classify requestors' proficiency in that domain. As shown in Figure 12.4(a) and 12.4(b), Bloom taxonomy is a matrix consisting of two dimensions: Knowledge dimension and Cognitive Process dimension. Knowledge dimension indicates the types of knowledge; Cognitive Process dimension indicates cognitive processing of knowledge. Each cell in the matrix is associated with a value ranging from 0 to 1, representing the level of proficiency. For example, let Figure 12.4(a) and Figure 12.4(b) indicate two of *Steve*'s partners *Albert* and *Christ*'s knowledge proficiency regarding the knowledge domain of "Software Engineering," respectively. In Figure 12.4(a), the cell (Factual knowledge, Remember) has a value of (0.9), which indicates Albert is good at memorizing factual knowledge about "Software Engineering." In Figure 12.4(b), the cell (Conceptual knowledge, Apply) has a value of (1.0), which indicates Chris is excellent at applying conceptual knowledge of "Software Engineering."

	Cognitive Process Dimensions					
Knowledge dimension	**Remember**	**Understand**	**Apply**	**Analyze**	**Evaluate**	**Create**
Factual knowledge	0.9	0.8	0.4	0.4	0	0
Conceptual knowledge	0.3	0.3	0.3	0.1	0	0
Procedural knowledge	0.6	0.5	0.3	0.2	0	0
Metacognitive knowledge	0	0	0	0	0	0

Fig. 12.4(a) Example of Albert's Bloom taxonomy matrix

	Cognitive Process Dimensions					
Knowledge dimension	**Remember**	**Understand**	**Apply**	**Analyze**	**Evaluate**	**Create**
Factual knowledge	0.2	0.7	0.7	0.7	0.8	0.8
Conceptual knowledge	0.8	0.7	1.0	0.8	0.7	0.7
Procedural knowledge	0.2	0.2	0.3	0.2	0	0
Metacognitive knowledge	0	0	0	0	0	0

Fig.12. 4(b) Example of Chris's Bloom taxonomy matrix

We will first discuss how to calculate requestor's knowledge association. Consider a requestor i who requests for a *partner* possessing specific piece of knowledge k with proficiency denoted by $BT_{(k)}$. The computation to decide whether a partner j conforms to the request is as follows:

$$KA_k(i,j) = KP_k(j) \bullet (BT_{(k)}(i))^T$$

where

$KA_k(i,j)$: indicates the knowledge association from a requestor i to a partner j, with respect to a certain domain of knowledge k. The higher the value is, the stronger *association* it is. $KA_k(i,j)$ is a Bloom taxonomy matrix.

$KP_k(j)$: The knowledge proficiency of the partner j with respect to a certain domain of knowledge k. $KP_k(j)$ is a Bloom taxonomy matrix.

$BT_{(k)}(i)$: indicates a requestor i requesting for a specific piece of knowledge k with proficiency $BT_{(k)}$. $BT_{(k)}(i)$ is a Bloom taxonomy matrix.

Using the motivation example presented in the Introduction Section, a requestor, Steve, requests for partners who have the knowledge proficiency of applying conceptual knowledge of Software Engineering to solve problems. Based on the aforementioned equation, we found two colleagues, Albert and Chris, whose $KA_{SE}(Steve, Albert)$ and $KA_{SE}(Steve, Chris)$ are non-zero, meaning both Albert and Chris conform to Steve's request in terms of knowledge association. The detailed calculation of knowledge association is as follows.

$$\text{Let } BT_{SE}(Steve) = \begin{bmatrix} 0 & 0 & 0 & 0 & 0 & 0 \\ 0 & 0 & 1 & 0 & 0 & 0 \\ 0 & 0 & 0 & 0 & 0 & 0 \\ 0 & 0 & 0 & 0 & 0 & 0 \end{bmatrix}$$

Let $KP_{SE}(Albert)$ be the matrix as shown in Figure 12.4(a) and $KP_{SE}(Chris)$ be the matrix as shown in Figure 12.4(b):

$$KA_{SE}(Steve, Albert) = KP_{SE}(Albert) \bullet (BT_{(SE)}(Steve))^T$$

$$KA_{SE}(Steve, Albert) =$$

$$\begin{bmatrix} 0.9 & 0.8 & 0.4 & 0.4 & 0 & 0 \\ 0.3 & 0.3 & 0.2 & 0.1 & 0 & 0 \\ 0.6 & 0.5 & 0.3 & 0.2 & 0 & 0 \\ 0 & 0 & 0 & 0 & 0 & 0 \end{bmatrix} \bullet \begin{bmatrix} 0 & 0 & 0 & 0 & 0 & 0 \\ 0 & 0 & 1 & 0 & 0 & 0 \\ 0 & 0 & 0 & 0 & 0 & 0 \\ 0 & 0 & 0 & 0 & 0 & 0 \end{bmatrix}^T$$

$$= \begin{bmatrix} 0 & 0 & 0 & 0 & 0 & 0 \\ 0 & 0 & 0.2 & 0 & 0 & 0 \\ 0 & 0 & 0 & 0 & 0 & 0 \\ 0 & 0 & 0 & 0 & 0 & 0 \end{bmatrix}$$

After *serialization*, $KA_{SE}(Steve, Albert) = 0.2$. Similarly, we can compute that $KA_{SE}(Steve, Chris) = 1$, as shown below.

$$KA_{SE}(Steve, Chris) =$$

$$\begin{bmatrix} 0.2 & 0.7 & 0.7 & 0.7 & 0.8 & 0.8 \\ 0.8 & 0.7 & 1.0 & 0.8 & 0.7 & 0.7 \\ 0.2 & 0.2 & 0.3 & 0.2 & 0 & 0 \\ 0 & 0 & 0 & 0 & 0 & 0 \end{bmatrix} \bullet \begin{bmatrix} 0 & 0 & 0 & 0 & 0 & 0 \\ 0 & 0 & 1 & 0 & 0 & 0 \\ 0 & 0 & 0 & 0 & 0 & 0 \\ 0 & 0 & 0 & 0 & 0 & 0 \end{bmatrix}^T$$

$$= \begin{bmatrix} 0 & 0 & 0 & 0 & 0 & 0 \\ 0 & 0 & 1 & 0 & 0 & 0 \\ 0 & 0 & 0 & 0 & 0 & 0 \\ 0 & 0 & 0 & 0 & 0 & 0 \end{bmatrix}$$

We will now discuss how to calculate *social association*. The social association indicates how a requestor is associated with her partners described in her social profiles. For the requestor and her partner, as denoted by *i* and *j* respectively, the social association between them is the product of the social relationship tie and the social reputation:

$$\mathrm{SA(i,j)} = SRT(i,j) \times SR(i,j)$$

where
SA: social association between requestor i and partner j.
SRT: social relationship tie indicates how the requestor and her partner treat each other.
SR: social reputation is confidence indicating the degree of trust from requestor i to partner j.

Social relationship tie (SRT) indicates how the requestor and her partners treat each other. The requestor needs to specify her SRT with each of her partners by filling forms and answering questionnaires. If the requestor does not specify SRT with one partner explicitly, the default value of their SRT is zero, meaning that there is no relationship between the requestor and the partner. SRT can exhibit different levels of social relationships, such as friend, team member, organization member, or community member. Meanwhile, SRT can be either positive or negative values ranging between [-1,1], indicating the relationship is good or bad. To perform quantitative analysis, we identify nine SRTs between the requestor *i* and her partner *j* as follows.

SRT(i,j) = 0, if there is no relationship between requestor i and her partner j.

SRT(i,j) = 0.8~1.0, if requestor i treats her partner j as a friend with positive relationship

SRT(i,j) = 0.5~0.7, if requestor i treats her partner j as a team member with positive relationship

SRT(i,j) = 0.3~0.4, if requestor i treats her partner j as an organization member with positive relationship

SRT(i,j) = 0~0.2, if requestor i treats her partner j as a virtual community member with positive relationship

SRT(i,j) = -0.8~-1.0, if requestor i treats her partner j as a friend with negative relationship

SRT(i,j) = -0.5~-0.7, if requestor i treats her partner j as a team member with negative relationship

SRT(i,j) = -0.3~-0.4, if requestor i treats her partner j as an organization member with negative relationship

SRT(i,j) = 0~-0.2, if requestor i treat her partner j as a virtual community member with negative relationship

The social reputation between requestor *i* and her partner *j* is denoted by *SR(i,j), which* is confidence indicating the degree of trust from requestor *i* to the requested partner *j*. *SR(i,j)* is used to determine whether the requested partner conforms to the requestor requirements of trust. The value of *SR(i,j)* is a percentage: the higher the percentage, the higher the confidence. For example, if the value of *SR(Steve, Albert)* is 78%, it means requestor *Steve* has 78% confidence that the requested partner *Albert* is trustworthy.

We utilize binomial probability's sampling to calculate *SR(i,j)*, based on a 95% confidence interval in terms of probability [23]. Here we first define the following terms:

- S is a set of interaction instances representing the samples of the requested partner's past interactions, $S = \{s_1, s_2, \ldots s_n\}$.
- Tr is a set of trust evaluation values containing past experience instances, and is denoted by $Tr = \{tr_1, tr_2, \ldots tr_n\}$.
- $Rating : S \rightarrow Tr$ $Rating(s)$: The Rating function maps the interaction instance s to past experience instance tr. In other words, the function associates past service instances with past experience instances, both being collected by the requestor's partners other than the requestor.
- $Accpet : Tr \rightarrow \{0,1\}$ A requirement hypothesis can be denoted as an $Accpet$ function.

The output of an Accept function is a "1" when past experience instance is accepted by the requestor, and a "0" if otherwise:

$$\text{Accept}(\text{tr}) \equiv \begin{cases} 1 & \text{Accept} \\ 0 & \text{otherwise} \end{cases}$$

Based on the usage of Large-Sample of Hypothesis for a Binomial Proportion to evaluate the simple error and true error of a hypothesis [23][24], the result of the hypothesis assessing the sample is a Boolean value (true or false). Thus, we can see that the hypothesis assessing the sample is a Bernoulli trial and the distribution of Bernoulli trial is a binomial distribution. The *binomial distribution* approximates the normal distribution when the number of sample is big enough. Simple error refers to correct rate in samples; true error refers to correct rate in a population. We can get a confidence interval according to the simple error and the area of confidence interval, representing a probability whose true error falls in the interval. In a normal distribution, the true error is 95% probabilities falling within the range of mean $\pm 1.96 \times \mathrm{SD}$ (Standard Deviation) in compliance with the experience rule. In other words, we can utilize the confidence interval to evaluate the lowest true error of the evaluating hypotheses.

Let *Accpet* function be the hypothesis and then we can evaluate the possible true error of the hypothesis based on the past instances S according to the *Evaluating Hypotheses theory* [23]. Whether tr ($tr \in E$) is accepted by *Accept* is a binomial distribution that approximates the normal distribution when the number of samples is large enough. Thus, we can utilize the normal distribution to calculate that the sample error closest with the true error. The true error is of 95% probabilities falling within a confidence interval, which will be approved as a trustworthy partner in the general application.

For example, let Steve be a requestor, and let Albert is a requested partner. We define the confidence symbol as the lowest bound of the true error. The trust of service conforms to the request's requirement when the confidence is higher.

$$\hat{p} = \frac{1}{n}\sum_{s \square S} Accpet(Rating(s)),\ SD = \sqrt{\frac{\hat{p} \times (1-\hat{p})}{n}},\ z_{95\%} = 1.96$$

$$Confidence \equiv \max\{\hat{p} - z_{95\%} \times SD,\ 0\}$$

As the number of samples increases, the standard deviation decreases relatively and the confidence will be closer to the true error. For example, assume Albert's past instances is denoted as S, and let $|S| = 256$. Steve proposes a Requirement Hypothesis *Accept*. If the result of calculation is $\hat{p} = 0.6$, the confidence can be calculated from the following equation:

$$\hat{p} = \frac{1}{256}\sum_{s \in S} Accpet(\mathrm{Rating(s)}) = 0.6,\ z_{95\%} = 1.96$$

$$Confidence = \hat{p} - z_{95\%} \times \sqrt{\frac{\hat{p} \times (1-\hat{p})}{256}} \cong 0.6 - 0.060012 = 0.539987$$

The calculated confidence *SR(Steve,Albert)* is 53.99%, which means Steve has 53.99% confidence that Albert can meet Steve's trustworthy requirement based on 95% confidence interval. In other words, we can assert that the Albert's degree of trust is 56.83% (53.99% over 95%) conforming to Steve's requirements.

Let Albert is one of Steve's team members with positive relationship (SRT = 0.7), then:

$$
\begin{aligned}
SA(i,j) &= SRT(i,j) \times SR(i,j) \\
&= 0.7 \times 0.5683 \\
&= 0.3978
\end{aligned}
$$

12.4.3 Context-aware content adaptation for finding right content presentation

In mobile environments, people mostly work with portable devices along with limited computing powers, small-size screens and changing conditions. Portable devices have distinct capabilities compared with desktop computers, which divergence increases the difficulty of presenting friendly contents for all kinds of devices universally. Besides, the condition of people's content access is more complicated in mobile environments. For example, people may need to access content while they stay outside or on a move. In addition, most current Web contents are designed for desktop computers only. The default settings and style-sheets, such as image size, font size, and layout structure, are not suitable to be presented on portable devices. As a result, a technique is needed to compose and deliver adaptive content from any platform in any format to any device through any network at anytime and at anywhere. To achieve this ultimate goal, one needs to know people's computing context (environment) and provides *content adaptation* based on such context.

Using the motivation example presented in Introduction Section, content need to be adapted before it can be delivered and presented to the users when a Web meeting switches from PC-based to PDA-based due to the contextual changes of device and network. In such a context-aware application to meet the changing computing environment, content adaptation is a technique to provide the most suitable content presentation according to corresponding computing context. Our context-aware content adaptation is designed to automatically adapt Web content to various formats and to enhance Web *accessibility* based on users' surrounding context, especially when requestors are using portable devices in a mobile environment. For example, if a requestor is accessing a film while she is in a meeting, then the context-aware content adaptation service should automatically turn off the sound to avoid from making noise because it is important to remain quiet in a public situation. For another example, if the requestor is accessing a film while driving, then the context-aware content adaptation service should automatically turn off the video, and leave audio only for the reason of safety. In this chapter, we focus on the design of content adaptation by providing rules for transforming objects' modality and fidelity to fit users' situated environment.

A Web page comprises a set of medium objects or simply objects, which are characterized with modality indicating their types such as text, video, audio, and image. Each modality is associated with fidelity indicating objects' quality such as image resolution, color depth, and video bit-rate. In order to render the same object in various devices, content adaptation needs to perform *transcoding* and changing object's modality and fidelity. For example, if a mobile phone can only play image with low resolution, the fidelity of an image needs to be turned to low.

We introduce a concept of *Unit of Display* (UOD) to define a smallest unit that can be displayed and adapted in a Web page. Environment profile contains the facts described by user's current contextual constraints, and the adaptation rule base contains the patterns of rules for context-oriented content adaptation. For example, if the context indicating user's situation is in a public situation, UOD with audio and video modality should be adapted by changing its fidelity to mute to remain quite. If the context indicating user's situation is driving, then the UOD with image modality should be adapted by changing its fidelity to blank for the reason of safety.

Content adaptation rules can be derived from the environment profiles defined in the context model. The rules are designed as patterns for transforming objects' modality and fidelity. Table 12.2 shows three categories of adaptation rules we have derived based on users' situated environment. For ease of representation, we present these rules in a lookup table format. Rows in each table are numbered by S, N, and D, which represent Situation, Network, and Device, respectively. As appeared on column head, the original object describes object's modality and fidelity before adaptation, while the adapted object describes object's modality and fidelity after adaptation. Object's fidelity before adaptation is defined to be original depending on how the objects are originally designed.

Please be noted that the assumed information listed in Table 12.2(a), 1(b), 1(c) is selected for ease of illustration. The contextual information could be much more complicated in a real world. Situation information can be further described by when, where, and what activities the person is involved, as well as location identified by GPS, sensor networks, RFID, and so on. Network information can be further described by communication protocols such as GPRS, 3G, VoIP, and WiFi. Device information can be further described with CC/PP, UAProf, and so on. We also simplified the description of modality and fidelity in this chapter. In a real world, modality information can be further classified into flash and streaming media, for example. *Fidelity* information can be further described with image types (e.g. BMP, JPEG), image resolution (e.g. 24 bit), color depth (e.g. 32bit), video bit-rate, and so on.

As shown in Table 12.2(a), situation lookup table is based on users' access situation in terms of lighting, noise level, visibility, and access situation. The underlying concept of this lookup table is that removing unnecessary medium can save transmission time. For example, it is obvious that one cannot read video in a dark situation; nor is allowed to listen to audio in a quiet public situation. Therefore, object fidelity can be adapted to blank (by removing video and leave audio only) for darkness, or adapted to mute (by removing audio and leave video only) for quietness, as shown in S2 and S4 in Table 12.2(a), respectively. Similarly, when a user's access situation is "meeting," an object should be adapted by changing its fidelity to mute to remain quiet (as shown in S21 and S23). If a user's access situation is "driving," an object should be adapted by changing its fidelity to blank for preventing the driver from browsing content (as shown in S22, S26, and S28).

Table 12.2(a) Situation lookup table

Rule#	original object	situation	adapted object
S1	Object(video,original)	Situation(normal)	Object(video,original)
S2	Object(video,original)	Situation(dark)	Object(video,blank)
S3	Object(video,original)	Situation(blurred)	Object(video,bright)
S4	Object(video,original)	Situation(quiet)	Object(video,mute)
S5	Object(video,original)	Situation(noisy)	Object(video,loud)
S6	Object(audio,original)	Situation(normal)	Object(audio,original)
S7	Object(audio,original)	Situation(dark)	Object(audio,original)
S8	Object(audio,original)	Situation(blurred)	Object(audio,original)
S9	Object(audio,original)	Situation(quiet)	Object(audio,mute)
S10	Object(audio,original)	Situation(noisy)	Object(audio,loud)
S11	Object(image,original)	Situation(normal)	Object(image,original)
S12	Object(image,original)	Situation(dark)	Object(image,blank)
S13	Object(image,original)	Situation(blurred)	Object(image,bright)
S14	Object(image,original)	Situation(quiet)	Object(image,original)
S15	Object(image, original)	Situation(noisy)	Object(image,original)
S16	Object(text,original)	Situation(normal)	Object(text,original)
S17	Object(text,original)	Situation(dark)	Object(audio,original)
S18	Object(text,original)	Situation(blurred)	Object(text,bright)
S19	Object(text,original)	Situation(quiet)	Object(text,original)
S20	Object(text,original)	Situation(noisy)	Object(text,original)
S21	Object(video,original)	Situation(meeting)	Object(video,mute)
S22	Object(video,original)	Situation(driving)	Object(video,blank)
S23	Object(audio,original)	Situation(meeting)	Object(audio,mute)
S24	Object(audio,original)	Situation(driving)	Object(audio,original)
S25	Object(image,original)	Situation(meeting)	Object(image,original)
S26	Object(image,original)	Situation(driving)	Object(image,blank)
S27	Object(text,original)	Situation(meeting)	Object(text,original)
S28	Object(text,original)	Situation(driving)	Object(text,blabk)

As shown in Table 12.2(b), network lookup table is based upon network bandwidth and object size. The idea behind this lookup table is to set up a threshold of network bandwidth and object's size. The object's fidelity will be adapted to lower resolution if the transmission bandwidth is less than 2M bps and the object's size is larger than 2M bytes. For a network with downloading bandwidth less than 2M bps, content adaptation needs to be done before any object over 2M (as shown in rules N2, N4, N6, and N8 in Table 12.2(b)) can be downloaded. We use Network(2M-) to denote the network bandwidth less than 2M bps, ObjectSize(2M+) to denote object size greater than 2M bytes.

Table 12.2(b) Network lookup table

Rule#	original object	network	adapted object
N1	Object(video,original)	Network(2M+)	Object(video,original)
N2	Object(video,original)	Network(2M-) and ObjectSize(2M+)	Object(video,low_resolution)
N3	Object(audio,original)	Network(2M+)	Object(audio,original)
N4	Object(audio,original)	Network(2M-) and ObjectSize(2M+)	Object(audio,low_resolution)
N5	Object(image,original)	Network(2M+)	Object(image,original)
N6	Object(image,original)	Network(2M-) and ObjectSize(2M+)	Object(image,low_resolution)
N7	Object(text,original)	Network(2M+)	Object(text,original)
N8	Object(text,original)	Network(2M-) and ObjectSize(2M+)	Object(text,low_ resolution)

As shown in Table 12.2(c), device lookup table is based on the types of users' devices. The idea driving this lookup table is that not every handheld device can play all types of rich media; it will save unnecessary transmission bandwidth by removing rich medium if it is un-playable on certain devices. For example, for some mobile phones with limited computing powers and rendering capability, they might not be able to play *video clips*. Besides, rich media take longer time to be delivered over wireless communication. As a result, if an object's modality is video, audio, or image, its fidelity will be adapted to lower resolution to be played on mobile phones as shown in rules D3, D6, and D9 in Table 2(c).

Table 12.2(c) Device lookup table

Rule#	original object	devices	adapted object
D1	Object(video,original)	Device(NB)	Object(video,original)
D2	Object(video,original)	Device(PDA)	Object(video,original)
D3	Object(video,original)	Device(Phone)	Object(video,low_resolution)
D4	Object(audio,original)	Device(NB)	Object(audio,original)
D5	Object(audio,original)	Device(PDA)	Object(audio,original)
D6	Object(audio,original)	Device(Phone)	Object(audio,low_resolution)
D7	Object(image,original)	Device(NB)	Object(image,original)
D8	Object(image,original)	Device(PDA)	Object(image,original)
D9	Object(image,original)	Device(Phone)	Object(image,low_resolution)
D10	Object(text,original)	Device(NB)	Object(text,original)
D11	Object(text,original)	Device(PDA)	Object(text,original)
D12	Object(text,original)	Device(Phone)	Object(text,original)

The results of context-aware content adaptation are shown in Figure 12.5. The left-hand side is Web content shown in a *mobile phone* without adaptation, where the requestors can only see a part of the content. In contrast, the picture shown at the right-hand side is the adapted version through context-aware content adaptation service, which provides better readability and loses no information compared with normal Web browsing.

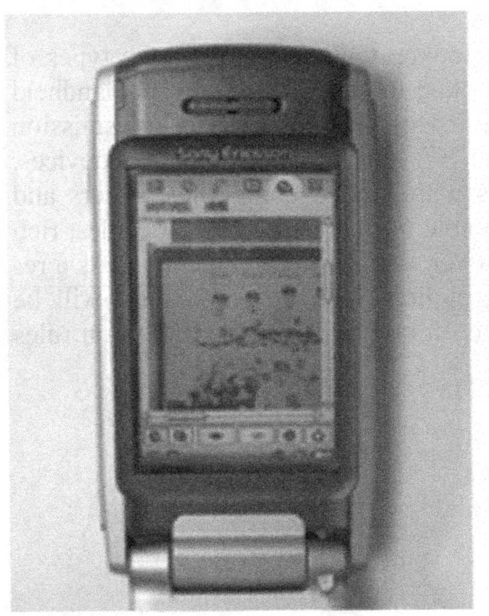

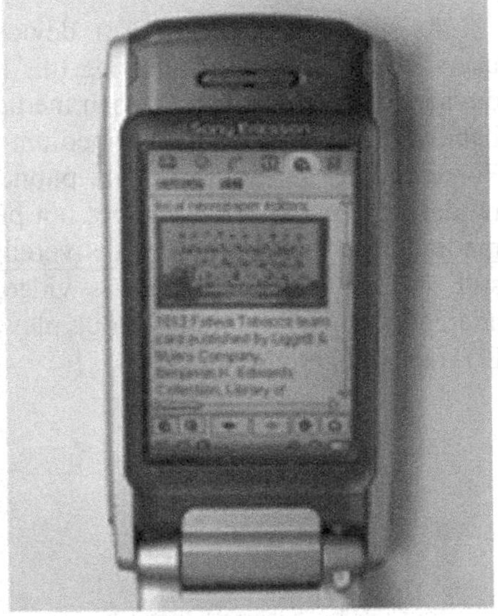

Fig. 12.5 A comparison of Web content access and adaptation

In this section, we proposed some practical context-aware application including context-aware services discovery for finding right services, context-aware social collaborators discovery for finding right partners, and context-aware content adaptation for finding right content presentation. Among the service computing, there are some important issues for achieving these goals such as OWL-S authoring, Web service discovery, Web service invocation, and Web service composition. There are many existing technologies can assist in building the foundation. In order to make available OWL-S, some *authoring* tools had been proposed, such as CMU WSDL2OWL-S, KSL OWL-S Editor, and Mind-Swap Ontolink. Web service *discovery* mechanism can help users to search services; some popular tools can assist in the building of *matchmaking*, such as CMU OWL-S Broker, CMU OWL-S for P2P, CMU OWL-S/UDDI Matchmaker, and KSL Semantic Discovery Service. Efficient service invocation ensure the efficiency service process, CMU OWL-S Virtual Machine is suggested for this concern. While one service cannot satisfy user's requirements, composition is an alternative way to get specific service for necessaries. Some existing tools can assist context-aware services in constructing the fundamental components, such as CMU Computer Buyer, KSL Composition Tool, and Mind-Swap Composer.

12.5 Conclusions

In this chapter, we have presented our context model to formally define context description pertaining to service requestors and services. We designed a context acquisition mechanism with tailored environment for collecting contextual information in three phases: form-filling, context detection, and context extraction. Based on the context model, we have presented three context-aware applications for demonstrating how to find right services, right partners (collaborators), and right information (content presentation).

For finding right services, our *CA-SOA* differentiates from the current SOA model in several significant ways. First, ontology is enabled to bridge between service requestors, service brokers, and services. Second, ontology-based contextual repository extends the current UDDI mechanism to define services in a more comprehensive manner. Third, context-based semantic matchmaking enables more precise services discovery and access. The way how we implemented our Context Acquisition Engine using Web services leads to four major advantages: (1) the engine itself reflects the SOA concept; (2) parts or whole of the engine can be reused with flexibility and extensibility; (3) our engine has potential to be published on the Internet to facilitate context-enabled services discovery and integration; and (4) our engine can be utilized as an extension to the current UDDI engine for better service matchmaking.

For finding right partners, we use social profiles in our context model to derive personalized social networks. Based on Bloom Taxonomy matrix, we proposed the algorithms of calculating knowledge association and social association. From the construction of the context-aware collaborator application, we conclude that, (1) the found partners are more relevant and they are more willing to share and collaborate because of higher social relationships; and (2) they are more likely to collaborate in the same social group.

For context-aware content adaptation, we use environment profiles in our context model to derive content adaptation rules into various content lookup tables, such as situation lookup table, network lookup table, and device lookup table. From the construction of the context-aware content adaptation application, we conclude that when people use PDAs, most of them can correctly explain the original meaning of an adapted image. When it comes to text and hyperlink, we found that most of the users have no difficulties in understanding the meaning of adapted images. When people use mobile phones, they have difficulties in browsing content due to the smaller screen size and less computing capability, which result in lower comprehension especially in understanding the meaning of an adapted image.

In our future research, we plan to enhance our CA-SOA in the categories of rule-base *truth maintenance* [25], request decomposition, service planning, services composition and verification. We also plan to conduct more experiments to examine performance metrics including precision and recall ratio of services discovery, efficiency of services composition, effectiveness of service verification, and reliability of service execution. Finally, we will explore using Web Services Agreement Specification (WS-Agreement) [26] from Global Grid Forum or Web Service Level Agreement (WSLA) [27] from IBM to carry the knowledge.

References

[1]. V. Poladian, J.P. Sousa, D. Garlan, B. Schmerl, and M. Shaw, "Task-based Adaptation for Ubiquitous Computing," *IEEE Transactions on Systems, Man, and Cybernetics, Part C: Applications and Reviews*, vol. 36, no. 3, May 2006, pp. 328-340

[2]. A.K., Dey and G.D. Abowd, "Toward A Better Understanding of Context and Context-Awareness," Technical Report GIT-GVU-99-22, Georgia Institute of Technology, 1999.

[3]. B.N. Schilit, N.I. Adams, and R. Want, "Context-Aware Computing Applications," *Proc. of the Workshop on Mobile Computing Systems and Applications*, Santa Cruz, CA, USA, Dec. 1994, pp. 85-90.

[4]. T. Berners-Lee, J. Hendler, and O. Lassila, "The Semantic Web: A New Form of Web Content That Is Meaningful to Computers Will Unleash A Revolution of New Possibilities," *Science and Technology at Scientific American.com*. Available from http://www.scientificamerican.com/article.cfm?articleID=00048144-10D2-1C70-84A9809EC588EF21&catID=2

[5]. K. Sycara, M. Klusch, and J. Lu, "LARKS: Dynamic Matchmaking Among Heterogeneous Software Agents in Cyberspace," *Kluwer Academia Publishers: Autonomous Agents and Multi-Agent Systems*, vol. 5, 2002, pp. 173-203.

[6]. K. Sycara, M. Paolucci, A. Ankolekar, and N. Srinivasan, "Automated Discovery, Interaction and Composition of Semantic Web Services," *Journal of Web Semantics: Science, Services and Agents on the WWW*, vol. 1, no. 1, 2003, pp. 27-46.

[7]. M. Korkea-aho, "Context-Aware Applications Survey," Available from *http://users.tkk.fi/~mkorkeaa/doc/context-aware.html, 2000.*

[8]. G. Chen and D. Kotz, "A Survey of Context-Aware Mobile Computing Research," *Dartmouth Computer Science Technical Report TR2000-381*, 2002, pp. 1-16.

[9]. Hewlet Packard, "Cooltown Project," Available from *http://www.cooltown.com/cooltown/index.asp*

[10]. A.K. Dey, "The Context Toolkit," Available from *http://www.cs.berkeley.edu/~dey/context.html*

[11]. S.K. Mostefaoui, A. Tafat-Bouzid, and B. Hirsbrunner, "Using Context Information for Service Discovery and Composition," *Proc. of 5th International Conference on Information Integration and Web-based Applications and Services (iiWAS 2003)*, Jakarta, Indonesia, Sep. 15-17, 2003, pp. 129-138.

[12]. M. Roman, C.K. Hess, R. Cerqueira, A. Ranganathan, R.H. Campbell, and K. Nahrstedt, "Gaia: A Middleware Infrastructure to Enable Active Spaces," *IEEE Pervasive Computing*, vol. 1, no. 4, Oct.-Dec. 2002, pp.74-83.

[13]. S.K. Mostefaoui and B. Hirsbrunner, "Context Aware Service Provisioning," *Proc. of the IEEE International Conference on Pervasive Services (ICPS)*, Jul. 19-23, 2004, Beirut, Lebanon, pp.71-80, 2004.

[14]. T. Lemlouma and N. Layaida, "The Negotiation of Multimedia Content Services in Heterogeneous Environments," *Proc. of The 8th International Conference on Multimedia Modeling(MMM 2001)*, Amsterdam, The Netherlands, Nov. 5-7, 2001, pp. 187-206.

[15]. J. Zhang, L.-J. Zhang, F. Quck, and J.-Y. Chung, "A Service-Oriented Multimedia Componentization Model," *International Journal of Web Services Research* (JWSR), vol. 2, no. 1, Jan-Mar. 2005, pp. 54-76.

[16]. T. Broens, S. Pokraev, M.v. Sinderen, J. Koolwaaij, and P.D. Costa, "Context-aware, Ontology-based, Service Discovery," *Proc. of 2nd European Symposium on Ambient Intelligence (EUSAI 2004)*, Nov. 8-10, 2004, Eindhoven, The Netherlands, pp. 72-83.

[17]. M. Khedr and A. Karmouch, "Negotiating Context Information in Context-Aware Systems," *IEEE Intelligent Systems*, vol. 19, no. 6, pp. 21-29. 2004.

[18]. M. Khedr, "A Semantic-Based, Context-Aware Approach for Service-Oriented Infrastructures," *Proc. of 2nd IFIP International Conference on Wireless and Optical Communications Networks (WOCN 2005)*, 2005, United Arab Emirates, pp. 584-588.

[19]. S.J.H. Yang, "Context Aware Ubiquitous Learning Environments for Peer-to-Peer Collaborative Learning," *Journal of Educational Technology and Society*, vol. 9, no. 1, Jan, 2006.

[20]. F.H. Ernest, "Jess in Action: Java Rule-Based Systems," 2003, *Manning Publications.*

[21]. R. Cross, R.E. Rice, and A. Parker, "Information Seeking in Social Context: Structural Influences and Receipt of Information Benefits," *IEEE Transactions on Systems Man and Cybernetics Part C-Applications and Reviews*, vol. 31, no. 4, Nov. 2001, pp. 438-448.

[22]. L.W. Anderson, D. R. Krathwohl, P. W. Airasian, K.A. Cruikshank, R. E. Mayer, P.R. Pintrich, J. Raths, M.C. Wittrock, "A Taxonomy For Learning, Teaching, And Assessing: A Revision Of Bloom's Taxonomy of Educational Objectives," 2001, New York: Longman.

[23]. T. Mitchell, "Machine Learning," *WCB McGraw-Hill*, 1997, pp. 128-141.

[24]. W. Mendenhall and R.J. Beaver, "Introduction to Probability and Statistics," *Duxbury Press*, 1999, pp. 442-446.

[25]. S.J.H. Yang, J.J.P.Tsai, and C.C. Chen, "Fuzzy rule Base Systems Verification using High Level Petri Nets," *IEEE Transactions on Knowledge and Data Engineering*, vol. 15, no. 2, 2003, pp. 457-473.

[26]. A. Andrieux, K. Czajkowski, A. Dan, K. Keahey, H. Ludwig, T. Nakata, J. Pruyne, J. Rofrano, S. Tuecke, and M. Xu, "Web Services Agreement Specification (WS-Agreement)," http://www.ggf.org/Public_Comment_Docs/Documents/Oct-2005/WS-AgreementSpecificationDraft050920.pdf.

[27]. H. Ludwig, A. Keller, A. Dan, R.P. King, and R. Franck, "Web Service Level Agreement (WSLA) Language Specification," Version 1.0, http://www.research.ibm.com/wsla/WSLASpecV1-20030128.pdf.

Index